Engineering Your Future

Launching a Successful Entry-Level Technical Career in Today's Business Environment

Stuart G. Walesh
College of Engineering
Valparaiso University

Prentice Hall PTR, Upper Saddle River, New Jersey 07458

Library of Congress Cataloging–in–Publication Data

Walesh, S. G.
Engineering your future : launching a successful entry-level technical career in today's business environment / Stuart G. Walesh
p. cm.
Includes bibliographical references.
ISBN (invalid) 0–13–221052–3
1. Industrial project management. I. Title
T56.8W35 1995
620',0068—dc20 94–34118
CIP

Editorial production: *bookworks*
Acquisitions editor: *Bernard Goodwin*
Manufacturing manager: *Alexis R. Heydt*
Cover designer: *Design Sources*
Cover credit: Salley Wern Comport/The Stock Illustration Source, Inc.

The publisher offers discounts on this book when ordered in bulk quantities. For more information, contact:

Corporate Sales Department
Prentice Hall PTR
Upper Saddle River, NJ 07458

Printed in the United States of America

10 9 8 7 6 5 4 3 2 1

ISBN 0–13–221052–5

Prentice-Hall International (UK) Limited, London
Prentice-Hall of Australia Pty. Limited, Sydney
Prentice-Hall Canada Inc., Toronto
Prentice-Hall Hispanoamericana, S.A., Mexico
Prentice-Hall of India Private Limited, New Delhi
Prentice-Hall of Japan, Inc., Tokyo
Pearson Education Asia Pte. Ltd., Singapore
Editoria Prentice-Hall do Brasil, Ltda., Rio De Janeiro

To Mom and Dad who always, somehow, managed
and who really cared for their customers.

Contents

Preface

Engineering Your Future: Launching a Successful Entry-Level Technical Career in Today's Business Environment provides the recent engineering or other technical graduate or the entry-level technical person with basic, pragmatic management and leadership concepts, knowledge, and skills. These results-oriented fundamentals should be immediately useful on the job and will also help the young professional learn more efficiently from day-to-day professional experiences.

Technical competency, although necessary, is not sufficient for the young engineer or other technically educated professional who wishes to quickly realize his or her potential in the consulting business, industry, or government. Technical competency must be supplemented with basic management proficiencies and leadership understanding if the entry-level professional is to be productive for his or her employer. Unfortunately, management concepts, knowledge, and skills are typically not introduced in undergraduate engineering and related curricula, and virtually nothing is taught about leadership. Accordingly, management and leadership must be learned by doing, often inefficiently at high monetary cost to the employer and at the risk of jeopardizing the young professional's career.

This is not a "transition from engineering to management" book. A premise of this book is that all engineers and other technical professionals are managers from day one—at least they are managers of their time, their assignments, their relationships with others, and their careers. The best leaders in technical organizations are most likely to be those professionals who began to develop management skill and leadership understanding very early in their careers—who knew, from the beginning, how to develop the "soft" as well as the "hard" side of their careers.

Career management is becoming increasingly important. The parents of today's young professionals often entered into unwritten but binding "contracts" with their employers. In that era, the young engineer or other technical professional would typically focus on technical matters and do them well, and the employer would, in turn, agree to provide long-term employment. Increasingly, such "employment contracts" are vanishing, average periods of employment with a given employer are diminishing, and major organizational upheavals caused by financial difficulties, acquisitions, and mergers are increasing. Perceptive young professionals will recognize these changes, anticipate employment problems, and prepare for employment opportunities. This book will help you engineer your career.

AUDIENCE

The book assumes that readers are, or soon will be, graduates (BS or MS) of an engineering or other technical program with little or no engineering, business, or management experience. The book further assumes that readers want to take a proactive approach and quickly build on their technically oriented education to become even more productive members of their organization.

Because of the intended audiences—senior and/or graduate student or recent graduates—the book is written to be used as either a textbook or a reference book for young professionals. It could also be used to support seminars and workshops directed at young engineers and other technical professionals. Portions of the book are intended to be of value as a textbook or reference book to young professionals outside of technical fields, that is, in business, government, and other areas.

Much of the material presented in this book will also be immediately useful to students. That is, while they are students, future technical professionals can utilize some of the tools and techniques presented in this book, such as, but not limited to, time management, delegating, managing meetings, project management, total quality management, business accounting, and marketing.

ORGANIZATION AND CONTENT

Many aspects of management and leadership are covered. Examples are self-management, management of others, understanding how organizations work and how to work in organizations, project management, total quality management, engineering or decision economics, basic business accounting methods, legal issues, ethics, the design function, the role and selection of consultants, marketing professional services, and shaping the future.

Chapters are arranged in accordance with the preceding order of topics. Although some sequencing of chapters should be followed for effectiveness, such as Chapter 10, Legal Framework, being followed by Chapter 11, Ethics, there is no compelling argument for reading or teaching all chapters in their order of presentation in the book.

A set of exercises is included at the end of most chapters to provide opportunities to further explore or apply ideas, information, and techniques presented in the chapter. Some exercises are well suited for modest to major team projects. Because effective teamwork is an important aspect of modern management, faculty members and others who use the book for teaching courses and for leading seminars and workshops are urged to assign some exercises to be done as team projects. By so doing, college students or seminar or workshop attendees will benefit in two ways. They will learn more about the subject matter, and they will learn more about being an effective member and occasional leader of a team.

ACKNOWLEDGEMENTS

Most of the material presented in this book was developed over a six-year period beginning in 1988 when I began teaching Engineering Management, a senior course in the Department of Civil Engineering at Valparaiso University. The material was further refined starting in 1990 when I initiated a two-day seminar titled "Management for the New Engineer" as part of the American Society of Civil Engineers' Continuing Education Program. This course was later offered in video-tape format to improve its accessibility by the entry-level technical professional community. Opportunities to conduct special in-house seminars and workshops and to provide management and leadership consulting for engineering organizations naturally grew out of the preceding teaching efforts and resulted in further expansion and refinement of what now is the content of this book. I clearly recognize and sincerely appreciate the many contributions made by former engineering students, seminar and workshop participants, and clients.

Other materials and ideas, reflecting primarily management and leadership applications, were obtained and developed over almost three decades while I was employed in the public and private sectors in engineering practice and education. Besides practicing engineering and being an educator, I administered and was administered to, managed and was managed, and led and was led. During that time, I witnessed some very enlightened and some very poor management of individuals, projects, and organizations. These were excellent learning experiences and constitute the personal experience base of this book.

I received a wealth of useful ideas and information from, and have been positively influenced by, numerous individuals. My debt to other professionals is suggested, in part, by the extensive list of references that appear at the end of each chapter. I drew ideas and information, and, therefore, reference materials from a wide range of sources. The resulting eclectic collection of cited and supplemental references indicates that technical professionals can learn much about all aspects of management and leadership by looking both within and outside of their fields.

Book writing labor and logistics are challenging and a major management effort. I gratefully acknowledge the crucial role of Vicki F. Farabaugh, formerly my administrative assistant in the College of Engineering at Valparaiso University, who supervised essentially all of and did much of the word processing, produced the graphics, and obtained reference citation information. I also appreciate the meticulous proofing of punctuation, grammar, and spelling performed by Camille Gudino, secretary in the college. Much of the writing of this book was completed while Jerrie, my wife, and I traveled and worked for six months on our vessel Sabbatical while on sabbatical from Valparaiso University. The university's support is appreciated. Finally, Jerrie provided source materials, constructively critiqued the entire text, and, as always, provided total support.

Stuart G. Walesh

1

Introduction

> *To make knowledge work productive will be the great management task of this century, just as to make manual work productive was the great management task of the last century.*
>
> (Drucker, 1978, p. 290)

The environment within which engineers and other technical professionals usually function is suggested by Figure 1–1. As shown, the diagram is most likely to apply to civil engineers, architects and other technical consultants. At any time in his or her career, a technical professional might be at any one of the three vertices of the triangle. During their career, technical professionals might move from employer to employer and eventually be employed within all three sectors represented by the triangle.

As suggested by Figure 1–1, the interaction process typically begins with the owner or client's retaining a consulting engineer, architect, or other technical professional to conduct a study, perform preliminary designs, or prepare a complete design and deliver a contract package consisting of plans and specifications to the owner. The owner then selects a contractor to construct the building, facility, or system. The consultant is sometimes retained by the owner to monitor the construction process so that the final structure, facility, or system is built in accordance with the original plans and specifications.

For a self-contained manufacturing organization, the bottom two vertices of the triangle shown in Figure 1–1 collapse into one point. That is, design and manufacturing or design and fabrication occur within a single organization. The owner or client shown at the upper vertex of the triangle is, in a manufacturing or fabrication setting, replaced by the customer or buyer.

The triangular model presented in Figure 1–1, and its manufacturing organi-

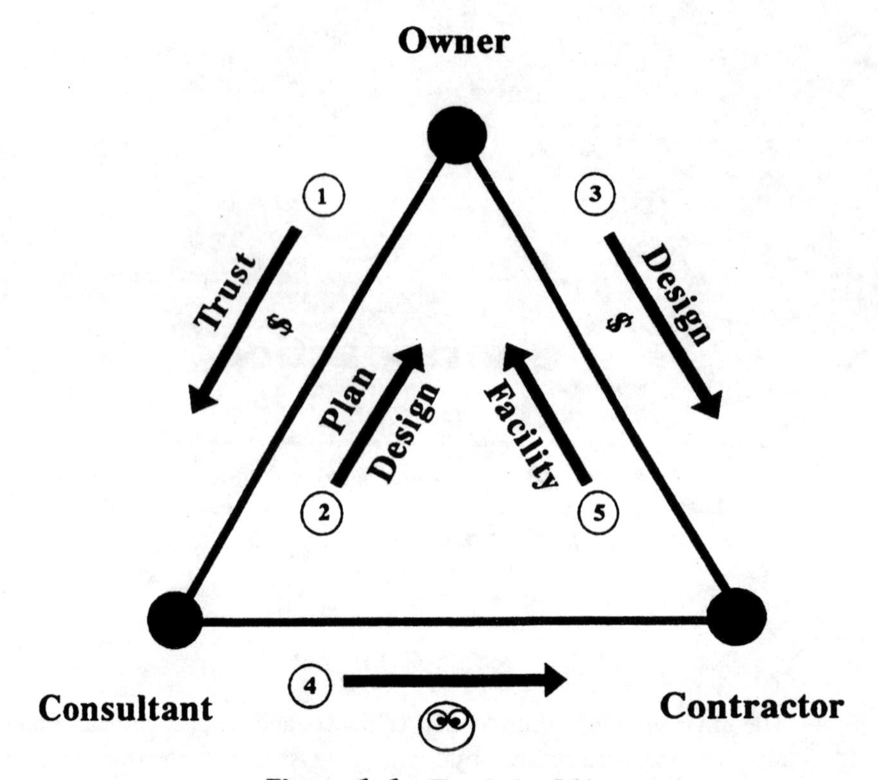

Figure 1–1 The playing field

zation variation, are frequently referred to in this book. In one sense, management is the process by which the various entities shown in Figure 1–1 interact with each other in the world of engineering and business.

DEFINITIONS OF ENGINEERING

The term *engineering/technology management* needs a definition. Accordingly, some definitions of engineering and of management are presented, similarities are noted, and then engineering/technology management is defined. The Engineers Council for Professional Development, now ABET, reflected the thinking of Theodore Von Karman, the aeronautical engineer, by saying "Scientists explore what is, engineers create what never has been" (ECPD, 1974). This succinct statement clearly shows how science and engineering differ; that is, creativity is essential in the latter. Later, ABET (1992) offered this definition of engineering which focuses on a science base, economic considerations, and the goal of benefiting society:

> *Engineering is the profession in which the knowledge of the mathematical and natural sciences gained by study, experience, and practice is applied with judgment to develop ways to utilize, economically, the materials and forces of nature for the benefit of mankind.*

The creative and humanist dimensions of engineering were captured by President Herbert Hoover (Fredrich, 1989, p. 546), who had a long and distinguished career as an engineer. He said:

> *It is a great profession. There is the fascination of watching a figment of the imagination emerge through the aid of science to a plan on paper. Then it brings jobs and homes to men. Then it elevates the standards of living and adds to the comforts of life. That is the engineer's high privilege.*

Beakley, Evans, and Keats (1986, p. 3) emphasized engineering's science base, creative dimension, timelessness, and human focus by saying:

> *The role of the engineer has not changed through the centuries. The primary task has always been to make practical use of converting scientific theory into useful application. In so doing, engineers help to provide for mankind's material needs and well-being. From era to era, only the objectives pursued, the techniques of solutions used, and the tools available for analysis have changed.*

Finally, Hardy Cross (1952, p. 141), using direct, plain words, clearly captures the central, people-serving goal of engineering when he writes:

> *It is not very important whether engineering is called a craft, a profession, or an art; under any name this study of man's needs and of God's gifts that they may be brought together is broad enough for a lifetime.*

Based in part on the preceding definitions, the following seven essential features of engineering are apparent. Engineering is

- Science-based.
- Systematic—However, except for trivial problems, judgment and other qualitative considerations always enter in.
- Synthesizing and creative.
- Goal-oriented—Get the job done on time and within budget.
- Dynamic—Technology changes, laws change, public values change, clients change, and the physical environment changes.

- People-oriented—Both in doing and in results, in that engineering is essential to the survival of human communities and to the quality of life.

DEFINITIONS OF MANAGEMENT

Sometimes, unfortunately, management is taken to mean a group of people at the top of an organization who direct the work. That group typically stands in strong contrast with a much larger group at the bottom of the organization who do the work. This is the "we versus they" concept of management with its emphasis on individuals or groups. This book does not subscribe to a polarizing model of management, but instead uses the word *management* to focus on the processes that should be at work.

According to Evans (1978), management is goal-oriented and "is the process of planning, organizing, directing, and controlling the task implemented to arrive at the objective." In a similar fashion, Allen (1969) defines management in terms of planning, organizing, leading, and controlling, and goes on to elaborate on each of these four activities.

A view of management that emphasizes synthesizing components is presented by Hicks (1966). He views management in terms of five components, namely, men (people), money, methods, machines, and materials. In a similar fashion, Coe (1988) views management as integration of various components. He defines management as "directing with skill" and goes on to say, "Managers read, write, think, confer with, and decide upon personnel, money, materials, methods and facilities in order to plan, organize, direct, coordinate and control their research, development, and production."

Following a more humanist approach, Madsen (1973, p. 5) writes that management is "getting things done through people" and "the efficient use and conservation of human and material resources in reaching predetermined objectives." Wortman (1981) suggests that management is achieving desired results through people.

Based in part on the preceding definitions of management, the five essential aspects of management are apparent. Management is

- Systematic
- Synthesizing and creative
- Goal-oriented
- Dynamic
- People-oriented in doing and people-serving in results

SIMILARITY OF ENGINEERING AND MANAGEMENT

The preceding definitions of engineering and management show the close similarities between the essence of engineering and essence of management. The engineering graduate or the senior engineering student is, in some ways, already a manager. He or she should be able to readily understand and identify with the key ideas of management. Similarly, students who are in or have graduated from other rigorous technically oriented programs are also predisposed by virtue of their education to be managers. Baird (1987) asserts, "Good managers are made, not born." He goes on to argue that engineering graduates are well positioned because their education helps them develop several characteristics that are essential in good management. These are, according to Baird, a problem-solving orientation; a scientific approach that is "unbiased, methodical and rational"; and skepticism that gets beyond the "how" of things, processes, and organizations and gets into the "why." In summary, students in or graduates of engineering and other demanding technically oriented programs are, by virtue of their education, very well positioned to become effective managers of engineering and other enterprises.

DEFINITION OF ENGINEERING/TECHNOLOGY MANAGEMENT

Knowing is not enough; we must apply. Willing is not enough; we must do.

(Goethe)

Engineering/technology management might be defined as the process used in the technically based arena for deciding *what* is going to be done, *how* it is going to be done, *who* is going to do it, *when* they are going to do it—and *how* are we doing? The essence of engineering/technology management might be most succinctly stated by defining it as making things happen through people and for people in the world of applied technology. What and who you know are secondary. What counts is how you utilize what and who you know to make good things happen.

STEPS OF ENGINEERING/TECHNOLOGY MANAGEMENT

The young professional's process of understanding and mastering management as defined in this chapter—making things happen through people and for people in the world of applied technology—may be viewed as moving up a staircase of proficiencies. The necessary steps are shown in Figure 1–2. As suggested by Figure 1–2, you must first learn how to manage yourself; then you can begin to manage relationships with others and move up the staircase to management of projects and eventually organizations, your profession, and perhaps a major segment of society. Chapters in this book generally correspond to and generally fall in the order of the steps shown in Figure 1–2.

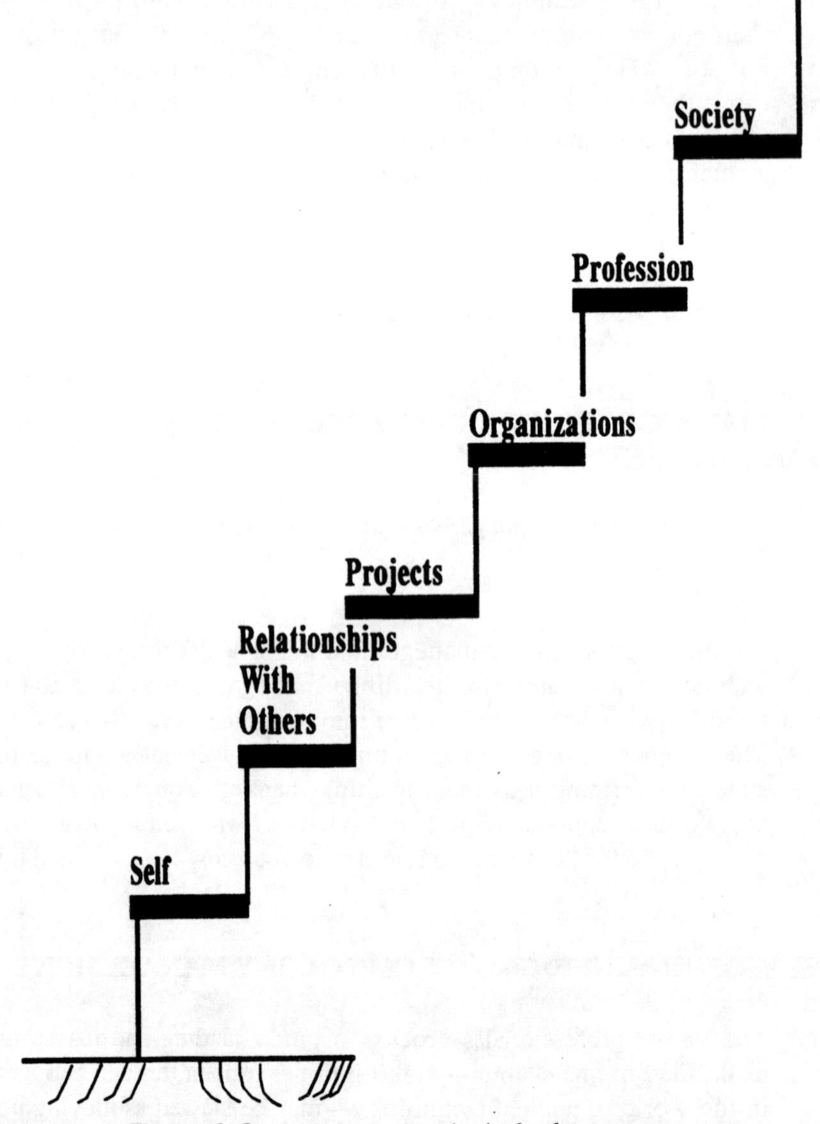

Figure 1–2 Steps for engineering/technology management

This is not a transition from engineering or technology to management book. This book assumes that all engineers and other technical professionals should be managers from day one—at least managers of their time, their assignments, their relationships with others, and their careers. The best leaders in the technical organizations of tomorrow are most likely to be those professionals who began to develop management skills very early in their careers.

LEADERSHIP, MANAGEMENT, AND PRODUCTION: DECIDING, DIRECTING, AND DOING

Leaders are people who do the right thing; managers are people who do things right. Both roles are crucial, but they differ profoundly.

(Bennis, 1989, p. 18)

Leadership, Management, and Production Defined

One paradigm for an organization, such as an engineering consulting firm, a manufacturing business, or a government agency, is that wholeness, vitality, and resiliency require attention to three different but inextricably related functions, namely leadership, management, and production. The meaning of each of these terms is suggested by the comparisons presented in Table 1–1. In a very simplified sense, the leading, managing, and producing functions can also be represented by the three D's—deciding, directing, and doing.

Figure 1–3 uses the metaphor of a three-legged stool to suggest how attention to the leadership, management, and production functions produces a stable organization—one that cannot easily be "knocked over." While an organization might temporarily survive, and perhaps even thrive, on two of the three legs, all three legs are needed for long-term survival. For example, a leaderless consulting engineering firm might do well for several years by balancing on two legs such as excellent management and production capabilities, but eventually be toppled because it lacked the third leg. That leg is leadership, especially the ability to see and act on changes in client needs and the means to serve those needs.

Even three legs are not always enough—each must carry its weight. For example, while leadership may be present, it may be weak. Bennis (1989, p. 17) says, "Many an institution is very well-managed and very poorly led. It may excel in the ability to handle each day all the routine inputs, yet may never ask whether the routine should be done at all."

TABLE 1–1 LEADERSHIP, MANAGEMENT, AND PRODUCTION DEFINED

Leadership	Management	Production
"Wrong jungle!" (Covey, 1990, p. 101)	". . .sharpening . . . machetes . . . writing policy and procedure manuals . . . setting up working schedules . . ." (Covey, 1990, p. 101)	". . . cutting through the undergrowth, clearing it out . . ." (Covey, 1990, p. 101)
"What are the things I (we) want to accomplish?" (Covey, 1990, p. 101)	"How can I (we) best accomplish certain things?" (Covey, 1990, p. 101)	—
"Leaders are people who do the right thing . . ." (Bennis, 1989, p. 18)	". . . managers are people who do things right." (Bennis, 1989, p. 18)	—
". . . leadership determines whether the ladder is leaning against the right wall." (Covey, 1990, p. 101)	"Management is efficiency in climbing the ladder . . ." (Covey, 1990, p. 101)	—
"Leadership is moving forward to create something new." (Kanter, 1993)	"Management is taking care of what's already been created." (Kanter, 1993)	—
". . . lead from the right (brain) . . ." (Covey, 1990, p. 147)	"Manage from the left (brain) . . ." (Covey, 1990, p. 147)	—
—	". . . management follows leadership." (Covey, 1990, p. 101)	—
"Concentrate on what is right rather than who is right." (Peyton, 1991, p. 73)	—	—
". . . having an opportunity to make a difference in the lives of those who permit leaders to lead" (Depree, 1989, p. 22)	—	—
Deciding what ought to be done.	*Directing* how things will be done, who will do them, and when.	*Doing* what we know has to be done or what we are told to do.

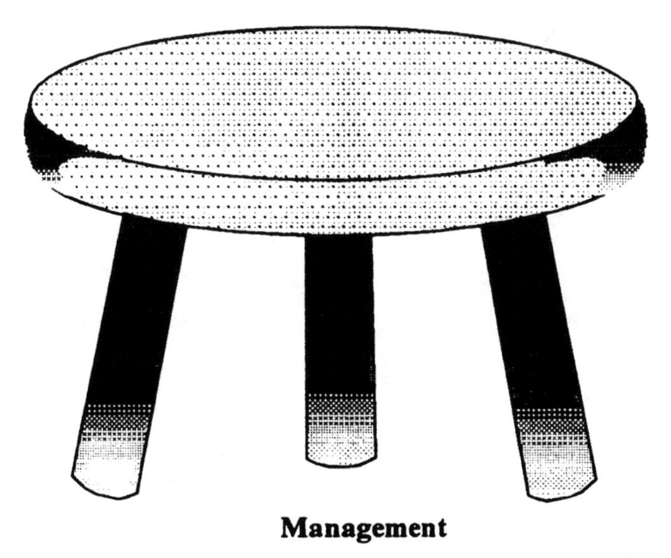

Figure 1-3 Essential functions of a vital organization

Traditional Pyramidal, Segregated Organizational Model

Assuming that you agree that each organization has leading (deciding), managing (directing), and producing (doing) responsibilities, consider the manner in which these corporate responsibilities might be met. More specifically, consider the matter of individual responsibility in achieving the three corporate responsibilities.

In what might be called the traditional pyramidal and segregated organizational model as illustrated in Figure 1–4, the three functions reside in three separate groups of personnel. The vast majority of employees are the doers or producers; a distinctly different and much smaller group of managers does the directing, and one person, or perhaps a very small group, does the leading.

Another way of viewing the production, management, and leadership functions in the traditional pyramidal mode is the serial or linear model illustrated in Figure 1–5. An aspiring and successful individual begins in a production mode and then passes serially or linearly through management and into leadership. Rather than being a trait that many can possess, albeit to different degrees, leadership is considered the end of the line or ultimate destination for a very few. But is this the optimum way for the modern or future organization to meet its leading, managing, and producing responsibilities? Probably not.

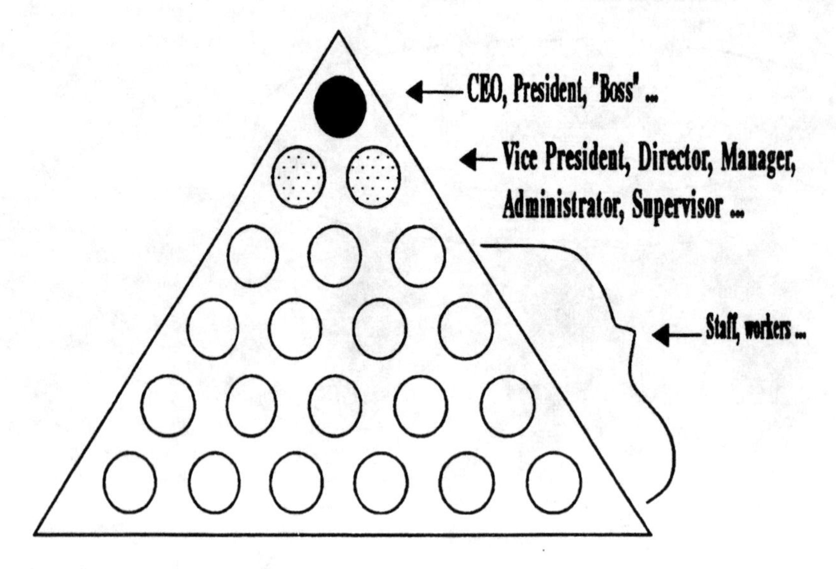

Key to Functions

- **Leading (Deciding)**
- **Managing (Directing)**
- **Producing (Doing)**

Figure 1–4 Traditional pyramidal and segregated organizational model

Figure 1–5 Traditional serial view of the production, management, and leadership functions

Shared Responsibility Organizational Model

But of a good leader,
who talks little,
When his work is done,
his aim fulfilled,
They will say, `We did
this ourselves'

(Lao Tzu)

An organization will be stronger if what used to be the three organizational responsibilities now also become individual responsibilities. The goal should be to enable each member of the organization to be a decider, a director, and a doer as illustrated by Figure 1–6.

While the relative "amounts" of leading, managing, and producing will vary markedly among individuals in the organization, everyone should be expected and enabled to do all three in accordance with his or her individual characteristics. This shared-responsibility organizational model, in contrast with the traditional segregated model, is much more likely to tap, draw on, and benefit from the diverse aspirations, talents, and skills that should be present within the organization. Because essentially all members are fully involved, the shared-responsibility organization is in a much better position to synergistically build on internal strengths, to cooperatively diminish internal weaknesses, and to learn about and be prepared to respond to external threats and opportunities.

Peyton (1991, p. 244) introduces the expression "cooperative individualism," which provides another means of describing the essence of the shared-responsibility organization model with emphasis on liberating and integrating the strengths of individuals. In cultures where individual freedom and initiative are

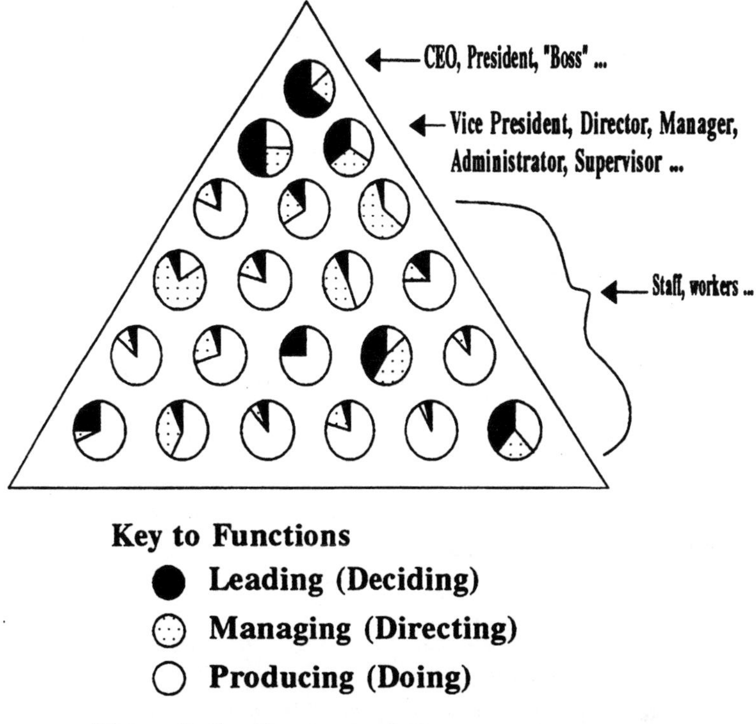

Figure 1–6 Shared-responsibility organizational model

valued, such as in the United States and many other countries around the globe, Peyton claims that the challenge is "to achieve organizational goals while allowing and encouraging each employee's individual freedom, initiative, creativity, productivity and responsibility." Many writers (e.g., DePree, 1989) emphasize the importance of identifying, freeing up, and focusing on the diverse individual aspirations, talents, and skills present in most organizations.

Focus of This Book: Management and Leadership

The entry-level engineer or other technical professional will, by definition, be well prepared for and will spend a vast majority of his or her time producing, that is, carrying out the production function of the organization. An undergraduate technically oriented education is typically a solid preparation for this function. The focus of this book is on the second and third of the three functions, that is, on managing and leading. Your education may not have prepared you well for these functions. Therefore, you need to assume major responsibility for developing your management proficiency and leadership understanding.

Incidentally, some engineering educators and practitioners claim that undergraduate engineering education actually drives youthful potential leaders away from the leadership function. The argument (e.g., IIT, 1991, pp. 2–5) is that undergraduate engineering educational programs increasingly emphasize, as students move through the program towards their bachelor's degrees, detail, deductive reasoning, subsystem analysis, determinate systems, and quantifiable uncertainty. This track, although it would not have to, generally ignores complementary leadership skills and characteristics, such as being cognizant of the big picture, integrative thinking, interpersonal relations, and dealing with uncertainty, diversity, chaos, and crises.

Professionals who are well beyond entry-level positions have typically acquired and demonstrated management skills. If so, they should increasingly become students of the leadership function. Additional ideas and information on leadership are presented in Chapter 15. Bennis (1989), Covey (1990), and Phillips (1992) offer useful introductions to this subject.

ENGINEER AS BUILDER

We recognize that we cannot survive on meditation, poems and sunsets. We are restless. We have an irresistible urge to dip our hands into the stuff of the earth and do something with it.

(Florman, 1976, p. 104)

Engineering is an old profession—some say it is the second oldest. Its roots can be traced back to the beginning of recorded history, when nomads first came together and formed communities along what are now the Nile river in Egypt, the Tigris and Euphrates rivers in Iraq, the Indus river in India, and the Yellow river in China. With the creation of communities came the need to provide basic infrastructure such as housing, transportation, defense, irrigation, water supply, and wastewater disposal. The work of the engineer had begun.

Besides being one of the oldest professions, engineering is one of the broadest. For example, ABET, the interengineering organization that accredits undergraduate engineering programs, recognizes 26 groups of engineering programs for purposes of undergraduate engineering accreditation (ABET, 1992). The groups are Aerospace, Agricultural, Architectural, Bio-Engineering, Ceramic, Chemical, Civil, Computer, Construction, Electrical and Electronic Engineering, Geological and Geophysical, Industrial, Manufacturing, Materials, Mechanical, Metallurgical, Mineral, Mining, Naval Architectural and Marine, Nuclear, Ocean, Petroleum, Plastics, Surveying, Systems, and Welding. Within any of the engineering groups a broad spectrum of functions is carried out by engineers, such as research and development; planning; design; construction, manufacturing, and fabrication; operations; marketing; and management.

Within and throughout the engineer's long history and great diversity, however, there is at least one widely shared interest and function: building. In the final analysis, whenever everything else is stripped away, the engineer is at the core a builder. When civil engineers "build," they usually call the process *construction.* When mechanical engineers "build," they routinely refer to it as *manufacturing.* From the perspective of electrical engineers, "building" is often referred to as *fabrication.* Call the process whatever you want, the ultimate end of the engineering process is to "build" something that never before existed that will meet human needs. Examples include the water supply system "built" by the civil engineer, the energy-efficient and safe automobile "built" by the mechanical engineer, and the electrical power distribution system "built" by the electrical engineer. Some engineers "build" less concrete but nevertheless important things such as computer programs, better ways to perform engineering functions, and improved ways to organize engineering organizations. Refer to *The Builders* (National Geographic Society, 1992) for a graphic interpretation of the historic and global role of the engineer as a builder.

Great responsibility and liability as well as great satisfaction accrue to "builders." President Herbert Hoover (Fredrich, 1989, p. 546) said this:

> *The great liability of the engineer compared to men of other professions is that his works are out in the open where all can see them. His acts, step-by-step, are in hard substance. He cannot bury his mistakes in the grave like the doctors. He can-*

not argue them into thin air and blame the judge like the lawyers. He cannot, like the architects, cover his failures with trees and vines. He cannot, like the politicians, screen his short-comings by blaming his opponents and hope the people will forget. The engineer simply cannot deny he did it. If his works do not work, he is damned . . . On the other hand, unlike the doctor, his is not a life among the weak. Unlike the soldier, destruction is not his purpose. Unlike the lawyer, quarrels are not his daily bread. To the engineer falls the job of clothing the bare bones of science with life, comfort, and hope.

As you begin your career, you will probably be increasingly cognizant of the diversity of engineers and the work they do. There is strength in diversity when that diversity is focused on a common and a meaningful interest. For engineering and other similar professions, that common bond is building for the benefit of society.

COMMON SENSE AND COMMON PRACTICE

Some of the content of this book might be correctly referred to as "common sense." An erroneous implication of this statement is that the material is obvious and, therefore, does not warrant study or explicit disciplined application. However, that which is common sense does not necessarily translate into common practice. Knowing something and using it are not the same. The entry-level engineer who is committed to high levels of achievement will take charge of his or her life. The young professional will understand the need to translate common sense into common practice through study and self-discipline. When common-sense ideas and approaches become common practice that is normal or habitual, the young professional will be well on the way to realizing his or her potential in the consulting business, industry, or government. Or, as stated by Mandino (1968, p. 54):

In truth, the only difference between those who have failed and those who have succeeded lies in the difference of their habits. Good habits are the key to all success . . . I will form good habits and become their slave.

REFERENCES

Accreditation Board for Engineering and Technology, *1992 ABET Accreditation Yearbook*, 1992.

ALLEN, L. A., *The Principles of Professional Management.* Palo Alto, Calif.: Louis A. Allen Associates, Inc., 1969.

BAIRD, B. F., "Moving Into Management," *MS/PhD*, 1987 ed., pp. 18–24.

BEAKLEY, G. C., D. L. EVANS, and J. B. KEATS. *Engineering: An Introduction to a Creative Profession*, 5th ed., New York: Macmillan, 1986.

BENNIS, W. G., *Why Leaders Can't Lead—The Unconscious Conspiracy Continues.* San Francisco, Calif.: Jossey-Bass Publishers, 1989.

COE, J. J., "Engineers as Managers and Some Do's and Don't's," *Journal of Management in Engineering- ASCE*, Vol. 3, No. 4 (October 1987), pp. 281–287.

COVEY, S. R., *The 7 Habits of Highly Effective People.* New York: Simon & Schuster, 1990.

DEPREE, M., *Leadership Is An Art.* New York: Dell Publishing Company, 1989.

CROSS, H. (R. C. Goodpasture, ed.), *Engineers and Ivory Towers.* New York: McGraw-Hill, 1952.

DRUCKER, P. F., *The Age of Discontinuity.* New York: Harper & Row, 1978.

Engineers Council for Professional Development, "Make Your Career Choice Engineering," 1974.

EVANS, H. G., "Management Careers for Engineers," *Engineering Issues—Journal of Professional Activities—ASCE*, Vol. 102, No. EI4 (October 1976), pp. 409–415.

FLORMAN, S. C., *The Existential Pleasures of Engineering.* New York: St. Martin's Press, 1976.

FREDRICH, A. J., ed., *Sons of Martha: Civil Engineering Readings in Modern Literature.* New York: American Society of Civil Engineers, 1989.

HICKS, T. G., *Successful Engineering Management.* New York: McGraw-Hill, 1966.

Illinois Institute of Technology, Association of Independent Technological Universities, and Historical Science Foundation, "Developing Leadership Through Engineering Education." In *Proceedings of the Conference New Challenges in Educating Engineers.* Chicago, Ill.: IIT, June 10–11, 1991, pp. 2–5.

KANTER, R. M., personal communication, February 22, 1993.

MADSEN, L. K., "Management Defined, Characteristics, and Human Behavior," *Effective Project Management Techniques,* ASCE, New York, 1973.

MANDINO, O., *The Greatest Salesman in the World.* New York: Bantam Books, 1968.

National Geographic Society, *The Builders: Marvels of Engineering.* Washington, D.C., 1992.

PEYTON, J. D., *The Leadership Way—Management for the Nineties.* Valparaiso, Ind.: Davidson Manors, Inc, 1991.

PHILLIPS, D. T., *Lincoln on Leadership—Executive Strategies for Tough Times.* New York: Warner Books, 1992.

TZU, LAO, *The Way of Life, According to Lao Tzu.* Translated by W. Bynner.

WORTMAN, L. A., *Effective Management for Engineers and Scientists.* New York: Wiley, 1981.

SUPPLEMENTAL REFERENCES

CROSBY, P. B., *Running Things—The Art of Making Things Happen.* New York: New American Library, 1986.

EDELHART, M., *Getting From Twenty to Thirty.* New York: M. Evans and Company, 1983.

KANTER, R. M., *The Change Masters.* New York: Simon & Schuster, 1983.

National Society of Professional Engineers, S. P. McCarthy, ed., *Engineer Your Way to Success.* Alexandria, Va., 1989.

EXERCISES

1.1 GOALS

Purpose

Motivate the young technical professional to think about his or her professional and other goals and how he or she plans to achieve them.

Tasks

1. Indicate your goals in each of the following areas for two and ten years from now.
   a. Annual salary and other income:
      i. Two years: ______________
      ii. Ten years: ______________
   b. Position (e.g., project engineer, project manager, owner):
      i. Two years: ______________
      ii. Ten years: ______________
   c. Function (e.g., design, marketing, field, general management):
      i. Two years: ______________
      ii. Ten years: ______________
   d. Other (e.g., international travel; present a paper; serve as officer in professional community or other organization; start own business; hold elective office; earn a Ph.D.):
      i. Two years: ______________
      ii. Ten years: ______________
2. List four things you will do within the next year to move you toward *one* of the above goals:
   a. ______________________________
   b. ______________________________

c. ______________________________

d. ______________________________

3. *Note:* You are in effect "planning a trip." How are you going to get to your destination? Do you have the necessary knowledge, skills, and experience or a way of obtaining them? Or are you going to let chance rule, perhaps using the rationale that everything will come to you if you "just work hard"?

1.2 ASSEMBLE PROJECT INFORMATION

Purpose

Assemble basic information on an engineering or engineering-related project, unique to each student, for subsequent use in one or more assignments.

Tasks

1. Select a technical "project" you worked on or are working on. Examples are as follows: a project you did during co-op; something you did during a summer job; your senior project; or a design project in one of your other courses. The project must have at least 15 different activities or tasks.
2. Prepare a brief, typed memorandum that includes:
   a. A paragraph describing your relationship to the project. That is, when, how, and why are or were you involved?
   b. A list of project activities, in approximate chronological order. Some activities may overlap, in this format:

Activity identification	Activity name or brief description
A	conduct field survey
B	plot field data
C	calculate loads
etc.	etc.

1.3 BOOK REVIEW

Purpose

1. Provide each student with an opportunity to study one professional/business author of his/her choice in depth and to critique the thesis of the book and the support for that thesis.
2. Introduce the student to the broad range of management books with the hope that the student will continue to read critically in this area.

Tasks

1. Select one business/professional book that treats some aspect of management. The book must be recent (published 1980 or later). Sources of books include:
   a. Books cited in references used in this course.
   b. Books reviewed in book sections of newspapers (e.g., see most major Sunday newspapers).
   c. Books cataloged under "Management" and similar terms in the university library, other libraries, and bookstores.
   d. Books recommended by others.
2. As soon as you tentatively select your book, advise the instructor via a memorandum and obtain formal (written) approval.
3. Read the book.
4. Prepare a report (5–10 double-spaced pages) in which you do the following:
   a. Cite your book (name, author, publisher, date, etc.).
   b. Describe the key ideas or theses presented in the book.
   c. Identify the author's evidence in support of the ideas/theses.
   d. Indicate whether or not you agree with the key ideas/theses.
   e. Is the author's support for the ideas/theses credible?
5. Use headings and subheadings in your report to aid the reader.
6. This assignment will be weighted as five times a normal assignment.

1.4 RESEARCH PAPER—INDIVIDUAL STUDENT VERSION

Purpose

1. Provide each student with an opportunity to study, in depth, an engineering/technology management topic of his or her choice.
2. Increase the student's awareness of the engineering/technology management literature.
3. Improve the student's ability to communicate in writing to a designated audience.
4. Enable all students who wish to do so to benefit from the results of the individual research efforts of others.

Tasks

1. Select one of the engineering/technology management topics listed below. Each topic may be selected by only one student, and selection will be on a first-come, first-served basis. Other topics of interest to a particular student may be selected if discussed with and approved by the instructor.

Brainstorming
Business organizational structures
Civil, electrical, mechanical . . . engineering in the twenty-first century
Challenges faced by the entry-level technical professional
Client–practitioner relationship
Communication
Cost control in engineering/construction
Creativity in planning/design
Dual ladder
Efficiency vs effectiveness
Engineering and/or other registration laws
Ethics
Expert systems
Failures and learning from them
History of some aspect of engineering/technology
Internationalization of engineering/technology
Leadership
Legal considerations
Liability
Marketing engineering or other technical services
Matrix organizations
Project management
Quality control in design/construction
Robotics in civil engineering
Spreadsheets in management
Teamwork
Time management
Total quality management

2. As soon as you select your topic, advise the instructor via a memorandum and obtain formal (written) approval.
3. Research the literature on the subject. Be sure to have some recent sources. (Read, use, and formally cite at least six sources in sufficient depth to describe the subject.)
4. You are encouraged, but not required, to use one or more personal contacts that you contact in person, by telephone, or by letter. If you use a personal contact, each of which can replace one of the required literature sources, cite them at the end of your paper using this format:

   Smith, J. A., Manager, XYZ Company, Chicago, IL, personal communication, January 28, 1988.
5. Write a paper on the subject. It must be typed, double-spaced, ten-page minimum, and written in the third person. References should be cited in text and listed at the end in alphabetical order by author, using format like that of reference lists distributed in class. Include tables and illustrations as needed.
6. Use headings in the paper such as Executive Summary (required, one-page maximum), Introduction, Conclusions, etc.
7. Assume that your reader is a senior engineering or other technical profession major who has not studied the topic in depth and knows little, if anything, about it.

8. Follow the report-writing tips presented in Chapter 3.
9. This assignment will be weighted ten times a normal assignment.

1.5 WIDESPREAD APPLICATION OF MANAGEMENT KNOWLEDGE

Purpose

Suggest to the student that an understanding of the management concepts, principles, methods, and so on presented in this book can help in understanding developments in the broader world of business.

Tasks

1. Select one issue (that is one day) of a major newspaper and clip at least four articles that relate to management topics presented in this book.
2. Tape the articles to or copy them onto one or more 8 1/2 × 11-in. sheets and number each article.
3. Prepare a memorandum (one page or less) in which you state and briefly discuss the management-related aspect of each article (using the numbers to identify the article).

1.6 RESEARCH PAPER—TEAM PROJECT VERSION

Purpose

1. Similar to purposes 1 through 4 in Exercise 1.4.
2. Simulate report preparation teamwork as typically done in offices of technically based organizations.

Tasks

1. Same as Task 1 in Exercise 1.4, except replace "one student" with "one project team."
2. Same as Task 2 in Exercise 1.4.
3. Same as Task 3 in Exercise 1.4, except replace "six sources" with "20 sources."
4. Same as Task 4 in Exercise 1.4.
5. Same as Task 5 in Exercise 1.4, except replace "ten-page" with "40-page."
6. Same as Task 6 in Exercise 1.4.
7. Same as Task 7 in Exercise 1.4.
8. Same as Task 8 in Exercise 1.4.
9. Same as Task 9 in Exercise 1.4.

1.7 RESEARCH PAPER TEAM PLAN

Purpose

Encourage proper planning of a team project.

Tasks

1. Assume that your team is doing Exercise 1.6.
2. Develop a team plan, that is, a plan that describes how your team will carry out Exercise 1.6. The team plan should be organized to address the points raised in the "what-how-who-when" definition of engineering/technology management presented in this chapter.
3. Submit a memorandum to the instructor that explicitly addresses the "what," "how," "who," and "when" of how your team will do Exercise 1.6.

1.8 RESEARCH PAPER—TEAM PROJECT VERSION

Purpose

1. Enhance the management ability of students by having them work together over an extended period of time on a major research and writing project.
2. Heighten student awareness of international differences in the approaches to engineering education and practice as well as variations in the status of engineers and the engineering profession, recognizing that many of today's engineering students will work in a global engineering and business setting.

Tasks

1. Select a country.
2. Research and otherwise gather information on the following aspects of engineering in the selected country.
   a. Overview of country (e.g., location, size, population, topography, economy, etc.).
   b. Engineering education.
   c. Engineering licensing.
   d. Engineering professional organizations.
   e. Area of science and technology for which the country is considered a leader.
   f. Image and/or social status of engineers.
   g. Any other aspect of engineering that interests one or more members of the team.
   h. If an engineer were to travel to this country for the first time to do professional work, what special considerations or approaches would be prudent?
3. Develop a team plan, that is, a plan that describes how the team will be structured and how it will function. Indicate the name of the team leader and describe what is

to be done when and by whom. The team plan should be organized to address the points raised in the "what-how-who-when" definition of engineering management presented in this chapter.

4. Submit a memorandum to the instructor that explicitly addresses how the team will do this exercise. This memorandum will be graded as one normal assignment for each team member.
5. Research the literature and other sources on the subject. Be sure to have some recent sources. Read, use, and formally cite at least ten sources in sufficient depth to describe the subject. Caution: Start searching early, because engineering-related sources are difficult to find.
6. You are *encouraged*, but not required, to use one or more personal contacts with whom you communicate in person, by telephone, or by letter. If you use a personal contact, each of which can replace one of the required literature sources, cite them at the end of your paper, using this format:

   Smith, J. A., Manager, XYZ Company, Paris, France, personal communication, September 19, 1992.

   Personal sources might include faculty or students with knowledge of other countries, faculty or students visiting from other countries, local or other engineers/managers who are natives of other countries or who have had international employment assignments, personnel at consulates of other countries, employees of travel agencies, and the like.
7. Write a paper on the subject. It must be typed, double-spaced, 30-page minimum, and written in the third person. References should be cited by author (not number) in text and listed at the end in alphabetical order by author, using a format like that of reference lists in this book. Include tables and illustrations as needed. Write the report in a consistent, formal style. Although the information for the report will be contributed by several individuals, it should "read" as though it were written by one person.
8. Use headings in the paper such as Executive Summary (required one-page maximum), Introduction, Overview of Country, Engineering Education, Engineering Licensing . . . Summary and/or Conclusions, Acknowledgements, Cited References, and so on.
9. Provide a title page followed by a table of contents that includes page numbers.
10. Assume that your reader is a senior engineering major who has not studied the topic in depth and knows little, if anything, about it.
11. Consider following the writing tips presented in Chapter 3 of this book.

The results of Tasks 5 through 11 will be weighted eight times a normal assignment for each person.

2

Management of Self

> *If it is to be it is up to me.*
>
> (Anonymous)

Before you, as an entry-level engineer or other technical professional, can effectively manage relationships with people, manage projects, and even think about managing organizations, you must learn to manage yourself. Most professional program students come a long way in self-management during their four or more years in college. The pattern of a whole new set of challenges that you faced on entering college is now repeated in that a different set of major challenges arises with entry into the world of professional work.

This chapter addresses many aspects of self-management. It begins by discussing the culture shock that the young technical professional is likely to encounter during the transition from education to practice. The employment versus full-time graduate school quandary is presented. Numerous time-management tools and techniques are discussed, followed by suggestions on how to get off to a good start in your first professional position. The chapter concludes with discussions of the importance of managing personal professional assets, participating in continuing education, being involved in professional organizations, and obtaining an engineering or other license.

THE NEW WORK ENVIRONMENT: CULTURE SHOCK?

As suggested by Figure 2–1, demands and expectations will change as you move from the world of study to the world of practice. Suddenly, the usual demands to get tasks and projects done correctly and on time are further complicated by the

- **No "partial credit"**
- **Little tolerance for tardiness**
- **Assignments are not graded**
- **Schedules are more complicated**
- **Higher grooming and dress expectations**
- **Teamwork is "SOP"**

Figure 2-1 The world of professional practice vs the university

expectations that the tasks and projects will be done within a budget. In addition to experiencing new demands and expectations, the intensity of those demands and expectations may change. You may experience new emphases, such as concern with budgets and production, but a lower level of intellectual challenge than that experienced at the university.

No Partial Credit

Although partial credit is routinely granted on examinations and assignments at the university, as it should be to encourage the learning process, the partial credit paradigm is much less applicable in engineering practice. Each of the structural, mechanical, aesthetic, operation and maintenance, and cost features of an engineered product, structure, facility, or system must meet the client's or customer's requirements if a project is to be successful. Neither the design professional nor his or her organization will receive partial credit if the product is lacking in any of the important attributes. The college strategy of banking quality points in easy courses to offset the poor performance in difficult courses simply won't work in engineering practice.

Doing especially well on one feature of a project, as admirable as that may be, is not likely to offset the failure to meet requirements on another part of the project. In fact, striving for perfection, that is, going well beyond the client's requirements, may result in excessive labor and expense charges against the project budget and cast you in a negative light.

Little Tolerance for Tardiness

There is little tolerance for tardiness. Although a proposal for an engineering or other professional services contract may be creative in its approach and handsome in its presentation, the proposal is likely to be rejected by the potential client if it is delivered after the stated deadline. Young professionals who frequently arrive late at meetings, even if they are well prepared, risk antagonizing their colleagues, many of whom place a high value on their time.

Assignments Are Not Graded

Depending on the mode of operation of the institution from which you graduated, you may expect all of your work to be reviewed and evaluated and perhaps even "graded." Many students grow accustomed to having professors read, critique, and grade their work. In the world of professional work, careful reading of all submitted materials is unusual, and written constructive critiques are rare. Teaching and mentoring are simply not the principal business of engineering and similar organizations.

A word of caution is in order. You may have derived considerable satisfaction from grades and other academic recognition received while in college rather than gaining satisfaction from the intrinsic value of the work. Accordingly, you may feel unappreciated as you begin employment because of the absence of continuous, positive feedback. Worse yet, your self-confidence may waver. You should recognize that many managers follow a "no news is good news" approach. That is, your good work may be appreciated, but you are not hearing about it.

Schedules Are More Complicated

The daily and weekly schedule of a typical university student is relatively simple in that so much of the time is blocked out for an entire semester or quarter by predetermined activities and events such as classes and meetings of professional, student government, and Greek and other campus organizations. Daily and weekly schedules in the engineering and business world are much less repetitive than those of the academic world and are much more likely to change quickly and dramatically in response to client and other needs. If you haven't already done so, you will need to develop a personal time-management system compatible with the erratic time dimension of the professional practice. Time management is discussed in detail in a later major section in this chapter.

Higher Grooming and Dress Expectations

University communities typically permit a wide range of dress and grooming. Engineering and other technical program faculty members should serve as role models and mentors for their students in all aspects of engineering and business, including grooming and dress. Unfortunately, the appearance of some professors and instructors falls far below the standard of the world of professional practice. This exacerbates the grooming and dress problem for some professional program students because they are not taught the importance of grooming and dress. The professional and business community has little tolerance for poor grooming and dress, although its expectations are rarely explicitly communicated. Accordingly, the young professional may lose opportunities and be relegated to secondary tasks because he or she does not understand the importance of personal appearance.

Teamwork Is Standard Operating Procedure

Teamwork is increasingly becoming standard operating procedure in professional practice. Except for the most trivial planning and design projects, interdisciplinary work and often formal interdisciplinary teams are required. Unfortunately, academia devotes relatively little attention to developing team skills. Students are often pitted against each other in competition for top grades and high grade point averages. The resulting highly individualistic paradigm, characteristic of many universities and their faculty members and students, is not likely to be well received in the professional practice community.

Expect and Embrace Change

Change is one of the few phenomena that you can count on throughout your career. The young professional must recognize the need to change and must evaluate his or her ability to move quickly and positively from the academic culture to the practice culture. Failure to make the necessary changes will frustrate desires to quickly realize one's potential in the consulting business, industry, or government.

EMPLOYMENT OR GRADUATE SCHOOL?

Assuming that you have a very solid academic record (B or better), consider the option of attending graduate school in engineering or a related technical field on a full-time basis immediately after earning a bachelor's degree. Most colleges and universities operate placement offices that help senior students secure full-time employment. Although not as evident, most colleges and universities also offer helpful resources to engineering and other students who wish to explore the graduate school option.

Young people interested in graduate study because it might lead to an academic career should read Hardy Cross's (1952) timeless book *Engineers and Ivory*

Towers and may also be interested in a helpful booklet available from the American Society for Engineering Education (Landis, 1989). Other graduate school aids, not necessarily focusing on an academic career, are also available (Adams, 1987; University of Illinois, 1982). The American Society for Engineering Education publishes a useful, periodically updated graduate school directory.

Figure 2–2 summarizes positive and negative aspects of full-time graduate school and full-time employment immediately after earning an undergraduate degree in engineering or other technical field. A mixed option, not discussed here in detail, is also available. This is full-time employment supplemented with part-time graduate study leading to a master's degree in engineering, some other technical discipline, or business. On the positive side, the cost of this form of graduate study is often reimbursed in whole or part by the young professional's employer. On the negative side, the work demands and schedules of professional

	Full-time graduate study	Full-time employment
Pro	■ In-depth study ■ Autonomy ■ Financial support ■ "Buy" time	■ Real-world applications ■ $
Con	■ Study burnout ■ Uncertainty of area of specialization ■ Short-term cost ■ Reduced number of positions	■ Technical burnout

Figure 2–2 Graduate study and employment options

practice may frequently conflict with classes and study. Even if the conflicts can be resolved, four or five years are likely to be required to earn a master's degree.

Full-Time Graduate Study

The best reason to enter full-time graduate study immediately upon completion of a B.S. degree in engineering or other technical field is the opportunity to do in-depth study in a field of interest. Many young people, as a result of the stimulation of their college education, learn to love their new-found discipline and want to continue to study the depth and breadth of that discipline. Full-time graduate study can help to satisfy that need.

Career autonomy and access to higher-level positions are other reasons to consider acquiring at least a master's degree and possibly a Ph.D. Graduate schools typically offer promising potential graduate students very attractive financial packages. While financial assistance should not be the primary reason for graduate study, it can help to offset the financial burden. Finally, some college seniors are simply uncertain as to what the future holds for them, and a year or so in graduate school during which they acquire a master's degree might be a productive way to "buy time" as they work through the decision-making process.

On the negative side, full-time graduate study immediately after undergraduate study might lead to, or raise the fear of leading to, a kind of "study burnout." By its very nature, graduate study in engineering, other technical disciplines, or business tends to be more specialized than undergraduate study and, therefore, is not advised for the recent graduate who is uncertain about his or her area of preferred specialization. Another negative aspect of full-time graduate study is that, even with excellent financial support, there will be a short-term cost and possibly a long-term cost when foregone wages are considered. Finally, graduate study leading to a terminal degree such as a Ph.D. might actually reduce the number of positions available to you. However, the opportunities offered by the smaller pool of positions may be clearly superior to opportunities available to holders of only B.S. degrees, and the number of professionals seeking the positions will be much smaller.

A recent study of United States and foreign student views of graduate engineering education (Barber et al., 1990, pp. 52–53) concludes that engineering students tend to be burned out by the end of their undergraduate programs. This discourages them from seriously thinking about and going directly to full-time graduate study. However, the same study also notes that undergraduate students who were involved in tutoring, professional organizations, undergraduate research, and cooperative education are more likely to go on to full-time graduate study.

Full-Time Employment

Just as the opportunity to continue in-depth study is probably the primary and highest attraction for full-time graduate study, the opportunity to participate in real-world engineering and other technical applications is the primary driving force for entering full-time employment immediately after graduation. As a new graduate, particularly if you have not been involved in cooperative education or summer employment related to your chosen technical discipline, you may have a strong desire to prove yourself in the technical practice world and see a product, structure, facility, or system come to fruition as a result of your studies.

The probability of high and secure salary is also a major attraction to the soon-to-graduate engineering or other technical profession student. Many students acquire significant debt as a part of their college education and, understandably, want to begin repaying that obligation. The young professional is advised, however, not to be too impressed by what appears to be high starting salaries. Your income over a several-decade career is not likely to be strongly correlated with a starting salary immediately after college. Other factors, including the holding of a graduate degree, may easily offset the apparent advantage of employment at a high starting salary immediately after earning a B.S. degree in engineering or other technical area.

Perhaps the principal negative aspect of full-time employment immediately after earning an engineering or similar degree, even when the salary is high, is the possibility of technical burnout. Unless the young person is able to continue, ideally with the assistance of the employer, a carefully crafted program of personal professional development, the person's value to the employing organization may gradually diminish. The young technical professional may be increasingly assigned routine work, while more sophisticated assignments are given to recent graduates who have just joined the firm and brought with them newer and higher technologies learned as part of their bachelor's or graduate programs. The crucial importance of managing your personal professional assets, of continuing your education, and of being involved in professional organizations is discussed at length later in this chapter.

Learn from Potential Employers

The employment search process, which most engineering and other technical program students participate in during their senior year, provides an excellent opportunity to learn about the value of graduate study. In their zeal to fill positions, recruitment representatives of private and public engineering and similar organizations may disparage full-time graduate study, arguing that advanced degrees are not really needed and, furthermore, you can always earn a master's degree on a part-time basis. In order that you may gain a comprehensive and balanced view of each recruiter's organization, request a list of top people or an organization chart showing who holds upper-level positions. Then find out how many of the organization's leaders hold advanced degrees. Draw your own conclusions regarding the importance of graduate education.

Much more insight into a particular organization's real position on postgraduate study can be obtained by the inquisitive student during cooperative education assignments and summer jobs. Note who holds graduate degrees and the disciplines represented by those degrees. As you get to know the holders of those degrees, ask them to share their experiences and views.

TIME MANAGEMENT

Time Is a Resource

Don't say you don't have enough time. You have exactly the same number of hours per day that were given to Helen Keller, Pasteur, Michelangelo, Mother Teresa, Leonardo da Vinci, Thomas Jefferson, and Albert Einstein.

(Anonymous)

Young, harried professionals often say things like "I don't have the time" as though they have less time than others. In fact, each person has all the time there is—no more, no less—24 hours a day and 365 days per year. Between graduation and retirement at about age 65, you have about 400,000 hours at your disposal. Some engineering and other technical program graduates will wisely use their allotment of time to achieve much in their personal life, family life, financial life, community life, and their professional life. Others will fill many of those 400,000 hours with lives of mediocrity. These individuals might live their last years regretting that they didn't do more with their gift of time. Time is a resource to be managed wisely. Without careful management, much of the time allotted to each professional is lost forever.

Time is often wasted unknowingly. One study (*Sales and Marketing Digest,* 1988) found that the average American, during his or her lifetime spends one year looking for misplaced objects, eights months opening junk mail, two years trying to return phone calls, and five years standing in line. Mackenzie (1990, p. 22) claims that "Most people work at about one third efficiency most of the time." He goes on to say (p. 201) that individuals can save two hours per day through improved time management.

Time Management: The Great Equalizer

While talent and intelligence will influence the entry-level technical professional's success or failure, how each uses his or her allotment of time will have a profound influence. As stated by management expert Peter Drucker, "Time is our scarcest resource and unless it is managed, nothing else can be managed" (Dorney, 1988). Unlike talent, intelligence, wealth, and other personal attributes and assets, time is distributed equally. Accordingly, effective use of time can be the great equalizer. You have the power, if not the responsibility, to carefully manage how you use your time. Lives are built on the use of minutes, hours, and days. No one can get more time, but anyone can do more with what he or she gets.

In a narrow sense, time management means getting more productivity out of a person's allotment of time through efficient telephone calling, correspondence handling, meeting attending, and task doing. In a broad sense, time management means going well beyond these practical tools and techniques and building a meaningful life. How the young professional decides to use his or her minutes, hours, and days determines how that young professional will spend his or her life. If you don't decide how to use at least some of your time resource, be assured that others will decide for you.

Time Management Tips

Time-management ABCs, that is, 26 time-management tools and techniques, are listed in Figure 2–3 and discussed here. Although some are self-explanatory, all are discussed because of the crucial importance of time management for the young professional. Tip A is fundamentally different from all other tips and much more important than any of them. Tip A urges you to decide where you want to be; the other 25 tips are suggestions on how to get there. Just as you must decide where you want to go on your vacation before you decide how you are going to travel, so you must decide where you want to go in your personal, family, financial, community, and professional affairs before you apply time-management tools and techniques. Tip A is the course you set on your ship's compass and tips B through Z are the skills you use at the helm. If you wish to go beyond the time-management tips presented in this chapter, attend a workshop on the subject or refer to the extensive literature on this topic (e.g., Brown, 1986; Covey, 1990; Day-Timers, Inc. (no date); Dorney, 1988; Engstrom and Mackenzie, 1967; Kelly (no date); Lock and Schwarz, 1976; Mackenzie, 1990; McDon-

A.	Written goals	N.	Discretionary time
B.	Plan each day	O.	Group similar activities
C.	Act immediately and constructively	P.	Telephone tag
D.	Bring solutions	Q.	Delegate
E.	Your best time	R.	Closed door
F.	Clean desk	S.	Write it down
G.	Organized space	T.	Write response on original document
H.	Efficiency vs effectiveness	U.	Travel and waiting time
I.	Professional files	V.	Word processing or dictation
J.	Keep related materials together	W.	Meet with yourself
K.	Meet only if necessary	X.	Log your time
L.	80/20 rule	Y.	Adopt a holistic philosophy
M.	Break projects into parts	Z.	Guarantee small successes

Figure 2–3 Time-management ABC's

ald, 1978; Oncken and Wass, 1974; Park, 1978; and *Sales and Marketing Digest,* 1988).

Tip A: Written goals—develop and be driven by them

... *If you don't know where you are going, any road will get you there.*

(Koran)

Conceptualize, refine, and write out monthly, annual, and multiyear goals for personal, family, financial, community, and professional areas and affairs. Clear goals, quantified to the extent feasible, are crucial to charting and navigating the seas of a business and professional career. Perhaps the acronym SMART will help you formulate your goals. It means that each goal should be

- *S*pecific
- *M*easurable, that is, the goal should be cast in quantitative terms if at all possible.
- *A*chievable, that is, while the goal will stretch you, you must be able to accomplish it.

- *R*elevant in that the goal is appropriate for your present situation and for your organization's current circumstances.
- *T*ime framed, that is, you establish a schedule for achieving the goal or its components.

Cypert (1993, pp. 118–119) offers a similar set of specific goal-setting suggestions. The process of writing, refining, and regularly examining goals will help to keep your career on course.

Consider communicating goals to family, friends, selected colleagues, your supervisor, and some subordinates as a means of enlisting their cooperation and support. Most professional and other acquaintances usually won't ask about an individual's goals because they think the topic is too personal or because they don't use a goal paradigm. But they are likely to be interested if the information is volunteered and they are often, with that knowledge in hand, in a position to assist in an achievement of your goals.

Assume, for example, that one of a young professional's goals is to present a technical paper at a regional or national conference. Obviously, the individual would scan various requests for papers to be published by professional organizations to find an appropriate opportunity. By sharing this goal with colleagues, particularly those who are involved in professional organization activities, the young engineer or other technical professional in effect enlists the potential help of many individuals and is more likely to achieve the goal.

Don't assume that colleagues, supervisors, supervisees, and others know your goals by virtue of working with or near you. Similarly, thinking that all good things will come your way if you simply put your nose to the grindstone is naive. Rather than being timid or overly modest about professional and other goals, adopt a proactive communicative approach and then be prepared to follow through as opportunities arise or are presented to you.

If the young professional has personal goals and understands and is comfortable with the employer's goals, he or she is in a good position to manage time. Discretionary time use should be gauged against personal and organizational goals and should favor important although not necessarily urgent activities over all other combinations. Figure 2–4 is an urgency versus importance matrix. Various activities on which one might choose to spend time are classified by their degree of urgency and degrees of importance; note that urgency does not necessarily mean importance. Figure 2–5, a variation on the preceding figure, suggests the concept of trying to focus your time resources on important-not urgent activities while relegating important-urgent activities to a secondary status and avoiding, to the extent possible, all nonimportant activities whether they are urgent or not urgent. For additional discussion of the emergency–importance matrix as it relates to utilization of time, refer to Covey (1990, pp. 149–156).

	Urgent	Not urgent
Important	I ACTIVITIES Crises Pressing problems Deadline-driven projects	II ACTIVITIES Prevention Building production capacity Relationship building Recognizing new opportunities Planning Recreation
Not Important	III ACTIVITIES Interruptions, some calls Some mail, some reports Some meetings Proximate, pressing matters Popular activities	IV ACTIVITIES Trivia, busy work Some mail Some phone calls Time wasters Pleasant activities

Figure 2–4 Time-management matrix—options (Source: Adapted with permission from Covey, S., 1990, p. 151. Copyright © 1989 by Stephen R. Covey. Reprinted by permission of Simon & Schuster, Inc.)

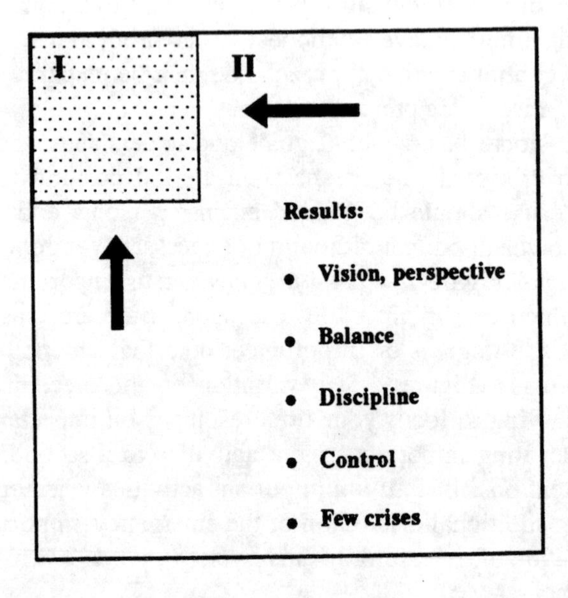

Figure 2–5 Time-management matrix—results (Source: Adapted with permission from Covey, S., 1990, p. 154. Copyright © 1989 by Stephen Covey. Reprinted by permission of Simon & Schuster, Inc.)

Tip B: Plan each day. Each day should be planned in writing, with tasks prioritized according to importance. Such planning, which might be done at the end of the preceding day or the beginning of the current day, will require less than 15 minutes of time but will be highly productive. Victor Hugo (1802–1884), the French writer and romantic who turned to politics in later life (Morse, 1973, pp. 54–55), said, "He who every morning plans the transactions of the day and follows that plan carries a thread that will guide him through the labyrinth of the most busy life. The orderly arrangement of his time is like a ray of light which darts itself through all his occupations. But where no plan is laid, where the disposal of time is surrendered merely to the chance of incidence, chaos will soon reign." In other words, if you fail to plan the use of your time, you probably plan to fail to use it effectively.

Others suggest (e.g., Covey, 1990, pp. 160–171) that, in addition to planning each day, the active professional should also plan each week. In his autobiography, Lee Iacocca, who rose to be the chief executive officer of the Chrysler Corporation, indicated that his practice was to devote Sunday evenings to planning his week's activities. He did this partly as a matter of organization and partly as a way of focusing his energies. He writes, "Every Sunday night I get the adrenalin going again by making an outline of what I want to accomplish during the upcoming week" (Iacocca, 1984, p. 20).

Tip C: Act immediately and constructively. After completing a transaction with one or more other individuals, immediately do something constructive. As a result of reading a piece of correspondence, hanging up after a telephone conversation, or saying good-bye to the person who visited you in your office, take a specific action. Examples include scheduling a meeting, drafting a letter, asking for a file, and requesting more information. Take one meaningful step rather than putting your notes on a pile with the good intention of getting to it later.

With specific reference to correspondence and other written and printed materials that come across your desk, immediately do one of three things with each item: discard it, file it, or act on it in accordance with the preceding suggestion. There are no other realistic options. Setting things aside somewhere in your office or work area until you "have the time" will simply result in a net accumulation of materials.

Tip D: Bring solutions—accompany each problem with at least one solution. Insist that subordinates bring solutions along with problems. This applies to the relationship between you and individuals who may be reporting to you on a permanent or project basis, and it applies to relationships between you and the people you report to on a project or organizational basis. Almost anyone who is perceptive enough to identify a problem is capable of suggesting at least one solution to the problem. And, as a result of their closeness to the problem, these individuals can often identify the best solution.

Writing out the problems is, for some professionals, a disciplining activity which tends to focus their thinking and enhance their understanding of all dimensions of the problem and, in turn, leads to solutions. Whatever your personal style, don't rush to your supervisor immediately upon discovering a problem. Think about it, perhaps write about it, and identify at least one feasible solution. Then talk to your supervisor. By so doing, you will enable both you and your supervisor to make the best use of your collective time. Expect the same of people who report to you.

Tip E: Your best time—identify and wisely use your best time of the day. Identify your best time of day, in terms of energy level and intellectual and creative ability, and try to schedule your most challenging tasks into that time period. Some individuals are "morning people," others are "night people," and others are most productive and creative at other times of the day. Recognize this in yourself and others, and use it to encourage effective time utilization.

Tip F: Clean desk—eliminate visual disruptions. Consider keeping your desk clean of all but your current project. This suggestion is offered because some people are easily distracted by materials in their area of peripheral vision, particularly if those materials remind them of other tasks or projects that need attention.

Tip G: Organized space—create an efficient work space. Items that are used frequently should be within easy reach of your desk, whereas less frequently used items can be more remotely located. Consider making improved use of drawers in desks and other office furniture. Instead of using them for bulk storage of infrequently used supplies and materials or to hold rarely used files, consider using some of the drawers in place of an "in basket," an "out basket," a "call back" tray or file and for other high-frequency uses. Such use of drawers keeps action items close, for convenience, but out of sight to avoid distractions.

Tip H: Efficiency vs effectiveness—distinguish between. Distinguish between efficiency and effectiveness, that is, between doing things right and doing the right things, giving preference to the latter. For example, a group of department heads in an organization may be very efficient in completing a weekly report and transmitting it to the corporate office, where no one reads it. The department heads are efficient, but the overall process is not an effective use of anybody's time. A situation like this often arises because the original legitimate need for the report has vanished but a process has been set in place and is blindly adhered to.

Ask this question of yourself and your colleagues: are we doing the right things or only doing things right? The ideal is to do both, but it is folly to do the wrong things right. Avoid the trap of doing, well or otherwise, that which should not be done at all.

Tip I: Professional files—create and maintain. Develop a set of personal professional files while in college or immediately upon beginning professional work. These files, which will of necessity start modestly, might be set up based on an initial set of technical and nontechnical categories. The existence of a systematic file system that can be expanded as needed is a partial answer to the problem of trying to read all interesting and potentially useful material that comes across one's desk. Remember the previous suggestion of discarding something, filing it, or acting on it; the availability of personal professional files provides a way to systematically file material that may have value in the future. Instead of trying to read and study every item, clip and file some of them for future retrieval and use.

Tip J: Keep related materials together. Keep all parts of the small project together as the project moves from person to person within an office environment. That is, to the extent feasible, when your responsibility for a small project such as writing a letter, doing additional work on a spreadsheet, or preparing a table for a report, is completed and the project responsibility shifts from you to someone else, try to keep the project intact. Large plastic sandwich or freezer bags provide an ideal container because they are slightly larger than the 8 1/2 x 11-in. paper so common in a U.S. office; they keep materials together, and their transparency allows the current user to quickly see what is available.

Tip K: Meet only if necessary—manage meetings. Meet only when necessary. To the extent feasible, ask or insist that meetings be carefully planned and conducted and that follow-up responsibilities be clearly assigned. Meeting management is so important to personal time management and overall organizational effectiveness that a major section of Chapter 4 is devoted to the topic.

Tip L: 20/80 rule—apply the vital few–trivial many rule. Respect the "vital few–trivial many" rule, "20/80" rule, or Pareto's law as described, for example, by Drucker (1963, p. 58) and Wortman (1981, p. 252). Most professional and other activities involve input and output, work and results, effort and accomplishments, or other cause and effect relationships. But each unit

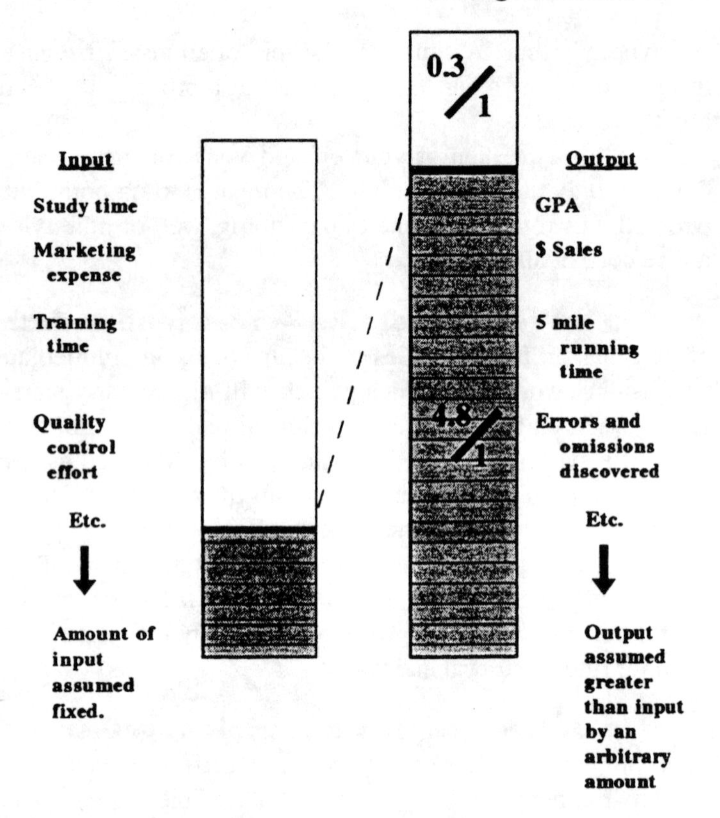

Figure 2–6 The 20/80 rule

of input, work, effort, and cause does not necessarily lead to the same relative result. Some kinds of input, work, effort, and cause are more productive than others. This is the concept portrayed by the 20/80 rule shown in Figure 2–6. Simply stated, 20 percent of the input produces 80 percent of the results. While there is absolutely no substantiation for the rule in a quantitative sense, the "20/80" rule, the "vital few–trivial many" rule or Pareto's law all make intuitive sense.

When an individual or group effort is undertaken, effective time management suggests that input, work, effort, or cause be partitioned and that each piece be examined for its individual contribution to the output, results, accomplishments, and effects. Input should not be considered just as a homogenous whole, but rather as a heterogenous mix of components, some of which are more effective than others. Less effective units of input should be replaced with more effective ones.

Recent college graduates should have an intuitive sense of the "vital few–trivial many" rule, the "20/80" rule, or Pareto's law on the basis of what they learned about studying. Each individual, for example, finds that certain types of studying are more effective than others. Some students learn best by reviewing and perhaps even rewriting class notes, whereas studying with friends is much less effective. These students have learned, over a period of time, that they can get more out of their studying time by spending most of it in reviewing class notes.

Tip M: Break projects into parts. Break major technical and nontechnical projects into pieces or components for planning and doing. As a result of an engineering or other technical education, most entry-level professionals have developed a powerful generic ability to examine complex mechanical, electrical, and other physical systems, to identify their components, and to understand the interrelationships between them. This analytic skill is transferrable and applicable to any apparently complex system regardless of whether it is primarily technical or nontechnical in nature.

As a result of visualizing the components of a major project or assignment and understanding the relationships between them, the entry-level technical professional is in a good position to apply the critical path method (discussed in Chapter 6) and other network methods for managing the time schedule of the major project or assignment. In addition, partitioning of a major project or assignment into its component parts provides the basis for effective delegation, which is discussed in Chapter 4, and is another way to make effective use of your time.

Tip N: Discretionary time—use it wisely. Use discretionary blocks of time wisely. Recognize that your work-related time can be viewed as falling in one of three categories: boss-imposed time, system-imposed time, and self-imposed time. Self-imposed time is discretionary time. Use it to move toward achievement of personal goals and organizational goals discussed under Tip A.

Tip O: Group similar activities. Group similar activities to achieve efficiency. For example, decide to return all of your telephone calls at the same time or to do a batch of dictation or word processing projects. Experience indicates that such grouping of similar activities is efficient. For example, it's generally easier to make the transition from calling one person to calling another person than it is to make the transition from calling one person to dictating a letter or working on a spreadsheet or to some other different activity.

Tip P: Telephone tag—don't play. Avoid "telephone tag," the wasteful practice by which two individuals repeatedly place calls to each other but do not connect or in any way exchange useful information. Use time-saving techniques such as making an appointment for a telephone meeting which you will initiate by calling the other party at the agreed-upon time, leaving a specific message with whomever you speak, asking to speak to someone else who might be able to assist, or leaving an intriguing message that might induce the elusive person at the other end of the line to call you. Given the difficulty of reaching some people by telephone, help to ensure the effectiveness of the conversation by first preparing a list of topics you would like to discuss, questions needing answers, and information you want to share. This written agenda could also serve as the basis for your written documentation of the telephone conversation.

Modern electronic devices and systems such as voice mail, electronic mail, facsimile machines, beepers, and cellular telephones are excellent ways to reduce the time required for effective telephone and other transactions. The young professional should determine which systems are available and become proficient in their use.

Tip Q: Delegate. Practice effective delegation of selected tasks, along with the necessary authority, to other capable individuals. Besides being a very productive time management tool, effective delegation offers many other organizational-strengthening benefits. Because of the overall importance of delegation to the entry-level professional and the organization, the subject of delegation is discussed at length in Chapter 4.

Tip R: Closed door—keep door closed but access open. Consider adopting, when the physical situation and the corporate culture allow, a closed-door policy with respect to your office or work space while implementing an "open-door" policy with respect to access to you. That is, you are less likely to be interrupted if your door is closed or the entry to your work space is arranged such that potential visitors realize that you are involved in some activity requiring concentration or privacy. On the other hand, for the sake of being an effective team player, establish a posture and procedure that clearly indicate your accessibility to colleagues and others.

Tip S: Write it down—document, document, document. Document everything unless you have a superb memory; even then, consider the consequences of your not being available when someone wants to make reference to your former projects. Time invested in taking notes at a meeting, writing a memorandum to file after the meeting, or writing a memorandum or letter to one or all of the meeting participants will often provide a significant return in terms of improved understanding by you and others of what was decided and why and, as a result, who will do what when. For many people, the process of capturing in one's own words the essence of a telephone conversation, a meeting, or a presentation adds immeasurably to understanding and retention. Beyond the immediate short-term needs, documentation is required as a reasonable safeguard against future litigation and potential personal or corporate liability. This aspect of documentation is discussed in Chapter 10.

Tip T: Write response on original document. Answer selected memoranda and letters by writing a response directly on the original document and sending it to the sender, perhaps with copies to others, while retaining a copy for the file. This practice requires much less time than creating a new document and also typically results in a quicker response. Exercise prudence in applying this time-management tip. While it may be appropriate for most internal documents, it is not likely to be suitable for most memoranda and letters received from outside your organization.

Tip U: Travel and waiting time—use productively. Many professionals, even entry-level engineers, spend considerable time traveling by rapid transit or air. They also often find themselves waiting for a meeting or a travel connection or having spare time between appointments. The travel and waiting time offers an excellent opportunity to be productive, provided that you have prepared by carrying work and materials with you. Besides taking materials to read or projects to work on, the traveling professional should also take appropriate basic supplies such as a dictaphone, computer, calculator, pens and pencils, a small stapler, and tape.

The potential for making effective use of travel and waiting time has been enhanced in recent years with the development of voice mail, electronic mail, facsimile machines, beepers, cellular telephones, and overnight mail, all of which provide means to obtain information from or send information to the home office or clients. The location of the professional person is, for most of the functions he or she carries out, increasingly becoming irrelevant because of the various communication devices and links that are now available.

Tip V: Word processing or dictation—don't write longhand. Use a word processor or dictation in lieu of writing longhand. Word processing and dictation are several orders of magnitude faster than hand writing and result in a product that can be used more effectively by others. An important exception to the general rule is the occasional sending of handwritten notes for special purposes such as thanking someone for a special favor or congratulating someone for an achievement. Given the ease with which words can be processed by modern electronic devices, the handwritten note has become a special, meaningful means of communication.

Tip W: Meet with yourself. Occasionally make appointments with yourself. To the extent that you can control at least part of your daily schedule, isolate blocks of time, preferably during the most productive time of your day, for work requiring higher levels of concentration. Like many time-management techniques, this one can be abused. However, if done in moderation, "making appointments with yourself" is a legitimate and effective means of managing your time.

Tip X: Log your time. Occasionally keep a time log, on about 15 to 30 minute intervals, for a period of several days to a week. Although this is a major effort, it may identify undesirable patterns or trends in the utilization of time and suggest ways to be more effective in meeting your personal goals and the goals of your organization. For example, in keeping with tip A, a time log may reveal that you are frustrating achievement of your goals by spending too much time on urgent but not important activities and not devoting enough time to not urgent but important tasks.

Tip Y: Adopt a holistic philosophy. Maintain the intellectual, physical, emotional, and spiritual dimensions of your being and the balance between them. The young professional is urged to adopt a holistic philosophy and program to defend against the strong tendency, in response to the pressures of the workplace, to focus excessively on intellectual activities.

Tip Z: Guarantee small successes. Arrange to have at least one success each day. The young professional should plan each day so that it includes one or more work-related activities that are both enjoyable and are likely to be accomplished. This may seem trivial. However, failures and other disappointments are inevitable. Their negative impact can be anticipated and to some extent ameliorated by giving yourself the opportunity to enjoy at least one success each day.

A Time-Management System

Assuming that you intend to practice some or most of the preceding time-management tips, how will you keep track of those things you plan to do and have done? A system is required, and many are available. Examples are those offered by Day Runners, Inc. of Fullerton, California and Day-Timers, Inc, of Allentown, Pennsylvania. Each system offers different features whose appropriateness must be determined by the individual. However, while adoption of a system is certainly needed, much more important aspects of your approach to time management are the commitment you make and the resulting extent to which the use of effective time-management tools and techniques becomes habitual.

Key Ideas about Time Management

Each young professional has all the time there is. The difference between the accomplishments of people, while it is certainly influenced by intelligence, talent, and other personal attributes, and by luck, is significantly influenced by how those people use their time. Time-management techniques and tools are intended to get more things done, and, even more importantly, to get the right things done. Your approach to time management should be guided by a compatible set of personal goals and organizational goals.

THE FIRST FEW MONTHS: MAKE OR BREAK TIME

As noted earlier in this chapter, demands and expectations will change abruptly as you move from the world of study to the world of practice. Success during the first few months of engineering or other technical practice is not likely to depend primarily on your technical knowledge. As part of the process of recruiting senior students to fill a particular position, the employer explicitly or implicitly defines the minimum range and level of technical competence. A candidate's ability to satisfy the technical requirements of an available position can be determined largely by examination of the candidate's resume and academic transcript. In other words, the selected candidate is assumed to have basic technical competency.

The degree of success enjoyed by a young entry-level staff member is likely to depend largely on nontechnical factors and performance. If you expect to be evaluated and to advance only on technical expertise, you will need to be a wizard or a genius.

Recognize and Draw on Generic Qualities and Characteristics

Fortunately, as part of his or her engineering or other technical education, the young professional is usually given ample opportunity to learn and develop generic, that is, widely applicable, qualities and characteristics. Hopefully, you have seen and seized that opportunity and can draw on those generic resources as you navigate the first few months of full-time employment. Consider adopting an attitude of quiet confidence based in part on what you have already accomplished by successfully completing, or almost completing, a demanding professional program. Generic qualities and characteristics that should have been developed as

- **Work hard**
- **Persistence**
- **Analytic approach**
- **Communication skills**

Figure 2–7 Generic qualities that should result from earning an engineering or other technical degree

part of the process of earning an engineering or other technical degree are listed in Figure 2–7 and explained as follows:

- The ability to work hard, that is, put in long hours of intense effort. For example, the typical engineering student is required to successfully complete significantly more courses for a bachelor's degree than the vast majority of students at a university. And most of those courses are very demanding. With the exception of the truly intellectually gifted, an engineering or similar rigorous degree certifies the holder's ability to work hard.

- Persistence, that is, carrying on in spite of difficulty, being resourceful and ingenious, and having the ability to see opportunities where others see problems.

- A high degree of analytic ability including skills such as understanding complex processes and systems, identifying components of those systems, understanding the relationships between components, determining the cause of problems or failures, conceptualizing and developing alternative solutions, comparing options and selecting the best course of action, and implementing the solution.

- Broad and effective communication skills, including listening, speaking, writing, mathematics, and the use of graphics.

Florman (1987, Chapter 5) also discusses values and widely applicable abilities that engineers, by virtue of education and experience, bring to their work and to society. He cites belief in scientific truth, ability to work hard, risk-taking, dependability, belief in order, pragmatic orientation, democratic tendency, creativity, and openness to change.

Incidently, most recent engineering and other technical program graduates will have opportunities to be involved in neighborhood, community, church, and synagogue activities such as service projects, recruitment efforts, political campaigns, and fund-raising campaigns. The generic qualities and characteristics that should have resulted from the technical education experience will be very valuable in these nontechnical endeavors.

Presented here are suggestions for enhancing your effectiveness during the first few months of full-time professional employment. This discussion assumes that you are adopting many of the time-management tools and techniques presented in the previous section of the chapter. Time-management skills, especially those that become habitual, provide one way to become productive during the first few months of employment.

Never Compromise Personal Reputation

> *Good character is more to be praised than outstanding talent. Most talents are, to some extent, a gift. Good character, by contrast, is not given to us. We have to build it piece by piece—by thought, choice, courage and determination.*
>
> (John Luther)

Engineers and other technical professionals provide advice and counsel—not a material product. The efforts of a craftsperson are judged primarily on the observed quality of the final product, such as a piece of pottery, a sculpture, jewelry, or a painting. In contrast, the credibility imparted to the advice given by an engineer is closely tied to the reputation of that engineer. The client of a technical organization or the individual served by a technical professional is usually not in a position to judge the advice or recommendations of the professional on the basis of the merit of that advice or recommendation. However, the user of the professional's service is quite capable of judging the quality of the professional. Ralph Waldo Emerson said, "What you are . . . thunders so that I cannot hear what you say . . ." Often, clients and the public cannot "hear" or fully understand what engineers and other technical professionals say. But they can and do judge character, and they use that to value and trust—or devalue and mistrust—the professional's conclusions and recommendations. Personal reputation, like a handcrafted crystal vase, takes a long time to create and once damaged, may never be repaired. Tell the truth. Keep your word. Give credit for ideas and information—do not use anything without crediting the source. Don't blame others—err on the side of taking blame on yourself when problems arise in your realm of responsibility. Give credit to others when significant accomplishments occur. The ethical aspects of engineering practice and, by extension, practice in similar technically based fields, are discussed in detail in Chapter 11.

One seemingly harmless way in which a young person's personal reputation may be tarnished is failing to keep what appear to be small promises. For example, you meet someone at a local meeting of your professional society, you exchange business cards, and you promise to send him or her a copy of your firm's brochure. But you forget. Or you run into an acquaintance whom you have not seen for some time, talk briefly, and the two of you agree you should get together for lunch. You offer to make arrangements. But you forget. Although any individual instance like the two cited here might be considered a harmless oversight, a series of them will damage your credibility. A time-management system, as discussed earlier in this chapter, is an effective mechanism for helping you keep your promises. By keeping small promises, you will build big relationships. For additional discussion of this topic refer to Benton (1992, pp. 97–98).

Learn and Respect Administrative Procedures and Structure

In the first few days of employment, you are likely to be deluged with forms, written procedures, policy statements, and information on the manner in which the organization is organized and does its business. There isn't a form that can't be refined or a procedure that can't be improved. Perhaps, after you are well established in the organization, you will want to make constructive comments about administrative policy, structure, and processes. At the outset, however, you should focus on learning and respecting the established policy, structure, and process. Focus on doing your assignments well.

Do All Assignments Well in Accordance with Expectations

> *. . . if you take care of your present job well, the future will take care of itself.*
>
> (King, 1944, p. 2)

Regardless of how unimportant or trivial initial assignments may seem to be, assume that your supervisor knows what he or she is doing. Some of the initial, simple mechanical types of tasks that you are asked to do may, in fact, be tasks typically assigned to technicians or other support staff. You may be given these assignments simply because the work has to be done and you are readily available to do it, or your supervisor may be assigning routine, simple tasks to help you develop a comprehensive understanding of the wide variety of work that is done in the organization. He or she may be grooming you to assume responsibility for managing that work.

During the first few months of employment, be very generous in giving your supervisor the benefit of the doubt with regard to the appropriateness of your assignments. Focus your energies on understanding the context of and expected results for each assignment, learning how to do the work, and appreciating the constraints under which you are to carry out each assignment. Do your assignments in accordance with those expectations and constraints and, in the process, learn all you can.

Get Things Done

The workplace places a premium on getting things done and making things happen. As noted in Chapter 1, with the definition of engineering/technology management, what and who you know are secondary; what counts is how you utilize what and who you know to make good things happen. Take the initiative to start an assignment and keep it going. Don't wait for someone to tell you what to do next—decide for yourself or ask. Be resourceful by seeing opportunities in problems and, at minimum, learning from them. Be persistent; that is, don't let setbacks become roadblocks.

Trim Your Hedges

Develop the habit of answering questions in a positive manner and stating your findings without excessive qualifications. What you write and say should be in the context of the expected or actual audience. For example, there is no excuse for giving an answer like "If I did the calculations correctly . . ." in answer to the question "What size electric motor do we need?" asked at a staff meeting. You are responsible to do the calculations correctly—or to find out how to do them. A qualified answer such as "Based on the limited field data, my guess is that there will be no foundation problems," might be an acceptable qualification in a conversation with professional peers who understand the complexity of your work, but is not likely to be appropriate in making a presentation to a client. Overly qualifying statements and responses on technical matters beyond the area of expertise of your audience is nonproductive. On hearing, while not understanding, the hedges, listeners may perceive a lack of competence, confidence, or commitment.

The tendency to overly qualify statements and responses suggests inadequate preparation, lack of ability, low self-confidence, or insensitivity to colleagues and clients. These traits detract significantly from the performance of the young engineer and, if not rectified, will interfere with advancement within the organization. The fact that you are well prepared, have ability, and are confident is irrelevant if you are perceived to be otherwise. Perception is fact.

Do all assignments well, in accordance with expectations as indicated earlier. When explaining or reporting the results to others, be very sensitive to the nature and interest of the audience. Speak in a simple, declarative, and brief fashion unencumbered with convoluted qualifications.

Incidentally, the preceding focuses on how what you say influences others.

What you say and how you say it will also influence you. Consider the more positive effect on your subsequent performance when you say, "I will get the draft report to you by Friday noon" rather than "I will try to get the report to you by Friday noon." Less hedging leads to more commitment. For numerous examples of how what you say and how you say it influence your performance and the attitude of others toward you, refer to Walther (1991).

Keep Your Supervisor Informed

Given the pressure of their responsibilities, many supervisors manage by exception. That is, you are unlikely to hear from them unless your performance on a task or project is unsatisfactory or is exemplary, or unless they have a new assignment for you. These types of management-by-exception individuals will probably expect you to function in a similar fashion, especially in your reporting to them. Determine their preferred mode of operation and function accordingly. If you are working with a management-by-exception individual, keep that person informed of the status of your assignments, particularly if the task or project is encountering problems that may have consequences for your supervisor and others.

Speak Up

Ask many questions, because you will have so much to learn about the organization you recently joined and about the effective completion of your assignments. Besides asking questions to help you quickly become a productive member of the organization, also offer suggestions when you see what appears to be a better way of doing things.

Compared to members of the organization who have been around for some time, you might be considered to have the disadvantage of having much to learn about the organization and the tasks you are given. But, in a sense, you have an advantage as a newcomer by being able to take a fresh, relatively unencumbered look at the organization and your assignments. This phenomenon has been referred to as the novice effect (Gross, 1991, pp. 125–163). Accordingly, you may have valuable insights that should be shared with your colleagues and supervisor.

When you speak up, whether it is in a conversation in the office or in a more formal setting such as a meeting, consider some of Benton's (1992, pp. 71–83) suggestions. Avoid a "bored room" voice, that is, a monotone. Do not speak too softly or raise your voice at the end of a declarative sentence, in effect, changing it to a question; women are more prone to do these things. Talk at a moderate speed and certainly not too fast. People tend to listen more carefully if they think you are thinking while you speak. Finally, talk a little but say a lot. Too many people talk a lot and say little.

Some final advice on the topic of speaking up: talk to strangers. This advice is probably contrary to that received from your parents when you were a child. However, you are no longer a child and presumably you can take care of yourself. Roane (1988, p. 10, pp. 185–190), in her book *How to Work a Room*, challenges her readers to "work the world." Adopt the philosophy that you are surrounded by opportunities to learn and make contacts. But you often have to take the initiative, whether you are at your place of work, doing personal errands in your community, or sitting in an airport between flights. Will talking to strangers always yield useful information or a new contact? Certainly not. However, Roane says, "That's not the point. The point is to extend yourself to people, be open to whatever comes your way, and have a good time in the process. One never knows. . . . The rewards go to the risk-takers, those who are willing to put their egos on line and reach out—to other people and to a richer and fuller life for themselves." As Wayne Gretzky said, "You miss 100 percent of the shots you never take."

Dress Appropriately

O wad some power the giftie gie us to see oursels as ithers see us!

(Robert Burns, 1759–1796, Scottish poet, in "To A Louse")

Your first impression is often your last chance.

(Anonymous)

Dress and grooming will significantly affect the professional success of the young engineer or other technical person. Appropriate dress and immaculate grooming are, of course, not sufficient, but they are absolutely necessary. While it is true that a person has a right to dress as he or she pleases, it is also true that others have a right to react as they wish to that person's dress and grooming.

The definition, usually unwritten, of what constitutes acceptable dress for advancing young professionals varies from organization to organization. The extremes range from dark-colored, traditional-style suits for men and women to jeans and sport shirts. There appears to be a trend to a more casual look (e.g., Snead, 1993) but, like many fashion trends, this one may be ephemeral. To determine the appropriate dress for your organization, observe individuals one or two levels above you in the organizational structure. Dressing in accordance with the position you want in your organization is rather bluntly labeled *aspiration dressing (Office Hours,* 1993).

If you work directly with people outside of your organization, you must also be sensitive to their perceptions of you and the organization you represent, especially early in a new relationship, based on your style of dress. Clothing that might be appropriate within your office may not be suitable within your client's or customer's workplace. Fitting clothing to occasions is referred to as *situational dressing* (*Office Hours,* 1993).

While dress styles will vary markedly from organization to organization, the range of acceptable grooming is likely to be much narrower. Strive to be clean and attractive at all times from the top of your head (e.g., clean, trimmed hair) to the bottom of your feet (e.g., clean, polished shoes). Specific grooming suggestions for women and men are presented in Figure 2–8.

MINOR DETAILS MAKE MAJOR DIFFERENCES

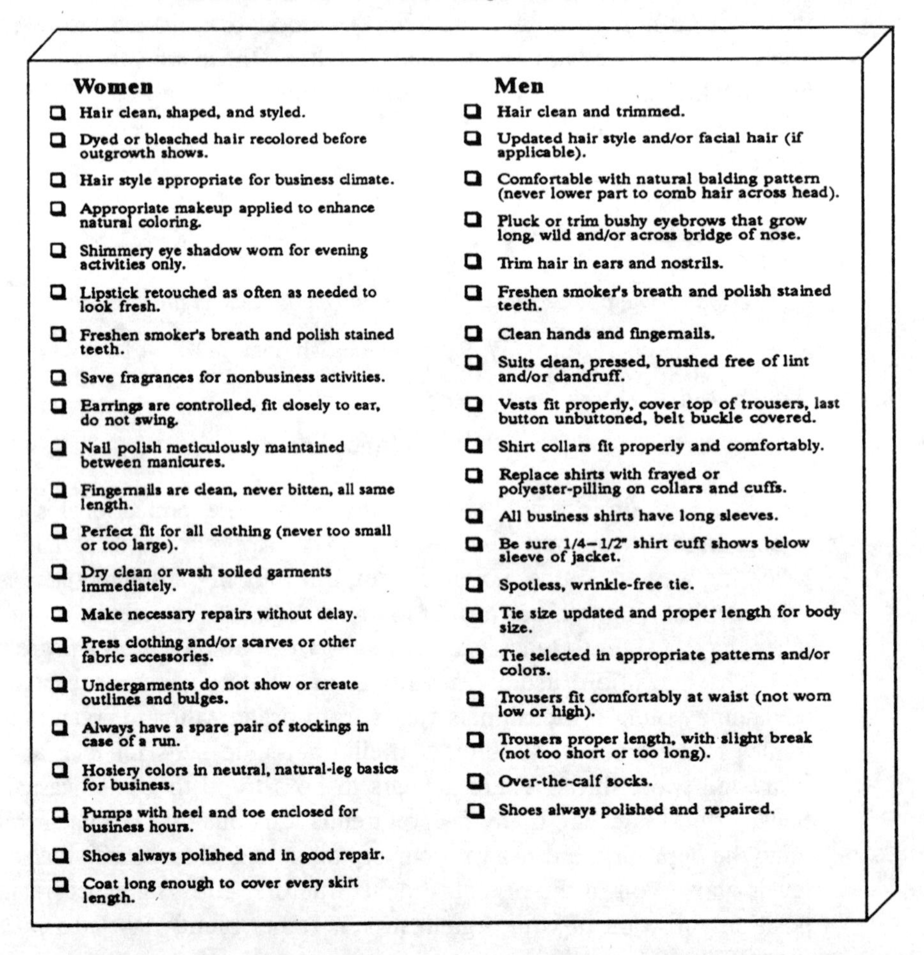

Women

- ❑ Hair clean, shaped, and styled.
- ❑ Dyed or bleached hair recolored before outgrowth shows.
- ❑ Hair style appropriate for business climate.
- ❑ Appropriate makeup applied to enhance natural coloring.
- ❑ Shimmery eye shadow worn for evening activities only.
- ❑ Lipstick retouched as often as needed to look fresh.
- ❑ Freshen smoker's breath and polish stained teeth.
- ❑ Save fragrances for nonbusiness activities.
- ❑ Earrings are controlled, fit closely to ear, do not swing.
- ❑ Nail polish meticulously maintained between manicures.
- ❑ Fingernails are clean, never bitten, all same length.
- ❑ Perfect fit for all clothing (never too small or too large).
- ❑ Dry clean or wash soiled garments immediately.
- ❑ Make necessary repairs without delay.
- ❑ Press clothing and/or scarves or other fabric accessories.
- ❑ Undergarments do not show or create outlines and bulges.
- ❑ Always have a spare pair of stockings in case of a run.
- ❑ Hosiery colors in neutral, natural-leg basics for business.
- ❑ Pumps with heel and toe enclosed for business hours.
- ❑ Shoes always polished and in good repair.
- ❑ Coat long enough to cover every skirt length.

Men

- ❑ Hair clean and trimmed.
- ❑ Updated hair style and/or facial hair (if applicable).
- ❑ Comfortable with natural balding pattern (never lower part to comb hair across head).
- ❑ Pluck or trim bushy eyebrows that grow long, wild and/or across bridge of nose.
- ❑ Trim hair in ears and nostrils.
- ❑ Freshen smoker's breath and polish stained teeth.
- ❑ Clean hands and fingernails.
- ❑ Suits clean, pressed, brushed free of lint and/or dandruff.
- ❑ Vests fit properly, cover top of trousers, last button unbuttoned, belt buckle covered.
- ❑ Shirt collars fit properly and comfortably.
- ❑ Replace shirts with frayed or polyester-pilling on collars and cuffs.
- ❑ All business shirts have long sleeves.
- ❑ Be sure 1/4–1/2" shirt cuff shows below sleeve of jacket.
- ❑ Spotless, wrinkle-free tie.
- ❑ Tie size updated and proper length for body size.
- ❑ Tie selected in appropriate patterns and/or colors.
- ❑ Trousers fit comfortably at waist (not worn low or high).
- ❑ Trousers proper length, with slight break (not too short or too long).
- ❑ Over-the-calf socks.
- ❑ Shoes always polished and repaired.

Figure 2–8 Grooming check list (Source: Used by permission of N. M. Grant and M. J. Cross of Grant Cross Communications Group, Inc., One Brighton, Oak Brook, IL 60521, 708–574–8323 and 219–923–4046, Copyright 1991.)

Incidentally, good posture is essential to achieving the full benefit of the professional clothing you choose to wear. As noted by Benton (1992, p. 54), "It's not what you wear but how you wear it that gets you to the top and keeps you there." Benton goes on to note that good posture also helps you appear energetic, improves health by reducing undue pressure on internal organs, and enhances voice quality. While on the subject of posture, Benton (1992, pp. 58–61) argues that all professionals should frequently stand up for a variety of reasons, including showing respect and signaling the end of a meeting.

Seek Opportunities to Develop Communication Skills

Look for opportunities to develop your communication ability. For example, offer to serve as secretary for a meeting. You will probably find, as a result of that effort, that you obtained the best understanding of the ideas and information exchanged at the meeting in addition to honing your writing skills. For similar reasons, offer to draft a letter, memorandum, or a report that will be eventually sent to one of your organization's clients or customers. Seek out ways to write papers on your work and then present them at conferences.

Volunteer to make oral presentations to colleagues, such as a summary of a seminar or workshop you attended; to clients; and to student, community, and professional groups. Every time you prepare for and deliver an oral presentation, you have the opportunity to improve on a skill that is highly and widely valued within the engineering and business world. Major sections of Chapter 3 are devoted to writing reports and making oral presentations.

Seize Opportunities for You and Your Organization

> *There is a tide in the affairs of men, which, taken at the flood, leads on to fortune; omitted, all the voyage of their life is bound in shallows and in miseries. On such a full sea are we now afloat; we must take the current when it serves or lose our ventures.*
>
> (Shakespeare, *Julius Caesar*)

> *Don't be afraid to go out on a limb. That's where the fruit is.*
>
> (Anonymous)

Luck is when opportunity meets preparation. You certainly will not be successful solely on the basis of luck. On the other hand, luck opens windows of opportunities, often for only a fleeting moment. If you see those opportunities and have the courage to seize them, you and your organization may benefit. For example, as a result of your initiative at a local meeting of an engineering society, you meet a representative of an organization and learn that the organization just happens to need specialized services provided by your firm. You immediately follow up by informing your company's marketing group, and your firm obtains a very attractive contract. For an expanded discussion of this topic refer to Benton (1992, pp. 216–228).

Choose to Be a Winner

Most externally imposed situations are neutral; that is, they are neither inherently "good" nor inherently "bad." They are what we make of them, as suggested by Figure 2–9. While we cannot control much of what happens to us, we can decide how we will respond and what we will do.

Many apparent problems are actually opportunities in disguise. For example, your failure to be selected to work on an exciting new project recently initiated in your office might prompt you to take the initiative to suggest that a similar project be undertaken with a different client. Or assume that you are asked to work long hours for many days to correct calculation errors made by a recently released employee so that a project report can be delivered on time. You could choose to view this as an unfair imposition. On the other hand, you could choose to develop a spreadsheet or computer program for use on this assignment which could be repeatedly used in the future to minimize errors and reduce the amount of time necessary to do calculations.

Recognize that individual and group attitudes, whether they are predominately positive or negative, are contagious in organizations. Unfortunately, negativism appears to move with even greater ease and speed than positivism through an organization. However, positivism can permeate an organization if a few people at all levels choose to take a winning, rather than losing, perspective and course of action. Harris (1968), in a little but useful book, effectively contrasts the approaches of winners and losers.

In a related manner, be known as an action-prone person rather than a talk-prone person. Experience suggests that for every ten to 100 people who say something like "You know what we ought to do?" no more than one will. Be that one.

Summing It Up

The world of professional practice is very different from the world of engineering and technical education. In the former, technical knowledge is assumed and success always depends partly and often heavily on performance in nontechnical areas. Forewarned is forearmed. Appreciate and use your generic qualities and

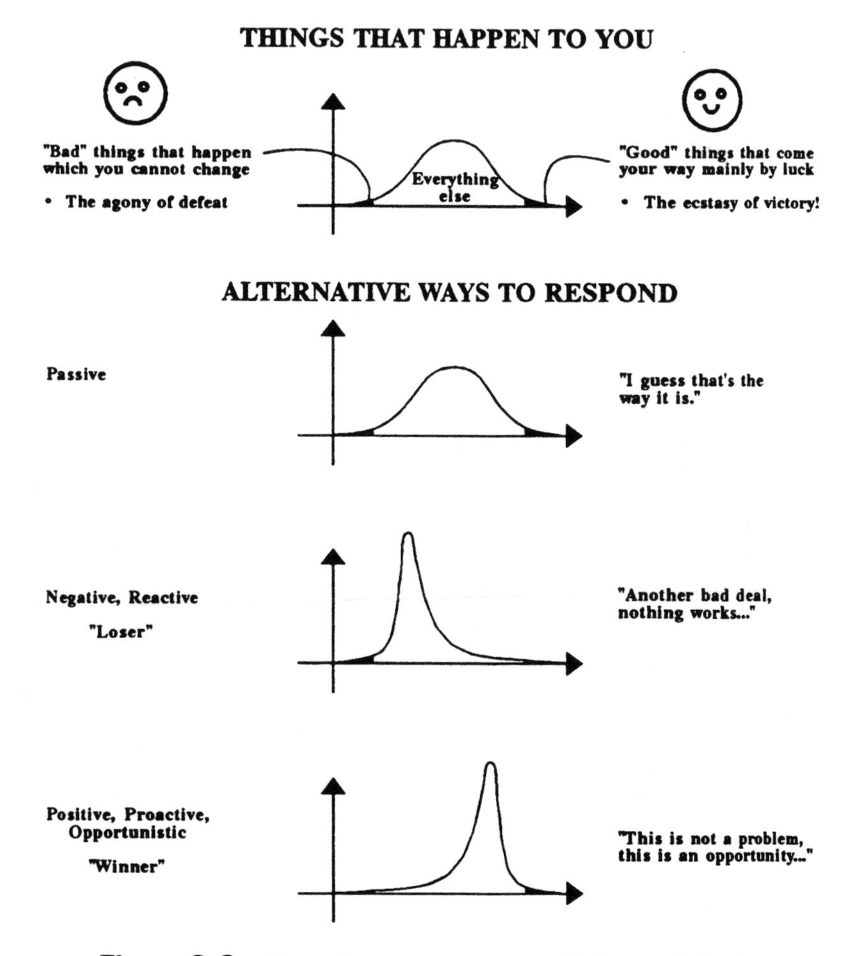

Figure 2–9 Alternative responses to externally imposed situations

characteristics. During the first few months of your first professional employment, guard your personal integrity, learn and respect administrative procedure and structure, do all assignments well, get things done, trim your hedges, keep your supervisor informed, speak up, dress appropriately, hone your communication skills, and be a winner.

MANAGING PERSONAL PROFESSIONAL ASSETS: BUILDING INDIVIDUAL EQUITY

A downturn in the stock market as indicated, for example, by a drop in the Dow Jones Index, is disconcerting for many individuals even the young professional who is just beginning to build his or her balance sheet. Suddenly, retirement accounts, mutual funds, and other investments plummet in value. Individual net worth may drop sharply. There is a gnawing fear of the negative long-term effect on the material well-being of a spouse and dependents. The young person vows to be more careful where they invest and to be more watchful of their investments. After all, prudence requires careful management of personal financial assets.

Personal Professional Assets

What about the status of and the attention to personal professional assets? By virtue of individual talents, education, and experience, each young engineer or other technical professional has significant value to society. In contrast with your financial assets, many of which can be measured to the penny, the value of your professional assets defies quantification. In a narrow sense, individual professional assets or personal professional equity might be valued as the present worth of the projected stream of future earnings. In a broader and more accurate sense, the young technical professional's assets are measured by the actual good accomplished and by all the good a person has the potential to accomplish through conscientious use of his or her talents, education, and experience.

Although the true value of personal professional assets defies quantification, it nevertheless is very real. Furthermore, the value of professional assets, even if narrowly defined, is likely to far exceed the value of the entry-level person's financial assets, except perhaps for those few fortunate people who are independently wealthy. Like personal financial assets, personal professional assets can appreciate, remain level, or decline.

Annual Accounting

The young professional should appraise his or her professional assets at least once a year, perhaps as part of a resume update exercise. What new areas of technology have been mastered? What new management techniques were used? What new concepts, ideas, or principles were studied? What new skills were acquired? What new challenges and responsibilities were accepted? What new opportunities were seized and new risks were taken? What knowledge was shared with professional colleagues? What new contributions were made? In what ways has the young professional been a "good and faithful servant" with his or her talents?

While experience is valuable, too much of one kind of experience can hamper your growth. As you review several annual accountings of your professional assets, will you find several years each filled with new experiences or one year of experience repeated several times? Resist the temptation to settle into the comfort of routine, rationalizing it in the name of gaining experience. Listen to Mandino's (1968, p. 53) warning about excessive experience:

I will commence my journey unencumbered with either the weight of unnecessary knowledge as the handicap of meaningless experience . . . In truth, experience teaches thoroughly yet her course of instruction devours men's years so the value of her lessons diminishes with the time necessary to acquire her special wisdom.

If any annual accounting of your personal professional assets reveals a loss or no increase in value, you have experienced a personally devastating "stock market crash." You have lost a year of growth and increased contribution, neither which can ever be redone or perhaps regained. As a result of mismanagement of your professional personal equity, you have failed yourself, employer, client, profession, and society.

Careful Management of Personal Professional Equity

Each young professional is gifted with a unique combination of talents. A challenge in the early years of professional life is to discover and develop, through reflection, education, and experience, that special set of talents and then to dedicate and direct those talents to meaningful professional work and service. You should commit to managing personal professional assets at least as well as you manage your financial assets. The following sections discuss how to manage your personal professional assets through continuing education, involvement in professional organizations, and licensing.

CONTINUING EDUCATION

Self-education is, I firmly believe, the only kind of education there is. The only function of a school is to make self-education easier; failing that, it does nothing.

(Isaac Asimov)

Continuing education is an important mechanism for maintaining your personal professional equity. A continuous series of challenging and varied work assignments is another important mechanism as suggested by a preceding section of this chapter. Many means are available to the entry-level engineer or other technical professional for immediately beginning his or her individualized continuing education and professional development program. Examples are internal and external workshops and seminars, university classes on a part-time basis possibly leading to one or more graduate degrees, correspondence courses such as those offered by various professional societies, remote learning via audio and video systems, and, for the very disciplined person, self-study. You have the primary responsibility for your continuing education and professional development.

Most technical personnel employers will support your continuing education by mechanisms such as offering some financial assistance and by providing some release time during normal office hours. However, to emphasize the need for you to take charge of this part of your professional life, note that "A recent survey of 165 A/E firms indicates that training is typically conducted on a hit-or-miss basis, is rarely documented, and is infrequently evaluated" (*Engineering Times*, 1992). The survey indicated that only 8 percent of the firms operated formal training programs and only 14 percent had a manager of in-house training. Furthermore, only about 36 percent of the firms responding to the survey provided special training for personnel moving into top positions. Chances are that if you are not in charge of your continuing education, no one will be.

Continuing education is vital to you for one or both of two reasons. First, rapid changes in technology require constant learning to be current. You probably noticed changes in technology during the short time you were in college. If the entry-level engineer expects to pursue a primarily technically oriented career, he or she must remain current.

A second reason to continue your education is to prepare yourself to function in areas other than the technical ones, such as research, marketing, administration, finance, and perhaps teaching. Continuing education is essential if you want to make one or more functional changes within your current employment situation or with a new employer. Think about your imminent commencement ceremony or about the commencement ceremony you recently enjoyed. On the surface, commencements celebrate the beginning of a career. Experience indicates, however, that such occasions also mark the commencement of the rest of your education. Although it may not be as formal as the first part, it will be just as important and it should last much longer—for the rest of your life.

INVOLVEMENT IN PROFESSIONAL ORGANIZATIONS: TAKING AND GIVING

I hold every man a debtor to this profession; from that which man has a course to seek countenance and profit, so ought they of duty to endeavor themselves, by way of amends, to be a help and ornament there unto.

(Francis Bacon, 1561–1626)

Active, as opposed to passive, involvement in professional organizations is the third way to continuously increase the value of one's personal professional equity. The other two principal means, as previously discussed, are varied and challenging work assignments and continuing education.

Beside the somewhat selfish concern of maintaining personal professional assets, entry-level professionals should realize that they will derive a satisfying and prosperous living from their profession and, accordingly, ought to give something back to their profession. The active professional uses the work of many predecessor professionals, many of whom produced the books, papers, conference proceedings, manuals of practice, computer software, and other valuable contributions for which they received little or no monetary compensation. As an indication of the contribution of others to the profession and of the corresponding obligation to do one's share, examine the personal library of a senior professional. Note the relatively large number of materials that were clearly produced, usually in the context of professional organizations, by largely volunteer labor.

The call to be actively involved in professional organizations goes beyond maintaining one's currency and meeting an obligation. Such participation provides an opportunity for you to enjoy and benefit from the company of leaders. Engineering and other technical professions and various subdivisions of them are like local congregations or synagogues—many members but very few doers. The doers are usually committed, creative, ambitious, and accomplished people. The young professional can learn much from associating with them, and the "ticket" is a commitment to being actively, as opposed to passively, involved in the work of professional organizations.

Once a commitment is made, the young professional faces the challenge of identifying the appropriate one or more technical and professional organizations. Beakley et al. (1986, pp. 128–140) lists about 60 engineering or engineering-related technical and professional societies from which you can choose.

Upon joining such an organization, or moving from the student membership status you had in college to the practicing professional status, you should select one or more types of activities for your involvement and contribution. Besides attending meetings, consider presenting and publishing papers, serving on and chairing technical and nontechnical committees, helping to arrange and run meetings and conferences, and serving as an officer. By investing your time and talent in one or more professional organizations, you will realize the significant return on your investment in terms of knowledge gained, satisfaction of contribution, and the association with the leaders of your profession.

As noted, presentation and publication of papers is one way to be actively involved in a professional organization. Personal and organizational benefits of individual or co-authored papers include:

- Improved writing and speaking ability, which is directly and immediately transferable to many aspects of your professional and personal life.
- Increased confidence as a result of interacting with peers.
- Expanded visibility of your organization with emphasis on its accomplishment and abilities.

- Earned membership in networks of leaders, which provides quick access to assistance when needed.
- Returning something to the profession.

Having considered some of the benefits of presenting and publishing papers, you may argue that you simply don't have the time. This may be a valid argument if you have to first do special work to serve as the subject for the paper. Instead, you and your potential co-authors, should seek to write papers about the good work that you have already done. The extra investment in time and effort, on top of what you have already done, will typically be very small compared to the total effort expended. Equally important, the extra investment in time and effort can be very small compared to the extra benefit that will accrue to you, your co-authors, and your organization. This concept of great return on the marginal investment is illustrated in Figure 2–10.

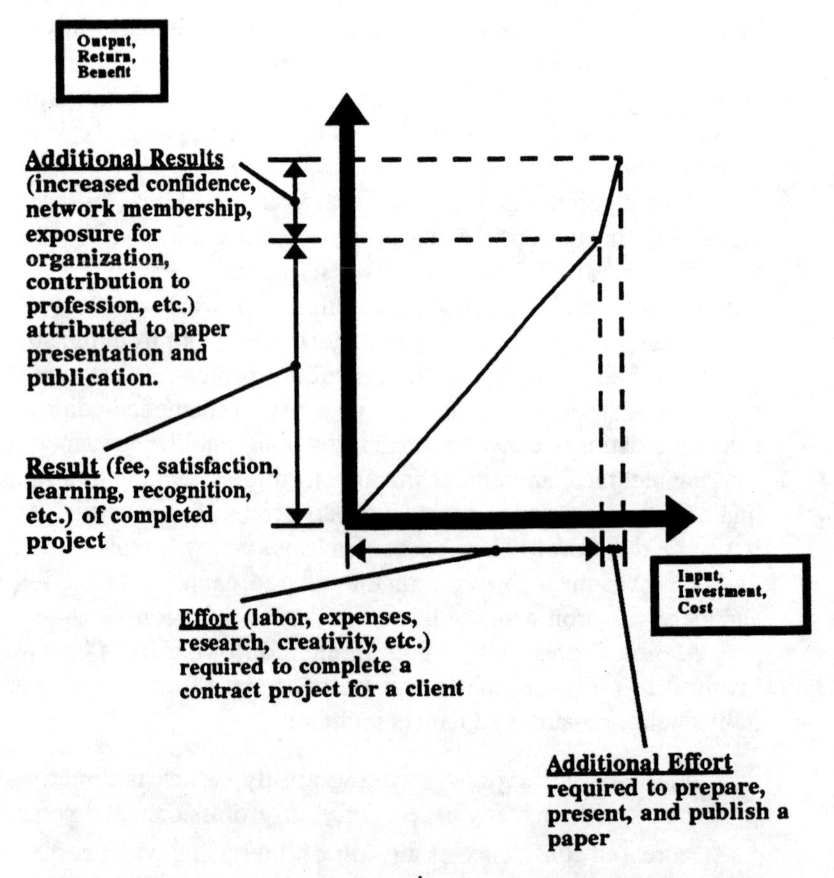

Figure 2–10 Great marginal benefit of papers, presentations, and publications

LICENSING

A system of licensing engineers and certain other technical professionals such as architects has been established in the United States and elsewhere primarily to protect the public by establishing minimum requirements for individuals who plan and design facilities used by the public. Various federal, state, local, and other laws and regulations specify when engineering work must be done under the direction of a licensed engineer (NCEE). Licensing laws focus on protecting the public. Because civil engineers, in comparison to most other engineers, work most closely with the public, licensing is a virtual necessity for civil engineers.

Accordingly, one benefit to the individual engineer of obtaining one or more engineering licenses is more engineering opportunities. Without the license, the engineer will usually be forced to always do engineering work for or under the direction of someone else. With the license, the engineer will be able to do higher-level work, be responsible for more engineering projects, have access to more favorable employment opportunities, and be in a position to someday own and operate his or her consulting engineering or other engineering-based business. To some extent, the holding of one or more engineering licenses is also a mark of achievement.

Licensing Process

The typical five-step licensing process as used in engineering is illustrated in Figure 2–11. The first step is usually to earn a four-year bachelor of science degree in an engineering program (e.g., civil engineering, computer engineering, electrical engineering, mechanical engineering, etc.) accredited by the Accreditation Board for Engineering and Technology. The second step is successfully completing the eight-hour Fundamentals of Engineering Examination typically taken on campus during the senior year of one's engineering education. As suggested by the title of this examination, it focuses on fundamentals. Therefore, the examination is most appropriately taken near the end of one's formal undergraduate education. Some engineering students fail to see the wisdom of taking the Fundamentals of Engineering Examination while they are at the university. Later, they realize the need for the examination and then have difficulty with the logistics of arranging to take the examination and often have considerable difficulty in passing it.

The third step in the engineering licensing process is to obtain four years of progressively responsible experience as an engineering intern. About one-half of the states in the United States allow some cooperative education time to count toward the four years of experience. Some graduate school time is also usually credited. The fourth step in the licensing process is successfully passing the

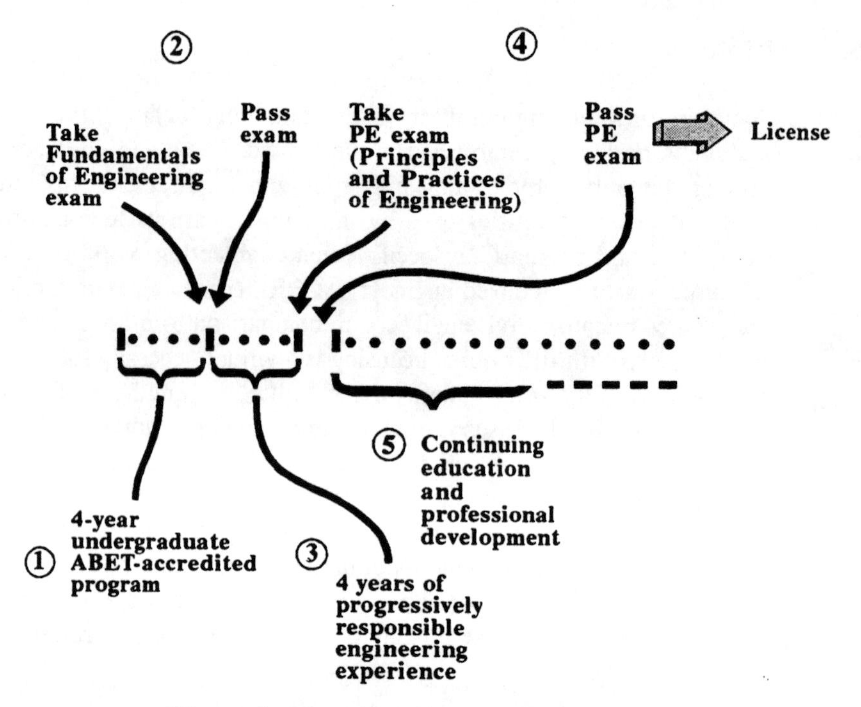

Figure 2-11 Licensing process for engineering

eight-hour Principles and Practices of Engineering Examination. Some states allow solving problems or answering questions in any area of engineering, and others require remaining in one or a few areas.

The final and longest-lasting step in the licensing process is continuing education and professional development. As indicated earlier in this chapter, this can be achieved through a combination of challenging and varied work assignments, continuing education, and active involvement in professional organizations.

Thoughts on Taking the Fundamentals Examination While in Engineering School

A few additional words are in order regarding the wisdom of taking the Fundamentals of Engineering Examination while in engineering school. Most civil engineering majors will take the examination at that time because they learn through faculty and others that holding an engineering license, or at least being in the licensing process, is expected by most civil engineering employers.

Engineering students in other disciplines, however, are likely to hear about what is called the "industrial exemption." On that basis, many will unwisely decide not to take the Fundamentals of Engineering Examination while at the university. Under the industrial exemption, which is in effect in most states in the United States, state registration laws indicate that an engineer does not necessarily need to have an engineering license. If the electrical, mechanical, or other industrially oriented engineer is absolutely certain that he or she will always be working in that situation, then the Fundamentals of Engineering Examination may not be warranted. But many changes occur during an engineering career, including moving into other employment sectors and deciding to begin one's own engineering business. Both of these changes are likely to require an engineering license. Acquiring such a license later in the career, particularly if an engineer has not successfully passed the Fundamentals of Engineering Examination, will be extremely difficult. Wise people keep options open.

Except for extreme circumstances, not taking the Fundamentals of Engineering Examination in college is like taking a driver's education course in high school but not getting a driver's license, running a marathon and quitting ten yards short of the finish line, or buying a red sports car and leaving it in the garage. You should think long and hard before you "buy" the industrial exemption argument.

While many engineering faculty will support the engineering licensing process, especially for the engineers who intend to practice engineering, some will be neutral to negative towards licensing. They may base this partly on the industrial exemption provision, as already discussed. They may also not value an engineering license because, in most situations, engineering faculty do not need licenses to practice their profession, that is, teaching and research. Accordingly, while an engineering license may not be appropriate for them, it may be very appropriate for you.

Comity

Comity is the process through which the state board of registration for engineers of one state may license a person to practice engineering in that state on the basis of a license issued by another state. Under comity, an engineer can usually have the Fundamentals of Engineering Examination results transferred from one state in the United States to another or obtain an engineering license in one state as a result of holding a valid license in another state. Although there are some exceptions (such as special structural engineering licensing procedures in some states), comity is common throughout the United States.

A problem does arise between the United States and Canada, however, because most Canadian engineering students obtain engineering licenses in their country without an examination. The basic requirement is two years of experience after completing their engineering education. As a result, U.S. engineering licenses are generally transferable to Canada, but Canadian engineering licenses are not readily transferable to the United States (Schwartz, 1988; Chapple, 1990).

There is also a comity problem between the United States and other countries. For example, unlike the United States, where engineering licensing is done on a state basis, European countries tend to license engineers at the national level. The European practice simplifies international comity agreements. U.S. and European professional organizations are working on ways to facilitate individual engineers becoming licensed to do engineering work in other countries (Stephens, 1989, Chapple, 1990).

License Renewal

While renewal of engineering licenses is required, retesting is not required as a condition of a license renewal. Furthermore, with the exception of a handful of states (e.g., Iowa, since 1978; Alabama, since 1991; and West Virginia, since 1992), documented continuing education is not a requirement of license renewal (Alabama, 1992; Escobedo, 1993; professional publications, 1988). Provisions of the state licensing laws for professional engineers are administered by state boards. Sanctions for violations range from letters of reprimand to permanent loss of a license to practice engineering in a given state.

REFERENCES

ADAMS, H. G., *Successfully Negotiating the Graduate School Process: A Guide for Minority Students*. Notre Dame, Ind.: National Consortium for Graduate Degrees for Minorities in Engineering, 1987, p. 39.

Alabama, State of, "Rule for Continuing Education for Professional Engineers and Land Surveyors," October, 1992.

BARBER, E. G., R. P. MORGAN, and W. P. DARBY, *Choosing Futures: U.S. and Foreign Student Views of Graduate Engineering Education*, IIE Research Report No. 21, Institute of International Education, New York, 1990.

BEAKLEY, G. C., D. L. EVANS, and J. B. KEATS, *Engineering: An Introduction to a Creative Profession*, 5th ed. New York: Macmillan, 1986.

BENTON, D. A., *Lions Don't Need to Roar: Using the Leadership Power of Professional Presence to Stand Out, Fit In, and Move Ahead*. New York: Warner Books, 1992.

BROWN, T. L., "Time To Diversify Your Life Portfolio," Inside Management Column, *Industry Week*, November 10, 1986, p. 13.

CHAPPLE, A., "Global Market Has Engineers Calling for Internationally Recognized PE," *Engineering Times*, October 1990.

COVEY, S. R., *The 7 Habits of Highly Effective People.* New York: Simon & Schuster, 1990.

CROSS, H. (R. C. Goodpasture, ed.), *Engineers and Ivory Towers.* New York: McGraw-Hill, 1952.

CYPERT, S. A., *The Success Breakthrough.* New York: Avon Books, 1993.

Day-Timers, Inc., "44 Telephone Tips," Allentown, Pa. (no date).

Day-Timers, Inc., "How to Organize Your Office," Allentown, Pa. (no date).

DORNEY, R. C., "How To Find the Time You Need Through Better Time-Management." Allentown, Pa.: Day-Timers, Inc., 1988.

DRUCKER, P. R., "Managing for Business Effectiveness," *Harvard Business Review*, Number 63303, May–June 1963.

Engineering Times, "A/E Firms Hit or Miss on Training," February, 1992.

ENGSTROM, T. W., and R. A. MACKENZIE, *Managing Your Time.* Grand Rapids, Mich: Zondervan Books, 1967.

ESCOBEDO, D., "Alabama Steps into Continuing Ed Spotlight," *Engineering Times*, December, 1993.

FLORMAN, S. C., *The Civilized Engineer*, Chapter 5, "The Engineering View." New York: St. Martin's Press, 1987.

GROSS, F., *Peak Learning.* Los Angeles: Jeremy P. Tarcher, Inc., 1991.

HARRIS, S. J., *Winners and Losers.* Allen, Tex.: Argus Communications, 1968.

IACOCCA, L., with William Novak, *Iacocca—An Autobiography.* Toronto, Can.: Bantam Books, 1984.

KELLY, M., "The Seven Secrets of Successful Time Management," *Administrator: Newsletter for Higher Education*, Madison, Wisc. (no date).

KING, W. J., *The Unwritten Laws of Engineering.* New York: American Society of Mechanical Engineers, 1944.

LANDIS, R. B., "An Academic Career: It Could Be For You," American Society for Engineering Education, Washington, D. C., 1989.

LOCK, M. C., and F. C. SCHWARZ, "Beat the Clock with Better Time Management," *May Trends*, Vol. 10, No. 3 (September 1976), pp. 7–10.

MACKENZIE, A., *The Time Trap.* New York: American Management Association, 1990.

MANDINO, O., *The Greatest Salesman in the World.* New York: Bantam Books, 1968.

MCDONALD, J. M., "Timely Tips on Managing Your Work," *Chemical Engineering*, November 20, 1978, pp. 197–200.

MORSE, J. L., ed., *Funk & Wagnalls New Encyclopedia.* New York: Funk & Wagnalls, 1973, Vol. 13.

National Council of Engineering Examiners, "Why Become a PE?—The NCEE Guide to Registration," NCEE, Clemson, S.C. (no date).

Office Hours, "Establishing a Professional Image," Vol. 93, No. 1, 1993.

ONCKEN, W., JR., W. and D. L. WASS, "Management Time: Who's Got the Monkey?" *Harvard Business Review*, November–December, 1974, pp. 75–80.

PARK, W. R., "Managing Your Time," Engineering Economics Column, *Consulting Engineer*, July 1978, pp. 16–17.

ROANE, S., *How to Work a Room: A Guide to Successfully Managing the Mingling*. New York: Shapolsky Publishers, 1988.

Sales and Marketing Digest, "Making the Most of Precious Time," October 1988.

SCHWARTZ, A., "Legal Corner," *Engineering Times*, Vol. 10, No. 7 (July 1988), p. 2.

SNEAD, E., "Casual Look Suits Today's Businessmen," *USA Today*, August 13, 1993.

STEPHENS, K., "U. S., Europe Draft Accord on Acceptance of Engineering Credentials," *Engineering Times*, March 1989.

University of Illinois, College of Engineering, "Advancing By Degrees," distributed by the American Society for Engineering Education, Washington, D.C., 1982.

WORTMAN, L. A., *Effective Management for Engineers and Scientists*. New York: Wiley, 1981.

WALTHER, G. R., *Power Talking*. New York: Berkley Books, 1991.

SUPPLEMENTAL REFERENCES

American Society of Civil Engineers, Committee on Employment Conditions, *Job Search Handbook for the Graduating Civil Engineer*. New York (no date).

BENNETT, F. L. and W. B. MCMULLEN, "What Are Engineering Employers Looking for in Engineering Management Graduates?" *Journal of Management in Engineering—ASCE*, Vol. 3, No. 4 (October 1987), pp. 267–274.

EDELHART, M., *Getting From Twenty to Thirty: Surviving Your First Decade in the Real World*. New York: M. Evans and Company, 1983.

ELSEA, J. G., *First Impression, Best Impression*. New York: Simon & Schuster, 1984.

FUTRELL, G. E., "What Do You Look for in a New Engineer?" *Journal of Management in Engineering—ASCE*, Vol. 1, No. 1 (January 1985), pp. 20–27.

GJOVIG, B., *Pardon Me, Your Manners are Showing*. Grand Forks, N.D.: Center for Innovation and Business Development, University of North Dakota, 1993.

Institute of Electrical and Electronics Engineers, Inc., "Professional Practices for Engineers, Scientists, and Their Employers," Task Force on Career Maintenance and Development, Washington, D.C., 1983, p. 12.

KLEIN, L., "Making Your First Job Count," *The Woman Engineer*, Fall 1988, pp. 50–51.

KOSIBA, R. J., "Entering Technical Management," *Journal of Management in Engineering—ASCE*, Vol. 1, No. 3 (July 1985), pp. 149–156.

MONTGOMERY, R. L., "Why Do So Many People Fail and So Few Succeed?" *Indiana Business*, November 1989, p. 89.

EXERCISES

2.1 PRODUCTIVITY TIPS

Purpose

Share personal productivity tips among members of the class and with the instructor.

Tasks

1. Identify one way in which you utilize time, handle paperwork, arrange workspace, keep alert, etc. (If done as a team assignment, identify several productivity tips.) Do not repeat specific productivity tips already included in this chapter.
2. Briefly describe the personal productivity tip(s) in a memorandum to the instructor.
3. *Note:* The instructor may share some of these with all members of the class but preserve the anonymity of contributor(s).

2.2 NONTECHNICAL FACTORS

Purpose

1. Emphasize the importance of nontechnical factors in the ultimate success of technical professionals and their projects.
2. Provide in-depth knowledge and understanding of one very accomplished "engineer."

Tasks

1. Select an accomplished engineer or other technical professional, or someone who made significant accomplishments in what is now considered a technical profession. To suggest the wide variety of individuals to choose from, acceptable examples are Sextus Julius Frontinus, A. G. Eiffel, Herbert Hoover, Joseph Strauss, and Orville and Wilbur Wright.
2. Write a memorandum to the instructor indicating your selection and the rationale.
3. Obtain the instructor's approval.
4. Research your topic with emphasis on the selected individual(s) nontechnical characteristics.
5. Write and submit a five- to ten-page paper presenting your findings. Use about one-tenth of the paper to summarize the engineering accomplishment(s) of your selected individual(s). Devote the remainder of the paper to positive and negative nontechnical characteristics of your selected individual(s). Explain the role of those factors in the technical professional's success.

3

Management of Communication Skills: Communicating to Make Things Happen

> *Seek first to understand, then to be understood.*
>
> (Covey, 1990, p. 239)

Communication skills are defined for the purposes of this book and illustrated in Figure 3–1 as consisting of listening, speaking, writing, and use of graphics and mathematics. The unity and completeness suggested by the circle indicate that a technical professional needs competence in all five skills to be a complete communicator.

If you agree that all five forms of communication are important, consider the instruction you received in each of them. Refer again to Figure 3–1, begin with mathematics, and proceed counterclockwise. Your engineering or other technically oriented education certainly included a strong emphasis on the use of mathematics and graphics, although neither may have been oriented towards communicating with largely nontechnical audiences. Writing instruction and critiquing typically receive moderate attention in technically oriented education programs. Even less emphasis is usually placed on developing speaking skills, and explicit instruction on building listening skills is rare. Accordingly, you may have some communication deficiencies and, listed in order of decreasing priority, they are likely to be in the areas of listening, speaking, and writing.

Recall the results-oriented definition of engineering/technology management presented in Chapter 1, namely making things happen through people in the world of applied technology. Your concepts, ideas, discoveries, creations, and opinions will contribute to making things happen only if they are effectively communicated to others. Effective communication, in the context of a particular situation, means using several or all of the five communication skills to understand colleagues and others and to accurately and convincingly convey your con-

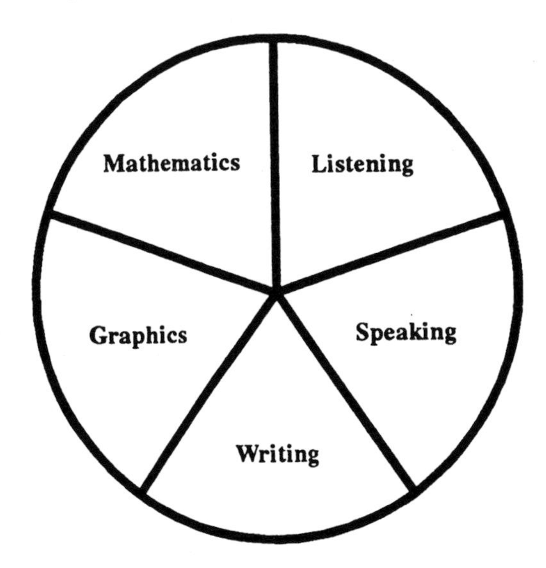

Figure 3–1 Components of communication

cepts, ideas, discoveries, creations, and opinions to them. The most exciting vision, the most thoughtful insight, the most elegant solution, and the most creative design are all for naught unless they are effectively communicated to others. Lacking such communication, the intellectual and other seeds that you plant within your organization and within your professional circles are not likely to sprout and bear fruit. You and your colleagues will be denied the bounty of your labors.

Stated differently, effective communication is necessary, but not sufficient, to success in engineering and other technical professions. There may be the rare exception, such as the noncommunicative genius tolerated by others because of his or her extraordinary intellectual or creative gifts. Then there is the occasional recent technical program graduate who also happens to be the boss's daughter or son and who lacks communication and other interpersonal skills, but is foisted on the other members of the organization. Unless you are a genius, are inextricably linked to the organization's ownership, or enjoy some other rare privilege, you will need effective communication skills to realize your potential.

You might be tempted to agree with the importance of communication, but argue that as an entry-level professional in a new environment, your plate is full and, therefore, you will defer developing effective communication skills for a few years. The fallacy of your position, if it persists, will gradually become evident as you find yourself thinking things such as:

- I know I had the best solution to the design problem but we decided on an inferior course of action.

- Mary and I started at about the same time, but she is increasingly having out-of-office contact with clients while "they" never let me out of this place.
- Juan was selected to attend a valuable two-day seminar, as indicated by the "brown bag" summary he gave the other day, but my request was turned down.
- I'm just not appreciated around here. I think I'll start looking around for other opportunities.

This chapter includes major sections on the three communication skills of listening, speaking, and writing. Uses of graphics and mathematics, the other two communication skills, are included in the listening, writing, and speaking sections. An introduction to body language concludes the chapter. The ideas and information presented here will allow you to build on a communication foundation provided as part of your formal education. If your undergraduate program did not stress the importance of communication and require you to practice communication skills, you probably have a serious liability. Ideas and information set forth in this chapter, coupled with your desire and self-discipline, should help you overcome that deficit.

LISTENING

Big thinkers monopolize the listening. Small thinkers monopolize the talking.

(David Schwartz)

Of the five types of communication noted in the introductory portion of this chapter, listening might appear to be the easiest. On the contrary, listening effectively, that is, listening to understand what others mean and how they feel is very difficult. Hearing, while it is necessary to listening, is not listening. Linver (1978, p. 115), says that hearing ". . . is a natural, passive, involuntary activity. Anyone with a normally functioning ear and brain will involuntarily hear sounds of certain intensity." Effective listening goes well beyond hearing by requiring being attentive, verifying understanding, and using what you learn.

Be Attentive

Covey (1990, p. 240) identifies five degrees or levels of attentiveness in how a person uses his or her hearing ability. As shown in Figure 3–2, they are ignoring, pretending to listen, selective listening, attentive listening, and empathetic listening. Ignoring the speaker, the lowest level of listening, may seem unusually rude,

- **Empathetic listening**
- **Attentive listening**
- **Selective listening**
- **Pretending to listen**
- **Ignoring**

Figure 3–2 Levels of listening

especially in a small group setting. However, most people have had the humbling experience of realizing that no one was listening to them.

While appearing to be attentive—that is, pretending to listen—may help, actually being attentive—through selective listening or attentive listening—will accomplish much more. Incidentally, Covey distinguishes between selective and attentive listening, levels three and four, by explaining that the former means hearing only parts of a person's message, perhaps by design, and the latter means hearing all of a person's words.

Empathetic listening, the highest and most desirable level according to Covey (1990, p. 240), means "listening with the intent to understand" or trying to see the situation as the other person sees it. Decker (1992, pp. 190–193) calls this "feeling listening" and distinguishes it from "fact listening." Empathetic listening is not necessarily sympathetic listening. "The essence of empathetic listening is not that you agree with someone; it's that you fully, deeply understand that person, emotionally as well as intellectually" (Covey, 1990, p. 240). Incidentally, empathetic listening is risky for the listener. If you truly achieve empathetic listening, then you have discovered how another person thinks and feels about something. As a result, you may be profoundly influenced and your thinking and feeling may change accordingly (Covey, 1990, p. 243).

Assuming that you want to achieve the highest level of listening, what techniques can you employ? Benton (1992, p. 81) suggests that you "Silence all internal dialogue. Let extraneous thoughts pass through your mind without dwelling on them or allowing them to take your attention away from the speaker." Comfortable eye contact is important, according to Roane (1988, p. 73), who suggests avoiding the extremes of glaring continuously at the speaker and looking everywhere but at the speaker. Other body language messages are also important, according to Benton (1992, p. 81), who cautions against looking bored, intimidated, or intimidating. She suggests encouraging the speaker by signaling "tell me more" in one way or another. Body language is discussed separately at the end of this chapter.

Verify Understanding

Recall that the goal is empathetic listening—understanding the other person intellectually and emotionally. You may need to verify your understanding of the speaker and his or her message. Obvious techniques include asking questions and paraphrasing or summarizing what you think the speaker is saying and seeking their confirmation. You may have to courteously interrupt the speaker to verify your understanding, but this will probably be viewed positively because it suggests that you are listening attentively and perhaps even with empathy. Engineers and other individuals with technical and scientific abilities sometimes use impromptu sketches as a way of verifying understanding.

Use What Is Learned

The most obvious way to demonstrate that you correctly and fully—intellectually and emotionally—heard what the speaker said is to reflect your new understanding in subsequent written and oral statements and in your actions. Your statements and actions won't necessarily reflect the speaker's wishes, but they will be informed by the speaker's perspective. No colleague can ask for more or should settle for less.

TWO CRITICAL DISTINCTIONS BETWEEN WRITING AND SPEAKING

Decker (1992, pp. 31–34) draws two distinctions between writing and speaking that are very valuable in preparing for and doing both. First, written communication is received by the reader as, to use Decker's words, "linear, single-channel input" as shown in Figure 3–3. Written words, one after another, are what the writer sends and the reader receives. Words only, with the possible exception of supporting tables or figures, must carry the written message, Therefore, the words must be carefully chosen and arranged.

In contrast, spoken communication is received by the listener—optimally an empathetic listener as defined earlier in this chapter—as "multichannel input"

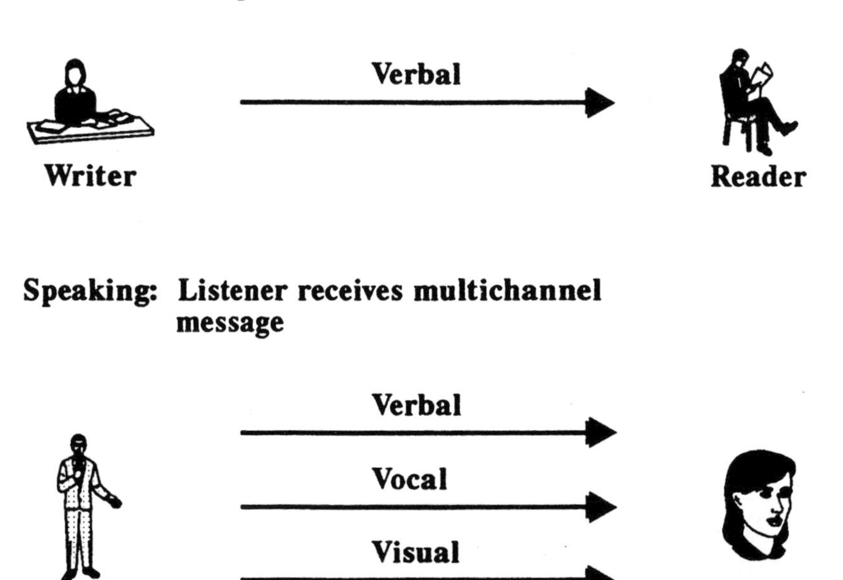

Figure 3–3 Writing and speaking as received by the reader and listener

as illustrated in Figure 3–3. The writer sends, intentionally or unintentionally, and the listener receives what Decker (1992, pp. 83–85) refers to as a three-component message. The components of your spoken message are verbal, the words you use; vocal, the way you use your voice to say the words; and visual, your facial and body expressions and motion and the visual aids you use as you speak. And what is the relative input of the three V's of spoken communication? Research indicates, as discussed further in the body language section near the end of this chapter, that visual is the most important and verbal is the least important! Forewarned is forearmed; plan your spoken presentations accordingly. Explicitly plan for and use the verbal, vocal, and visual components of spoken communication, and place appropriate emphasis on each.

Decker's (1992, pp. 31–34) second writing versus speaking distinction is that writing is much more effective in communicating "facts, data and details" than is speaking. In contrast, speaking clearly holds the power of persuasion. Again, forewarned is forearmed. Depending on your primary objective—information transfer or persuasion—wisely choose and carefully execute your medium.

REPORT WRITING TIPS: A CHANCE TO SHINE

I write to find out what I'm thinking.

(Edward Albee)

Hard writing makes for easier reading.

(Brown, 1988)

Very often, the most used and influential product of a technical investigation is a written report. That product, the "written report," is defined broadly in this book to include formal reports as well as memoranda and letters. Assuming that the data and information are available, the ideas or tips presented in Figure 3–4, are discussed to assist the entry-level professional to produce all or portions of an effective written report. The focus is on producing a report with minimal effort and maximum communication potential.

Define and Write to Likely Audience or Audiences

Engineers often write reports intended for technical and nontechnical individuals and organizations. The same report has to be intelligible and useful to multiple audiences. For example, a report produced by design engineers in a manufacturing organization may be directed to technical personnel such as engineers in manufacturing and to nontechnical personnel such as staff in marketing and finance departments. Similarly, a report produced by a civil engineering consulting firm may be directed to technical personnel such as the city engineer or director of public works and nontechnical individuals such as the mayor, city council members, and citizens.

Accordingly, the following report structure is suggested as a means of communicating with multiple, technical, and nontechnical individuals and audiences:

- Transmittal letter or memorandum (used for formal reports and largely nontechnical in content)
- Table of contents (what is in the report and where?)
- Executive summary (mostly nontechnical and completely self-contained in that there are no explicit references to other parts of the report)
- Chapters (mostly technical)
- Acknowledgements (say thank you)

- **Define and write to audience(s)**
- **Ask about written report-writing guidelines and standards**
- **Outline and incubate**
- **Retain some of the outline in the report**
- **Write "easy" parts first**
- **Write in third person**
- **Use a gender-neutral style**
- **Write in active, direct manner**
- **Adopt flexible format for identifying tables, figures, and references**
- **Design a standard base map or diagram**
- **Use format writing**
- **Establish milestones**
- **Produce an attractive, appealing report**
- **Cite sources**
- **One more time**

Figure 3–4 Report-writing tips

- Cited References (give credit)
- Appendices (highly technical or very detailed)

With this model, confidential or very large appendices may not be included with all copies of the report, but may be available on special request.

As a further expansion of the concept of a report structured to communicate with various audiences, refer to Figure 3–5. Note how various parts of the report

Transmittal Letter (as may be appropriate)

Title Page (title, date, author, etc.)

Table of Contents

Chapter 1 - Executive Summary (2 page maximum)

- **Introduction**
- **Methodology**
- **Conclusions**
- **Recommendations**

Chapter 2 - Introduction

- **Background**
- **Purpose**
- **Scope of Work (e.g., planning, design, research, etc.)**
- **Related Studies/Projects**

Chapter 3 - Description of the Project Area/Site/Topic/etc.

Chapter 4 - Evaluation of Alternatives and Recommendations

- **Design Criteria/Constraints**
- **Cost Estimation Procedure**
- **Alternative 1**
- **Alternative 2**
- **Etc.**
- **Recommendations (what, why, when, who)**

Acknowledgements

Cited References

Appendix A

Appendix B

etc.

Note: Appendices (sometimes called attachments or exhibits) could include items like cost estimates for alternatives, definitions of terms, list of abbreviations, derivations of equations, sample computer input and output, descriptions of equipment, maps, data tables, key letters/ memoranda, etc.

Figure 3–5 Generic report outline

will naturally be of interest to certain technical and nontechnical readers. McKenna (1990, p. 38) notes that most readers of a report will want to quickly know the "bottom line," that is, the most important information. Accordingly, he advocates the inverted pyramid approach, a news-writing method, that presents the most important information first. Note that the report structure suggested in Figure 3–5 does this.

Young engineers may be troubled by the inverted model. Presenting the conclusions and recommendations before describing the data collection and analysis procedures is certainly contrary to the way the work was actually done. But, to be blunt, most readers of the report will not care what was done, other than to have the confidence that it was done correctly. They will focus on the recommendations and implications for them, their area of interest, or their constituencies.

Ask about Written Report-Writing Guidelines and Standards

Some organizations have style guides and procedural manuals for writing of reports and other documents. These helpful guidelines might address topics such as abbreviations, capitalization, citation of references, gender-neutral writing, punctuation, type styles, graphics, and overall report format. Inasmuch as engineering reports are often written as a team effort, style guides and procedure manuals help to achieve internal consistency within any given report. They also contribute to interreport consistency for the organization. Finally, writing guides save time and minimize frustration by greatly reducing the need to make numerous decisions about the mechanics of a report. Determine if your organization has guides or manuals. If not, perhaps you will be in a position, after having written several reports, to suggest that such documents be prepared and to contribute to their preparation.

Outline and Incubate

Using the suggested overall structure, and report-writing style guides or procedure manuals provided by your employer, outline the report in ever-increasing detail. For example, within each chapter, identify likely first-, second-, and third-order headings. Prepare a preliminary list of likely appendices. Identify key figures and tables and determine approximately where they will appear.

While a formalized outline may not be your preferred approach, resist the temptation to write complete sentences and paragraphs until at least a crude outline is fully expanded and has been refined on several different occasions. Provide "incubation time," that is, an opportunity for your subconscious mind to work on the project. As nicely stated by Benton (1992, p. 200), "plant subjects you are considering in the garden of your subconscious." In other words, begin work on the outline of the report, set it aside for several hours or a day or so. Then pick up the project again and work further on the outline. You are likely to find that you have gained new insights and ideas by virtue of providing an opportunity for your subconscious mind to work on the project while you are doing other activities. Repeat this process several times. By evolving a detailed report outline, the principal writer of the report or part of a report is less likely to omit items of importance or to overemphasize one or more topics. The outline can also be a very effective tool for internal review of report structure and content prior to initiating the more time-consuming writing. Consider asking the client or ultimate recipient or recipients of the report to provide input based on the working outline.

Retain Some of the Outline in the Report

Carefully selected, placed, and spaced first-, second-, and perhaps third-order headings are useful to the reader who skims the report for topics or areas of particular interest. Those headings, perhaps down to the second-order level, can also be used to construct a table of contents for the report which further enables the selective reader to focus on critical parts of the report. Recognize that the author of the report may be asked, often on short notice or in an impromptu situation, to summarize the report. By "turning the pages" on a report that has carefully designed headings and subheadings, effective presentations can be made even though they are done on relatively short notice.

Write "Easy" Parts First

As the writer of the report, you will probably be very familiar with or most enthusiastic about certain aspects of the project that will appear in the report. For example, perhaps you developed a computer program and used it to do part of the design. Maybe you directly supervised data collection in a laboratory or carried out a literature search. While keeping the relative importance of your favorite parts of the report in perspective, do not hesitate to write those parts first. That is, a report does not have to be written in chronological fashion given the versatility of modern word processing equipment.

Write in the Third Person

Most engineering reports are written in the third person as a matter of convention. Exceptions are certainly appropriate where the overall effectiveness of the report will be improved. For example, you might decide to use a first-person structure for that part of the report where recommendations are presented. Each recommendation might be preceded with the first-person statement "We recommend . . . ." By switching abruptly from third person to first person, the report structure draws the reader's attention to the importance of the recommendations. The style change also reminds the reader that recommendations are based on the data and analysis presented in the report combined with the experience and judgment of the engineering group or organization that prepared the report.

Employ a Gender-Neutral Style

Unless the expected audience is entirely composed of one known gender, draft the report so as not to favor one gender over the other. If the basic style of your report, or the accompanying oral presentation, is offensive to part of the audience, your message may not be "heard."

Besides use of the now common "he or she" pronouns, Walters and Kern (1991) offer two useful suggestions. First, avoid or greatly reduce gender-specific pronouns. For example, instead of writing "When a person needs assistance with software, he or she should contact the Office of Electronic Information Services" use the gender-neutral pronoun "anyone." Rewrite the sentence to read "Anyone needing assistance with software should contact the Office of Electronic Information Services." The second approach suggested by Walters and Kern is to write in a plural format rather than a singular format. For example, instead of writing "If a member of the staff needs assistance in making travel arrangements, he or she should call the Human Resources Office," write "Staff members needing assistance with travel arrangements should call the Human Resources Office."

As you strive to write in a gender-neutral style, be careful to avoid the grammatically incorrect "anyone/everyone-their" construction. For example, the sentence, "Everyone should bring their project materials to the meeting," is grammatically erroneous because "everyone" is singular and "their" is plural. Unfortunately, their is no singular gender-neutral pronoun except "it." One way to correct the example sentence is to sacrifice some specificity by omitting "their."

Write in an Active, Direct Manner Rather than a Passive, Indirect Manner

Consider for example the passively structured sentence, "It was determined from application of computer modeling that the existing freeway system could not accommodate rush-hour traffic." A preferable active reconstruction of this sentence is "Computer modeling indicated that the freeway system will not carry rush-hour traffic." The active structure is shorter and more easily understood. Furthermore, the active structure eliminates the word "it," which defies any logical meaning. As another example, consider the passively constructed statement, "It should be noted that Type C soils are most common in the project area." An active, direct reformulation of this sentence is "Type C soils are most common in the project area."

Use Rhetorical Techniques

Rhetoric is the art of using words effectively. Conger (1991) effectively argues that the most effective presenters use carefully selected rhetorical techniques. He identifies and gives examples of metaphors, analogies, brief stories, repetition, rhythm, and alliteration. Martin Luther King's "I Have A Dream Speech," for example, is used to suggest the power of repetition and rhythm. You should incorporate carefully selected rhetorical techniques in your writing.

Adopt a Flexible Format for Identifying Tables, Figures, and References

As a report evolves, particularly in an interdisciplinary environment, numerous changes will be suggested and many of them will be implemented. The text will be expanded and deleted; tables and figures will be modified, expanded, or deleted; new chapters may be introduced; and additional reports, papers, books, and other information sources are likely to be cited in the report.

Recognizing the dynamic, evolving nature of the report, one should adopt certain techniques to minimize the effort required to incorporate the inevitable changes. For example, in the text, cite references to reports, papers, books, and other sources of information using a format like "Smith, 1984" rather than "Reference 3." The latter format usually requires frequent renumbering of the reference list and of referenced citations in the text as references are added. In contrast, the former format simply refers to an alphabetized list of references at the end of a chapter, or perhaps as an appendix to the report, that gradually expands, but requires no changes of numbers or of the citations in the text.

If tables and figures are to appear in chapters, consider numbering them within each chapter (e.g., Table 4–1, Table 4–2, etc.) rather than consecutively from the beginning of the report. With this approach, the addition of a table or figure in the chapter will limit the renumbering process to only that chapter. Pages can be numbered in a similar fashion. Again, the extent of renumbering will be less if this scheme is used.

All tables, figures, appendices, and attachments should be explicitly mentioned in the text with some explanation of their significance. These supporting items should appear immediately after they are mentioned, with the exception of appendices and other special supporting materials which appear at the end of the report, or perhaps in a separate document available on special request.

Use Lists

Add to the variety, clarity, and attractiveness of your report by occasionally using lists as an alternative to narrative paragraph form. An example is the list used in the earlier "Define and Write to Likely Audience or Audiences" section of this chapter. Chan and Lutovich (1994) offer the following suggestions on using lists:

- Provide ample white space by widening the margins and placing extra space between lines.
- Mark items with neutral symbols, that is, use numbers and letters only if necessary such as to indicate priority or to provide for ease of subsequent reference to selected items in the list. (This list provides an example of the use of symbols.)
- Use end punctuation only when one of the listed items contains more than one complete sentence. If punctuation is used at the end of any item, it should be used at the end of all items. (This suggestion is followed in this list.)
- Explain the list by introducing it to the reader with an appropriate statement.
- Use a similar construction for all items in the list. (For example, all items in this list begin with a verb.)

Design a Standard Base Map or Diagram

The effort required to prepare graphics will be minimized if base maps, plan sheets, diagrams, and other graphics are carefully designed at the beginning of the report preparation process. Computer-based maps, plan sheets, diagrams, and other graphics offer more flexibility and ease of modification than manually based materials. However, the anticipated use of computers does not diminish the need for careful design of graphics.

Assume that a report sets forth a rapid transit plan for a growing metropolitan area. The base map on which the various alternatives and recommendations will be shown should be carefully designed at the beginning of the report preparation process. In fact, base-map decisions should be made early in the project work. Similarly, assume that the engineering project involves the retrofitting of a heating, ventilating, and air conditioning system into a large, older building. A set of diagrams or plans and elevations of the building should be developed early in the project, or certainly at the beginning of the report writing phase, so that base plans and elevations can be repeatedly used in the display of alternatives and the recommended system.

Use Format Writing

Consider, for example, a report that will repeatedly describe similar things or activities, such as alternatives considered or computer simulations that were run. Develop a generic format for alternatives or simulations and use it repeatedly, with minor variations for interest, to describe all alternatives or simulations. The generic format to be followed for the description of each of a series of alternatives might include the following subtopics in the indicated order: location, function, cost, positive attributes, negative characteristics, and implementation sequence. The format writing helps the writer write and the reader read.

Establish Report Milestones

Milestones are dates by which key parts of the report writing project will be completed. For example, milestones might be established for completion of the outline of the report, each of the chapters, tables and graphics, and appendices. Such milestones are important because there is no limit to the improvements and refinements that can be made to any report. There are always better ways to present data, describe analysis, and present recommendations. Establishment of milestones will tend to offset the openendedness of report-writing projects and focus the project team on producing results.

Produce Attractive and Appealing Report

Modern word processing features such as underlining, right justification, boldfaced type, special fonts, color, and graphics provide an opportunity to produce an attractive and appealing report. In fact, failure to produce such a report suggests that the authoring organization is not well equipped in its report production capabilities and perhaps its other capabilities. Give special consideration to the cover of the report, because it is all that some potential readers may ever choose to look at. Rarely does anyone *have* to read your report—make them *want* to read it by virtue of its appearance and structure.

Cite All Sources

Carefully give credit for all ideas, information, and data used in preparing a report. Respect the intellectual and creative works of others. This is typically done by citing the references in the text and providing a list of the cited references at the end of chapters or at the end of the report.

An acknowledgements section is often very appropriate. It provides an opportunity to recognize the assistance of private- and public-sector individuals and organizations. That is, considerable help is often provided by such individuals and organizations, even though they have not authored a report or other document that can be cited as a formal reference. Be gracious—err on the side of acknowledging the assistance of individuals and organizations even though some of them may have been only marginally involved in the work culminating in the written report.

An acknowledgements section, typically placed near the end of a report, also provides a means by which the report author or authors indicate that, although they appreciate assistance provided by others, they are responsible for the essence of the report. The acknowledgement section of the report might read, in part, as follows: "The authors gratefully acknowledge the assistance received from the indicated individuals and organizations. However, the authors are solely responsible for the opinions expressed, conclusions drawn, and recommendations made in this report."

One More Time

The difference between the right word and the almost-right word is the difference between lightning and the lightning bug.

(Mark Twain)

As noted at the beginning of this report-writing section, the focus is on production—generating a report with minimal effort and maximum communication potential. A small diversion is in order at this point in the process. When your report is finished, take one last critical look at it in the privacy of your office or home. Or perhaps ask your spouse, a friend, or a colleague to critically read the report. Try to take a fresh perspective. Perhaps a sentence needs to be shortened, a table tightened up, or a figure fixed. Make whatever refinements are needed to yield a final polished product.

Concluding Thoughts

Frankly, excellent reports are few and far between. An excellent report is attractive, usable by various audiences, logically structured, well-written, informative, and convincing. The production of an excellent report about a project is a major project itself. Sadly, clients are often denied the value of an excellent engineering project and the participants in that project are denied the satisfaction of seeing their work come to fruition because the report, the crucial link between the engineers and the clients, is mediocre. The quality of the report, whether it's a letter, a memorandum, or a formal document, should be at least as high as the quality of the technical work being described. The report-writing tips presented here are intended to achieve that objective.

SPEAKING TIPS: HOW TO MAKE AN EFFECTIVE PRESENTATION

. . . we are all public speakers. There's no such thing as a private speaker—except a person who talks to himself.

(Decker, 1992, p. 27)

Numerous opportunities are available to professionals—even entry-level technical professionals—to make or help to make formal and semi-formal presentations. Typical audiences are colleagues within your organization, clients and potential clients, members of professional associations, service clubs, teachers, and students.

In some of these oral presentations, very much is at stake. For example, a technical professional may be trying to convince colleagues of the validity of his or her ideas in order to see them implemented. Or a young engineer may be trying to communicate with representatives of a client, urging them to accept a recommendation based on an engineering investigation. The decision made by a potential client to retain your firm may hinge on the effectiveness of the presentation made by a representative of your firm to the client. For other presentations, the stakes are moderate. Nevertheless, much good can be accomplished and much satisfaction achieved if the presentation is excellent.

Therefore, the community of technical professionals places a high premium on the ability to make effective presentations. Individuals who develop good to excellent speaking skills are well rewarded in terms of span of influence, promotion, compensation, perquisites, added opportunities, and perhaps most important of all, personal satisfaction for a difficult task well done.

Reluctance to Speak

Although numerous opportunities arise within the engineering and business world to make effective presentations and although effective presentations are needed and valued, relatively few professionals, young or otherwise, voluntarily seek out such opportunities. Furthermore, many of the presentations that do occur are mediocre to poor, even some of those made at regional and national conferences.

Effective speaking, like any other skill, can be learned through study and practice. However, because this particular skill must be practiced before other human beings and because of fear of failure, relatively few professionals make the commitment to become effective speakers. The speaking they do is usually infrequent, by request, and under duress. They react to the need to speak rather than seek the opportunity to do so.

As an aside, look for opportunities to test the hypothesis that people's impressions of organizations, private and public, are often based on only one contact with a representative of that organization. Engage people in conversation upon various organizations as opportunities arise. Ask them what they think about a particular organization, such as a consulting firm, a manufacturing company, a government agency, or an educational institution and then ask them how they know. You might be surprised to find the basis for their perspectives. Many times there will have been a contact with one member of that organization, and that contact will have been made in a public-speaking situation. The impression, whether it is positive or negative, will be based on very little evidence, namely, a professional presentation by one representative.

As an entry-level technical professional, you are urged to commit to continuously improve your presentation ability. The ideas presented in Figure 3–6 are offered to assist you to prepare for, present, and follow up on formal and semiformal presentations.

Define the Audience and the Setting

As with the audience for a written report, discussed earlier in this chapter, the audience to which you will speak may in fact be made up of many audiences. However, unlike a written report that can be written at various levels so that the multiple audiences can choose which portions they want to read, you must make the oral presentation to the entire audience. Each member of the audience is expected to listen to the entire speech. This is a major challenge for the speaker.

Accordingly, find out as much as you can about the audience. Inquire about age, gender, education level, the reason the group is assembled, and the common interest or interests among the expected attendees. If you are one among a group of speakers, find out as much as you can about the other speakers, their topics, and the chronology of the presentations. Also determine the total time allotted for your presentation and postpresentation questions and discussion and plan to respect those limits. Determine if you will be permitted to entertain the questions immediately after your presentation or if all questions will be deferred until all speakers have completed their formal remarks.

- **Define the audience and the setting** 
- **Prepare the script** 
- **Prepare the graphics** 
- **Practice out loud**
- **Arrange for and verify audiovisual Equipment** 
- **Suggest a proper introduction**
- **Deliver the speech** 
- **Prompt postspeech questions and answers** 

- **Follow up**
- **Take extra care with international audiences** 

Figure 3–6 Speaking tips

Prepare the Script

The process of preparing the material from which you will speak at a presentation is very similar to the previously presented process of writing a report. Accordingly, refer again to the following topics within the portion of this chapter titled Report-Writing Tips: A Chance to Shine; Outline and Incubate; Retain Some of the Outline in the Report; Write "Easy" Parts First; Employ a Gender-Neutral Style; Write in an Active, Direct Manner Rather than a Passive, Indirect Manner;

and Use Rhetorical Techniques. Assemble and organize, in your script, many more ideas and much more information than you will need during your presentation. Adopt the approach that you know or will learn much more about your topic than you will actually share during your formal remarks.

As you select material, particularly data, to tell or show your audience, use discretion. Don't, in your zeal to support your principal points, overwhelm your listeners with quantitative material so that they miss your message. Don't allow them to "drown in data while thirsting for knowledge" (Decker, 1992, p. 66).

Consider adopting the simple, but very effective, speech delivery technique that might be referred to as the "tell, tell, tell" approach. That is, tell your audience what you are going to tell them, tell them, and then tell them what you told them. Remember that individuals who hear your presentation, unlike individuals who read your report, cannot easily refer back to what you said earlier. Unless they are very conscientious listeners and note takers, you will need to reiterate your principal points. The "tell, tell, tell" approach is an effective model for meeting this need.

As already noted, if you are one of a group of speakers, you should become familiar with the general themes or content of the other presentations and the order in which the presentations will be made. Consider contacting one or more of the other speakers to learn more about their presentations and how you can coordinate your efforts. Craft your presentation so that it relates to, supports, or contrasts with the other presentations. Whenever you are part of a group of speakers, you are not speaking in a vacuum. Therefore, you have an obligation to try to coordinate your efforts with theirs.

Prepare the Graphics

When preparing graphics, whether they are 35-mm slides, overhead transparencies, or other images, do not indiscriminately make direct copies of tables, figures, and similar material found in published reports. The likelihood that such visuals will be viewable by an audience is very small. Recognize that tables, figures, and other graphics included in reports can be viewed at very close range by the reader and can be examined for an extended period of time or reviewed frequently.

In proportioning the graphics, use a height-to-width ratio consistent with your selected medium. For example, 35-mm slides have a long-to-short dimension ratio of 1.5:1.0 while the standard video screen has a height-to-width ratio of 3 to 4. Slides, relative to video, offer added formatting flexibility because they can be projected in portrait or landscape configuration. By respecting the proportions of your selected graphic medium, you will be able to fully utilize the available space and provide a more aesthetic image for your audience.

Utilize the 1/30th rule of thumb. Each alpha-numeric character should have a height that is at least 1/30th the total height of the projected image. Heights less than this are difficult for the audience to see even if the projected image is of very high quality.

Each projected image should clearly portray one or a few very simple ideas. Remember that your audience is simultaneously listening to you, looking at your visual aid, and presumably trying to understand new concepts or ideas or assimilate new information. Material that is familiar to you is very new to them. Use carefully designed graphics to help them understand what you are trying to say.

Practice Out Loud

A word fitly spoken is like apples of gold in a setting of silver.

(Bible, Proverbs 25.11)

Practice your presentation out loud, preferably in front of one or more people with whom you feel comfortable and who are likely to offer you constructive criticism. If you are not able or inclined to do that, still practice your presentation out loud in the privacy of your office or some other location. Never give a presentation without having practiced it out loud, even if only to yourself.

"Out loud" practices help to establish the actual time required to make your presentation. Reading or speaking the presentation "to yourself" will give you a poor measure of the actual presentation time. You are likely to greatly underestimate the delivery time. Furthermore, each time you practice your presentation "out loud," you become more versed in and more knowledgeable about the content of your speech. You will discover additional words and phrases to say what you want to say. You will be able to master words that you initially had difficulty pronouncing. When you actually make your presentation to the audience, you will, accordingly, have more experience to draw on. You will already have given a very similar speech several times. Many good speeches that have been well prepared in terms of content and organization are spoiled because the speaker seems to be searching for the right words. This probably reflects the lack of "out loud" practice, not a lack of substantive content or familiarity with material.

Another reason why you should practice your presentation aloud to trusted colleagues or to family members is that by so doing you can identify and eliminate distracting habits. Examples are: playing with change or keys in your pockets, saying obnoxious things such as "ah" and "you know," avoiding eye contact, looking at just a portion of the audience, mispronouncing words, speaking in a monotone, using unusually long sentences, rocking back and forth, and frequently taking your glasses off and putting them on. Again, if you are reluctant to practice your presentation aloud with a live audience, make an audio recording or, better yet, make an audio-video recording. As you practice your presentation, consider the techniques you have admired in accomplished speakers. Try mimicking some of them, with suitable modifications for your personality and presentation, to see if you can find more effective ways to communicate. Finally, consider asking a colleague or family member to attend the actual presentation and provide you with a postpresentation critique.

Do you know how your voice sounds to others? Unless you've listened to and studied recordings of your voice, you don't know how you sound to others. As noted by Decker (1992, p. 123–125), " . . . the voice on the tape is much closer to what others actually hear than the voice we ourselves hear as we speak." As explained by Decker, the voice you hear is conducted largely through the bones in your head while the voice others hear is transmitted through the air. He says that the "reel" voice is the "real" voice, at least as far as your audiences are concerned.

Unless you plan to talk only to yourself, you ought to know how you sound to others and, if you don't like it, change it. This is another reason to consider recording all or part of your presentation as part of your preparation to speak and when you speak to a group. A small microcassette recording device that you turned on and unobtrusively placed on the lectern or near you at the beginning of your talk will capture the verbal and vocal components of your presentation. Privately study the recording and be prepared to be shocked by the strengths and weakness of your spoken communication. Although much more difficult to arrange, an audio-video recording is even more valuable, because it will capture all three components of spoken communication—verbal, vocal, and visual.

Use your organization's voice-mail system to hear how you sound to others in the normal course of the business day (Decker, 1992, pp. 130–131). For example, when you send a voice-mail message to a colleague, send a copy to your voice mail box. Later, when you access your messages critically listen to the verbal and vocal components of the spoken messages from you. Identify strengths and weaknesses and explicitly build on the former while you diminish the latter.

As you identify areas needing additional work by you, prioritize them and tackle them one or perhaps two at a time as part of your next speaking opportunity. Assume, for example, that the audio recording of your last presentation revealed these three deficiencies: too many "aahs," voice intensity trailing off at the end of sentences, and poor pronunciation of certain words. If too many "aahs" is the most distracting of the three liabilities, explicitly work on it as part of your next presentation. Lay a large note on the lectern where only you can see it, that says "No aahs!" Think about this distracting habit throughout your presentation and consciously avoid saying "aah." Record all or part of your presentation and observe the progress you made. Work further on this problem or take on a new distraction at your next presentation. Use the same focused approach to build on your speaking strengths. Gradually you will replace bad habits with good ones.

After carefully preparing your speech, practicing in front of colleagues or family, and "getting it right," you may be tempted to plan on reading your perfected presentation to the audience. After all, by so doing, it will come out "perfect." Don't do this! It may come out "right," but will be extremely boring unless you are a very gifted orator. In fact, many members of the audience will be insulted because they could have read your presentation on their own time rather than coming to hear you in person.

A possible exception to the advice of not reading to an audience is a presentation that, in effect, is a formal testimony before a judge, jury, commission, or other legally constituted body. In such cases, reading from a manuscript may be advisable along with submission of the manuscript as support for your testimony. These situations are very rare.

The suggested practicing process is not meant to result in a memorized speech. Audiences tend to recognize and react negatively to memorized presentations. The goal of "out loud" practicing is to know your material very well, establish the major points you want to make and the order in which they will be presented, and to know how long your presentation will take so that you can speak in a largely extemporaneous, natural manner.

You are likely to need some general notes or an outline to guide your presentation. This script should be minimal—just enough to remind you of the next topic, point, or idea. If you are using an overhead projector and need notes, consider writing them on opaque material along one edge of the transparencies where they can be readily seen by you, as you look at the audience, but not seen by the audience. Perhaps your visual aids will be sufficient, without notes, to "drive" your presentation. As an alternative to using a conventional written outline or list, Hoff (1988, pp. 14–19) suggests trying a series of cartoons or symbols. Readily seen only by you as you speak to the audience, the cartoons or symbols will guide your presentation.

Arrange for and Verify Audiovisual Equipment

Well ahead of time, and perhaps as a condition of accepting the invitation to speak, determine your minimal audiovisual needs and communicate them to whomever is making the arrangements for the presentation. These needs might include one or more of the following: lectern; lectern light; speaker's table; electronic or other pointer; screen; 35-mm projector; overhead projector; movie projector; video cassette player and monitor; sound system with fixed or lapel microphone; special seating arrangements such as classroom style, U-shaped, and round tables; flip chart; blackboard; chalk; colored markers; and extra transparencies. Do not assume anything—determine your needs and communicate them in writing to whomever is making arrangements.

If you are to make a presentation at a regular meeting of a professional organization, the organizers of the meeting will often inquire about your audiovisual and other needs. However, if such inquiries are not directed to you, take the initiative to make absolutely sure that plans are in place to provide everything you will need. Your preparation for audiovisual equipment and other support does not, however, end at this point in the process.

Immediately prior to your presentation, say about an hour, but no less than one-half hour before the meeting or session will begin, visit the presentation room and check the availability and the functioning of all requested audiovisual equipment and other facilities. Operate on the assumption that if something can go wrong with audiovisual equipment before or during your presentation, it probably will. If something you requested is missing, immediately ask for assistance. Do not assume that something is functioning simply because it is there. Test the equipment or have someone test it and, if it is not functioning properly, immediately request a replacement or repair.

If you are using transparencies, a 35-mm projector, a video cassette player, or some other imaging device, project some samples and make sure that everyone in the audience, regardless of where they are seated, will be able to see the images you project. Go to various vantage points in the room, sit at those locations, and account for the fact that the room will be full or partially so. Determine if attendees will be able to see what you project, taking into account where you will be standing when you speak. Recognize that some personnel who are responsible for setting up audiovisual equipment, even though they do it professionally, do not take into consideration the conditions that will exist in the room at the time of your presentation. Although audiovisual personnel are very familiar with setting up equipment, they may not ever give presentations using such equipment. Try to project images as high as possible so your audience will have an unobstructed view.

As you check out the visibility of 35-mm slides, remember that some slides are displayed with the long dimension horizontal and others with the long dimension vertical. Make sure that you project both types during your preview so that your images will be contained on the screen. Do everything you can to make the images as large as possible. For example, consider moving the screen farther back or the projector farther in the opposite direction. Some projectors have adjustments for establishing the size of the image. Use those adjustments and get things just right.

Locate the controls for room lighting and experiment with various lighting levels to determine the optimum conditions for your presentations. Arrange to have someone make the adjustments as needed during your presentation, or do it yourself. Do not allow the lighting to be turned too low during your presentation. Overhead projectors can typically be used with full lighting conditions. Thirty-five-mm slides and other imaging systems do not require an extremely low level of lighting if the images are carefully prepared. If the lighting level is too low, the audience is deprived of seeing your gestures and facial expressions. With low room illumination, some members of the audience may doze off or fail to concentrate on your presentation, particularly if your presentation is scheduled after lunch or in the evening.

At regional and national conferences, a session assistant is often available to provide help with visual aids, lighting, distribution of handouts, and other special needs. Determine if an assistant has been assigned to your session and, if so, introduce yourself to that person. Share your needs with the assistant and seek his or her support. Typically, these people are volunteers or have been volunteered and often have no expertise in supporting presentations and speeches. However, they are usually very willing to assist provided you give direction by indicating your needs and suggest ways in which they can be satisfied.

Suggest a Proper Introduction

To the extent feasible, ask the host or session chair to give you a brief but proper professional introduction. Audiences like to know whom they are hearing from. The introduction for a young professional would typically include a brief summary of your post-high-school education, the naming of your employer and perhaps reference to some of your project experience. Even the very young professional has a background that is of interest to the audience. In anticipation of being properly introduced, prepare a short (one or two short paragraphs) narrative that hits the high points and offer this to the person who will introduce you.

Deliver the Speech

Begin by being gracious. Thank the chair of the session or the person in charge who introduced you for the introduction. Consider expressing your appreciation for the opportunity to speak to the audience. Incidently, at the end of your formal remarks, consider thanking the audience for their attention.

As you speak, remember where your audience is located. Your audience is not the wall to your left or to the right, they are not on the ceiling, they are not in your notes, and they are certainly not in the images projected on the screen behind you. Your audience consists of those people seated directly in front of you. Speak to them—each and every one of them. Do this explicitly by gradually moving your gaze around the room. Make eye contact of up to five seconds with specific individuals. These "eye-fives" will make your message more personal and help you avoid speaking to just one portion of the audience.

With the exception of occasionally glancing at the images that are being projected on the screen or on the monitors, to make sure that items are in order and focused and visible, do not spend any significant time looking at those images. You already know what they look like—you prepared them. A common failure of many speakers is a tendency to look at, and actually speak to, that which is projected on the screen, forgetting about the audience. Speaking in the direction of the images, rather than at the audience, makes it difficult for the audience to understand what you are saying.

Do not apologize for any aspect of your presentation including, but not limited to, the amount of time you spent preparing for it, the content, the organization, the visibility of visual aids, your experience or lack of it. All of these things are within your realm of control and, out of respect for your audience, you should have taken care of all of them before you were introduced. A possible exception is something that happens at the last minute for which, in spite of all of your and the organizers' planning, you cannot control. A simple apology might be appropriate and then you need to take feasible corrective or offsetting action and get on with your presentation.

Recognize that for many members of your audience, especially if you are speaking at a regional, national, or international conference, you will be all that many members will ever know about your organization. As suggested earlier in this chapter, this is a fantastic opportunity to faithfully represent the values and capabilities of your organization. What you wear, how you are groomed, how you speak, what you have to say, and how you interact with the audience will define what many members of your audience will think about your organization. You are, in the eyes of many members of the audience, your entire organization. Rather than viewing this as a tremendous burden that has to be carried by you against all odds, view it as a very special communication opportunity.

As you proceed with your presentation, use gestures and facial expression to support what you are saying. Consider the use of props and display items to make a particular point. While an image of an object is better than just a word description, the object itself may be best. Many times, engineers and other professionals are discussing objects that can be simply brought, in whole or in part, to the site of the presentation and displayed including possibly passing them around the room. If you can, bring the actual object. Depending on its size or weight, consider bringing a simplified model of it. Prop examples include a scale model of a proposed or constructed building, selected components from an automobile engine, or a sample of a new material being used in the manufacture of a product.

Never exceed your allotted time. To do so is to show complete disregard for other speakers on the program and for the audience. Conclude your speech with a definite ending both for effect and to effect a transition to either the next speaker or to a question and answer session stimulated by your presentation. An example of a definite ending is: "Thank you for your attention. I would be pleased to respond to your questions or suggestions."

Prompt Postspeech Questions and Answers

Probably the only thing more satisfying than delivering an effective speech is having it followed by a stimulating exchange of ideas and information with active participation by many members of the audience. However, getting that exchange started is often difficult.

If there is a lull after conclusion of your presentation, consider asking a general question of the audience such as "Would anyone care to share a similar or different experience?" Or try to find someone in the audience who has an inquisitive or doubtful expression on his or her face and diplomatically "challenge" the person on the chance that he or she will be compelled to speak. For example, look at the person and say, "You look very skeptical about the usefulness of what I just described. How do you feel about it?" Then remain silent. Most people will feel compelled to fill the silence with a comment. Their comment is very likely to "break the ice" for other members of the audience. You might even consider "planting" a person in the audience who agrees to ask a question of his or her preference if no one else does. This may be the catalyst needed to stimulate participation by many members of the audience.

During the question and answer session, consider repeating the question or perhaps rephrase it for clarity. Although you may be able to hear the question clearly because the questioner is facing you, members of the audience, especially those behind the questioner, may not understand the question. Therefore, they may not be able to benefit from your answer. If during the discussion, you simply do not know the answer to the question, consider referring the question to the audience—perhaps they can help. Another approach is to say that although you do not know the answer, you would be pleased to get back to the questioner if they would give you a business card at the end of the session.

Follow Up

Usually, after you have given an effective presentation, members of the audience will engage you in private or small group conversations at the conclusion of the session or meeting. They may ask you for a copy of your paper, related information, or information about your organization. Perhaps they will refer you to a book, paper, or other publication that may be of value to you, or they may mention one or more individuals who may be able to assist you in further study of your topic.

By all means, follow through on anything you promise to do. For example, if you promise to send a copy of your paper, do so shortly after the meeting, transmitting it with the letter that references the presentation, and reminds the reader that he or she asked for a copy of the paper; consider asking the person to share the paper with others and to report any critiques back to you. If you were referred to a paper, some other publication, or another person, try to follow up. Consider writing to the referrer, thanking them for the contact. By properly following up on postspeech conversations, you learn more about your subject and widen your network of professional contacts. Learning more about your subject and getting to know more people are two very significant personal and organizational benefits derived from making effective presentations. Make sure you get this return on your investment.

When people ask you to follow up after your presentation, ask for their business cards so that you have the necessary information. Afterwards, examine the business card closely to determine the person's organization and to deduce their responsibility in that organization. This may prompt you to send that individual items of interest, such as your organization's brochure, copies of materials that may be in your personal professional files, and names of individuals that you think the person might want to know. Consider adding the individual to your organization's newsletter or other mailing list.

Take Extra Care with International Audiences

Seek opportunities to make presentations to international audiences. This may seem somewhat ambitious to you, a young professional. However, the world of engineering and business is changing rapidly as a result of demographic, technologic, and economic forces. Engineering is being internationalized and, therefore, you are likely to have global opportunities including speaking to international groups assembled in your country or abroad.

If you are an American or from another English-speaking country (e.g., Canada, U. K.), you face special challenges when speaking at international conferences. Such gatherings typically use English as the official language simply because it is the most practical. This is not done because the English language is the "best," but because of a series of historic decisions that have resulted in, among other things, many people in western, eastern, and other countries having studied English.

Frankly, when you stand in front of that international group, two strong feelings are likely to be held by some members of the audience. First, some listeners will resent the fact that your language, not their language, is the official language and that resentment may be targeted at you because you represent a country that uses the official language. Second, and more positively, people in other nations will expect you, as one who normally speaks English, to set a high standard. English is usually their second language and they see, or more precisely "hear," your presentation as an opportunity to improve their use of the English language. Choose your words carefully, avoid slang, and be sensitive to ethnic, religious, economic, and political differences.

Bottom line: If you take seriously the advice presented in this chapter about making oral presentations, take it even more seriously when you make a presentation at an international conference. Such a conference provides you with an opportunity to sharpen your speaking skills, set a fine example for your country and its language, and earn the respect and friendship of people from other nations.

BODY LANGUAGE

Communication effectiveness, especially listening and speaking, are markedly enhanced or diminished by body language—yours and his, hers, or theirs. Body language is nonverbal communication such as posture, facial expression, the position of arms, handshake, eye contact, and dress.

One study (Decker, 1992, pp. 83–85; see also NIBM, 1988, p. 1) concluded that the words a person uses in speaking account for only 7 percent of that person's influence on others! Voice tone explains 38 percent of the impact, and visual impression 55 percent. Riggenbach (1986) discusses the role of body language in various types of negotiations and claims that 95 percent of the nonverbal gestures being received during negotiations are not used by the receiver. According to the author, the nature and interpretation of body language vary among countries and cultures. Finally, Riggenbach says, "Body language shows the inner feelings and attitudes of a person—actions do speak louder than words!"

As extreme and irrational as these results and claims may seem, review some of your recent positive and negative transactions with one or more other people. To what extent was your speaking effectiveness influenced by your body language? How much were you affected, as you spoke and as you listened, by theirs? Figure 3–7 lists many and varied types of body language—the "vocabulary" of body language—in the right column and possible interpretations—how the "vocabulary" is "heard"—in the left column. The creators (NIBM, 1988, pp. 28–29) of the tabulation emphasize that a single type of body language at an instant or during a short period of time should not be interpreted out of context

LIKELY INTERPRETATION	TYPE OF BODY LANGUAGE
Happiness or satisfaction	Smiling Enlarged pupils Relaxed posture Serene facial expression Free, unrestrained movement
Unhappiness or dissatisfaction	Frowning Constricted pupils Tense posture Pursed lip, furrowed brow, flared nostrils Rigid body, lack of movement, or nervous movement
Agreement	Nodding Winking Smiling Relaxation following concentration Continuation of serene eye contact
Disagreement	Shaking the head side to side Frowning Crossing the arms Pursing the lips
Interest or receptiveness	Serene eye contact Stillness of body Even breathing Arms folded loosely over lower body
Disinterest or distraction	Looking away Hunched shoulders Arms folded on chest, face placid Vacant eyes Sighing Finger drumming or desk tapping
Anger or irritation	Accentuated breathing Intense, aggressive eye contact Arms folded on chest, face taut Clenched fists, or hands gripping desktop
Disbelief	One raised eyebrow Crooked smile Head shaking side to side Tilted head
Surprise	Two raised eyebrows Enlarged pupils Sudden attention directed to speaker
Decision-making in progress	Eyes directed to ceiling, blinking rapidly Turning away and looking steadily at nothing Standing and walking back and forth
Decision has been made	Deep breath followed by a sigh and relaxation End of facial tension followed by smiling or earnest eye contact
Superior status	Takes central spot in meetings Speaks without seeking permission Initiates and terminates most transactions Exhibits dominant behavior, such as standing taller in confrontations Pats people on the back
Subordinate status	Takes peripheral spot in meetings Seeks permission before speaking Waits for dominant individual to initiate or terminate most transactions Exhibits submissive behavior, such as curling shoulders forward, in confrontations Seeks pats on the back

Figure 3–7 Body language types and interpretations (Source: Adapted with permission from NIBM, "Body Language for Business Success," New York: 1988, pp. 28–29.)

with other types of body language. This is one reason for the grouping of types of body language in Figure 3–7.

REFERENCES

BENTON, D. A., *Lions Don't Need to Roar: Using the Leadership Power of Professional Presence to Stand Out, Fit In, and Move Ahead.* New York: Warner Books, 1992.

BROWN, T., "Communications Overload," *Industry Week*, October 3, 1988, p. 13.

CHAN, J. F., and D. LUTOVICH, "Using Lists . . . A Simple Solution to a Complex Problem," *HYDATA*, American Water Resources Association, March 1994, pp. 9–10.

CONGER, J. A., "Communicating a Vision that Inspires," *Engineering Management Review*, Institute of Electrical and Electronics Engineers, Vol. 19, No. 4 (Winter 1991), pp. 69–77.

COVEY, S. R., *The 7 Habits of Highly Effective People.* New York: Simon & Schuster, 1990.

DECKER, B., *You've Got to Be Believed to Be Heard.* New York: St. Martins Press, 1992.

HOFF, R., *I Can See You Naked: A Fearless Guide to Making Great Presentations.* Kansas City, Missouri: Andrews and McMeel, 1988.

LINVER, S., *Speak Easy.* New York: Summit Books, 1978.

MCKENNA, J. F., "Tales From the Circular File," *Industry Week*, March 19, 1990, p. 38.

National Institute of Business Management, "Body Language for Business Success," New York, April 1988.

RIGGENBACH, J. A., "Silent Negotiations: Listen With Your Eyes," *Journal of Management in Engineering—ASCE*, Vol. 2, No. 2 (April 1986), pp. 91–100.

ROANE, S., *How to Work a Room: A Guide to Successfully Managing the Mingling*. New York: Shapolsky Publishers, 1988.

WALTERS, R., and T. H. KERN, "How to Eschew Weasel Words . . . ," *Johns Hopkins Magazine*, December 1991, pp. 25–32.

SUPPLEMENTAL REFERENCE

TUFTE, E., *The Visual Display of Quantitative Information.* Cheshire, Conn.: Graphics Press, 1983.

4

Management of Relationships with Others

> *I'll tell you what makes a great manager: a great manager has a knack for making ball players think they are better than they think they are. He forces you to have a good opinion of yourself. He lets you know he believes in you. He makes you get more out of yourself. And once you learn how good you really are, you never settle for playing anything less than your very best.*
>
> (Reggie Jackson)

Assume that you have mastered—or almost so—self-management. You have decided on full-time professional employment rather than full-time graduate school, made the initial transition from the world of study to the world of practice, and you are in your first full-time employment situation. You have an effective time-management system; you are conscious of the importance of personal integrity; you function well within the administrative structure of your organization; your assignments are being completed well and on time. Although you watch what you say and how you say it, you are not reluctant to speak up when you have a question or suggestion; you dress appropriately, are developing communication skills, and have a positive outlook. In addition, you passed the Fundamentals of Engineering Examination and are beginning to gain valuable experience toward your engineering license. Finally, you are involved in continuing education and are active in at least one professional organization.

Having managed to do all of those things, you notice that you are beginning to have opportunities to "manage" other people, or at least some of the efforts of other people. As suggested by Figure 4–1, you are increasingly interacting with clerical staff, technicians, draftpersons, subcontractors, data processing staff, surveyors, and vendors. You are also being asked to assist with the orientation of the

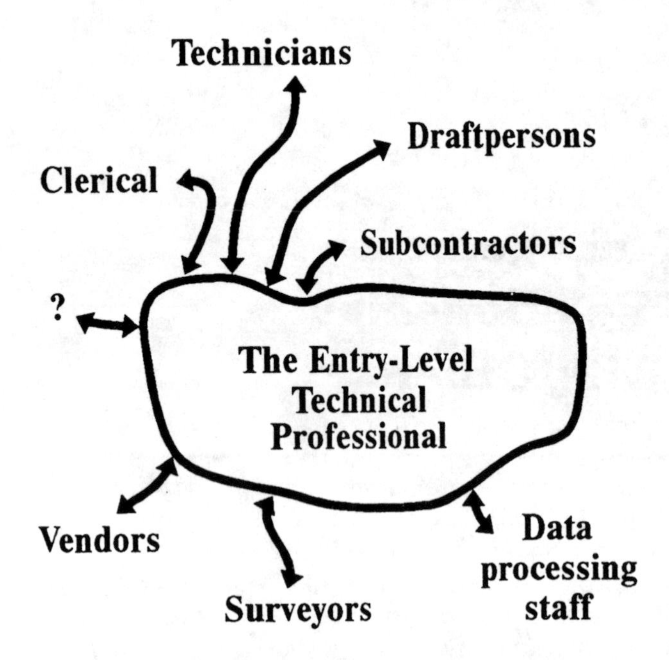

Figure 4–1 Early opportunities to interact with others

new entry-level professional staff and to direct some of the efforts of those staff on technical projects.

You are now ready to focus more of your time and energy on managing relationships with others. Early success in professional practice, as satisfying as it can be, may result in young technical professionals' failing to grasp the importance of the next step in their professional development. That step is the further development of interpersonal skills. Technically and scientifically oriented people, especially young people, are probably most prone to think, to put it bluntly, that "it's what you know that counts." While "what you know" is certainly necessary, it is not sufficient in the world of engineering and other technical practice. The need to develop interpersonal skills is nicely stated by Benton (1992, p. 88):

> *No matter how ambitious, capable, clear-thinking, competent, decisive, dependable, educated, energetic, responsible, serious, shrewd, sophisticated, wise, and witty you are, if you don't relate well to other people, you won't make it. No matter how professionally competent, financially adept, and physically solid you are, without an understanding of human nature, a genuine interest in the people around you, and the ability to establish personal bonds with them, you are severely limited in what you can achieve.*

Patterson (1991) advises you, the young professional, to "Take time to smell the social roses." He goes on to observe, "It's just that the real shots are called by those who find a way to fit within the social scheme."

This chapter offers guidance on how to effectively manage your relationships with the wide variety of people you are likely to encounter within and outside of your organization as a result of the practice of engineering or some other technical profession. The chapter begins by discussing various types of people, or more precisely, ranges of attitudes and perspectives that you are likely to encounter. Maslow's hierarchy of needs and theories X and Y are presented. Major sections of the chapter are devoted to the art and science of delegation and of managing meetings. The chapter concludes with a discussion of how to work with support personnel and some thoughts on really caring for colleagues.

TYPES OF PEOPLE

At the risk of an oversimplified categorization of people, or at least of the attitudes that people exhibit, a brief review of a variety of simple models is useful. Within your organization and among the people you serve as a result of your professional work, you are likely to encounter:

- People who make things happen, who watch things happen, and who ask what happened.
- Risk takers, caretakers, and undertakers.
- Individuals who are on top or on tap.
- Those who see "it" as an opportunity or see "it" as a problem.
- Winners and losers (Harris, 1968).
- Proactive people and reactive people. These two approaches are exemplified by typical expressions used by proactive and reactive individuals as shown in Figure 4–2.

The preceding simple models are essentially ways of suggesting that individuals vary markedly in their outlook and perspective. These wide variations occur even in presumably homogeneous groups such as civil engineers employed in a small consulting firm or mechanical engineers involved in design and manufacturing in an automobile plant. Of course, individuals can change. Although each person has natural tendencies, he or she can recognize the available range of outlooks and perspectives and work towards those that are more likely to be productive.

You, as the entry-level technical professional, must try to understand the needs and motivations of the people around you, particularly those with whom you must work to carry out your responsibilities. Although you will not share much of what you hear and see in terms of personal philosophy, try to understand

Reactive Language	Proactive Language
• There's nothing I can do.	• Let's look at our alternatives.
• That's just the way I am.	• I can choose a different approach.
• He makes me so mad.	• I control my own feelings.
• They won't allow that.	• I can create an effective presentation.
• I have to do that.	• I will choose an appropriate response.
• I can't.	• I choose.
• I must.	• I prefer.
• If only	• I will.

Figure 4–2 Reactive and proactive language (Source: Covey, 1990, p. 78. COPYRIGHT © 1989 by Stephen R. Covey. Reprinted by permission of Simon & Schuster, Inc.)

and respect the individuals who hold those philosophies. Recall the powerful idea of empathetic listening discussed in Chapter 3. One of Covey's (1990, p. 235) seven habits of highly effective people is appropriate. He says, "Seek first to understand, then to be understood."

MASLOW'S HIERARCHY OF NEEDS

One way to get a better understanding of why the people around us and with whom we work and interact exhibit such great variation in outlook and perspective is to consider the work of psychologist Abraham Maslow. He developed one of the first models of people's needs (Martin, 1986; McQuillen, 1986; Schermerhorn, 1984, Chapter 12; Wortman, 1981, Chapter 2). Maslow's work, which was done in the 1950s, followed Frederick W. Taylor's "scientific management," which emphasized training workers for efficiency and essentially ignored human factors. Maslow's work also followed Mayo's human relations approach, which was published in 1945, and pointed to the uniqueness of individuals and suggested that workers should have some degree of group control over their work (Martin, 1986).

Maslow's hierarchical model helps managers understand the basic drive or motivation of people around them, including those with whom they have supervision responsibilities and those who supervise them. As shown in Figure 4–3, Maslow envisioned five levels of needs, any one of which may apply to any individual at any time. The most basic needs are physiological: basics such as food and shelter. The second level has to do with safety and security: being physically, psychologically, and economically safe and secure. Assuming that the first two levels are satisfied, Maslow identified belonging as the third level. Individuals want to be accepted as part of a group, feel wanted and appreciated, and give and receive affection.

Maslow referred to the first three levels as *lower-order needs.* The upper two levels he called the *ego needs.* The first of the upper needs is esteem: having self-respect and the respect and recognition from others. The highest-level need is self-fulfillment or self-actualization—the idea of thinking and feeling that one has fully realized one's potential. Awareness of needs and the satisfaction of those needs do not necessarily occur in sequence, and rapid, temporary changes can occur. An example is an unexpected life-threatening situation, which could take someone you normally work with from a level four down to a level one.

Another way of explaining the possible relevance of Maslow's hierarchy to the entry-level professional is to consider Figure 4–4. After several months of employment, the young professional's physiological needs are certainly met and the individual now feels safe and secure, certainly in an economic sense. But there may be a strong desire to have more indications that you are accepted as part of a group and that you are wanted and appreciated.

Just as you may be at one point in Maslow's hierarchy, imagine that any one of your colleagues and co-workers could be at any one of the other levels at any given time. Maslow's needs model helps the entry-level technical profes-

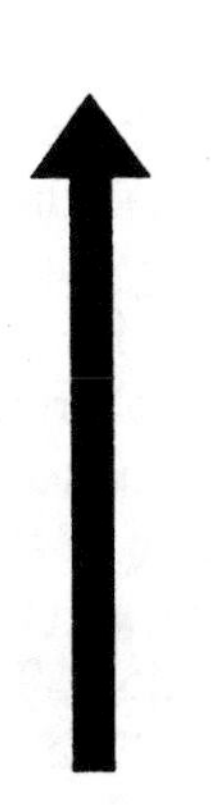

Self-fulfillment

Esteem

Belonging

Safety

Physiological

Figure 4–3 Maslow's hierarchy of needs

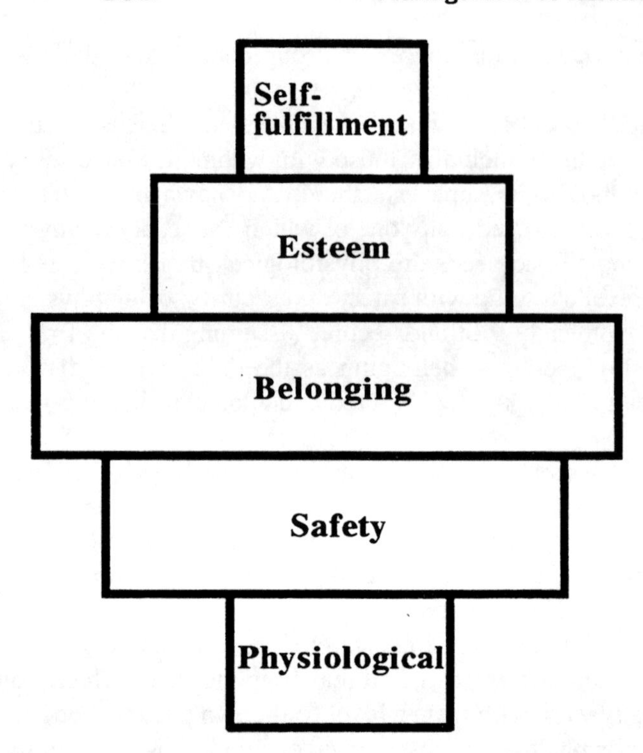

Figure 4–4 Possible Maslow's hierarchy for the young professional

sional understand the actions and reactions of others by providing insight into what motivates them. Consider, for example, the technician who does what he or she is told and appears to have no other aspirations. Perhaps this person is under extreme financial pressure and is focusing primarily on economic security. Having employment is more important than advancing up the organizational ladder.

You may know of an upper-level manager or executive who works long hours, appears "driven," and is involved in many extracurricular projects. This individual may be focusing on level four with emphasis on seeking the respect and recognition from others. Consider also one or two other young professionals in the office who are in the process of seeking other employment opportunities. While they feel safe and secure in their present positions, and the work is certainly challenging, they are at level three; they are not finding, within the present employment situation, the feelings of acceptance and appreciation that they need.

The discussion in the preceding section of types of people might be understood in the context of Maslow's hierarchy of needs. Maslow helps the young engineer or other technical professional understand why people in the work environment vary markedly in their outlook and perspective. Martin (1986) suggests that Maslow's model can also be applied to entire organizations. That is, it may help to explain the culture of one organization vis-à-vis the cultures of others.

THEORIES X AND Y

In understanding and honing one's attitude toward work and in interacting with others and their work attitudes, the young professional should be aware of two fundamentally different perspectives. Referred to as theory X and theory Y, these two perspectives bracket the range of attitudes towards work. They were developed by Douglas McGregor (1960) and are widely discussed in the literature (e. g., Brown, 1988; Martin, 1986; McQuillen, 1986; and Wortman, 1981, Chapter 2). Most college students or recent college graduates have enough part-time or full-time work experience to recognize these two work styles, although they may not know them by the names theory X and theory Y.

Definitions

Theory X people dislike and avoid physical and mental work. They hate work and want mainly security—the kind of security represented by Maslow's level one and level two needs. Theory X people will work in response to threats, coercion, control, and close direction, and they must be constantly reminded that their security is at risk. Theory X people must be watched and controlled.

In stark contrast, theory Y assumes that physical and mental work are natural—as natural as play and leisure. Theory Y people want to contribute by assuming responsibility, exercising self-discipline, working hard and smart, and putting something of themselves into the product or service. They want to understand, support, and become active participants in the mission, objectives, and efforts of the group. Accordingly, theory Y people respond very positively to rewards for and recognition of achievement. Theory Y assumes that most people possess a wealth of intelligence, creativity, imagination, and energy. When released in the work environment, these resources bring individual satisfaction and organizational success.

Perspective

As stated by H. S. Geneen, former president and chief executive officer of ITT and author of the book *Managing* (1984), "You cannot run a business, or anything else, on a theory." Or can you? More specifically, does the theory X–theory Y model make sense and have value? Experience suggests that each person will tend to favor either theory X or theory Y; that is, an individual is likely to be closer to one of the two ends of the available spectrum. Furthermore, the majority of individuals probably prefer theory Y. The ultimate test of one theory over the other for a particular work environment is the product produced. Idealistic notions will not ultimately prevail if the results fall short of the organizational or corporate mission.

A theory Y environment will tend to be supportive of two-way communications, whereas a theory X environment will tend toward one-directional (top down) communication. Incidentally, a theory Y manager may appear to be a theory X manager because he or she is a poor communicator. That is, a theory Y manager with poor communication skills could be viewed as a theory X manager and in turn fail to effectively utilize the resources, particularly the theory Y adherents, within his or her part of the organization. In this environment, the professional staff receives very heavy assignments often at odds with their basic interests and talents because the manager assumes that everyone is totally engrossed in all aspects of the organization and fails to obtain input from the people who make up the organization. Furthermore, such a manager typically fails to communicate appreciation for efforts expended and tasks completed.

Applications of Theory X and Theory Y Knowledge

Assuming that you see some validity in the theory X–theory Y model, consider some hypothetical examples of its usefulness as applied to individuals and even organizations. Theories X and Y can help you understand the responsiveness or nonresponsiveness of others to typical interpersonal situations. For example, a theory X manager, because he or she is not satisfied with the progress being made by a supervisee who happens to be a theory Y adherent, threatens to discharge the supervisee. The theory Y individual does not respond with the expected fear, but instead is insulted and disappointed.

Or a theory Y-oriented young engineer, seeking a first permanent employment opportunity, interviews with what he or she immediately determines to be a theory X-run organization. The young person is perceptive enough to not pursue the position further, even though an attractive salary and benefit package is offered. Finally, consider this situation. A newly arrived theory Y manager is meeting individually with all of his or her supervisees. While meeting with one of the supervisees, a Theory X adherent, the manager asks for ideas and inquires about the supervisee's aspirations. Later, after the employee leaves the manager's office, the employee tells a co-worker to "watch out for the new manager," noting that the manager talks strange and must "have something up his or her sleeve."

Probable dominance of theory Ys. Mr. John Mole, an educated and experienced manager, became successful writing novels, quit his management position, and went underground for two years from 1985 to 1987 as a "management mole." During that time, Mr. Mole had temporary jobs in 11 organizations—jobs he obtained without revealing his educational and business background. He wrote about his experience in the book *Management Mole: Lessons from Office Life* (Bredin, 1988; Mole, 1988).

Mr. Mole reports in his book that the vast majority of junior-level staff he encountered wanted to learn, work, contribute, and succeed, that is, they were primarily adherents of theory Y. Unfortunately, much of what junior-level staff offer and aspire to is wasted because of poor management practices. On-the-job training and orientation were virtually nonexistent in the organizations on which the book is based. As a result, the "blind lead the blind," that is, young employees who have little or no training or orientation informally "train" and "orient" new personnel. Mr. Mole's book strongly suggests that there is a great unused resource out there—many theory Y people. This resource can be tapped and enabled through the process of enlightened, sensitive management.

DELEGATING

Good managers never put off until tomorrow what they can get someone else to do today.

(Anonymous)

Delegating means to legitimately assign part of your tasks to someone else. Properly done, the delegator retains overall responsibility, because the original assignment part of which has now been delegated was given to him or her, while giving up some of the authority. From the perspective of the delegator's "boss," the delegator still has all the responsibility. When called to task for a deficiency in the overall assignment, the delegator should never blame the delegatee. Delegation is not giving orders, that is, holding back authority while giving someone else responsibility. Delegating is also not dumping, that is, getting rid of responsibility, especially when the going gets tough. The treatment of delegation in this section is based in part on ideas derived from Covey (1990, pp. 171–179) and Day-Timers (1990).

Reasons to Practice Effective Delegation

There are many reasons to practice effective delegation. In spite of compelling arguments in support of delegation, there is considerable reluctance to do it; that reluctance is often evidenced by young engineers and other technical professionals. Before exploring resistance to delegation, however, consider the following reasons to practice effective delegation:

• Delegation frees up experienced people to take on new responsibilities, projects, and challenges—to do things they are better prepared for or would rather do. The idea here is to carefully leverage your experience, talents, intelligence, and interests, as suggested by Figure 4–5.

• Delegating gives other members of the organization an opportunity to learn, grow, and contribute in new ways to the work of the organization. Within a delegation environment, everyone, regardless of position, rank, or salary, or hourly rate is being challenged, stretched, and pushed to learn and contribute more.

• Delegation helps individuals learn from others with, surprisingly, the learning often flowing from the actions of the subordinates to their superiors. Because of the novice effect, which is discussed in Chapter 2, whenever an "old" or routine task is given to a new person, fresh and improved approaches are likely to result.

• Delegation reduces tasks' costs, that is, it tends to push the cost of each task to the lowest level possible consistent with the required results. Through effective delegation, typical organizational tasks such as drafting a letter, typing a memorandum, calling on a potential client, making a presentation to a potential

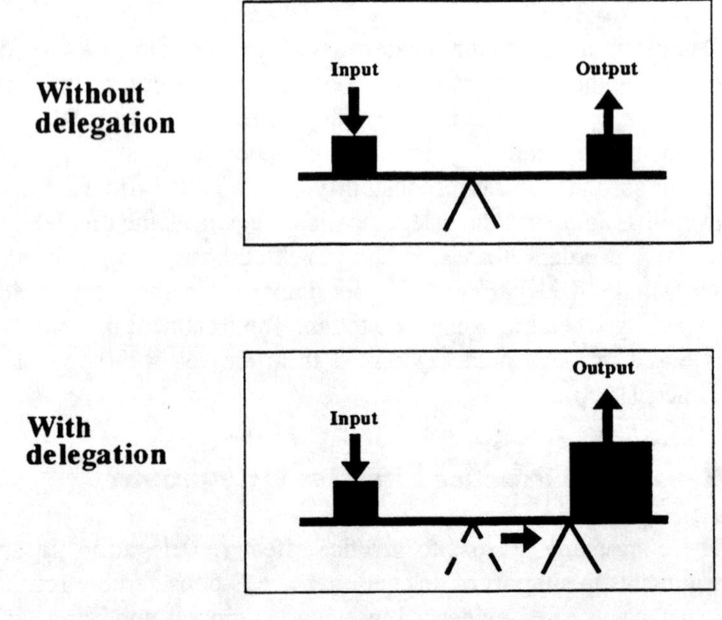

Figure 4–5 Delegating as leveraging (Source: Covey, 1990, p. 172. COPYRIGHT © 1989 Stephen R. Covey. Reprinted by permission of Simon & Schuster, Inc.)

client or customer, managing a project, preparing a poster, drafting a report, and preparing a proposal are all done at the lowest possible labor costs because they are all pushed down into the organization to be done in whole or part by individuals who have lower labor rates and have the ability to carry out the tasks. However, whenever an individual delegates a task that requires a person to learn new tools and techniques, the delegator will have to make an up-front investment of training or teaching time.

- Delegation builds resiliency into an organization. By spreading understanding of tools and techniques to other members of the organization, more people know how to do more things. This is analogous to the concept of "strengthening the bench" on an athletic team.

Reluctance to Delegate

In spite of all the good reasons to delegate, some young technical professionals never learn how to do it. This failure hampers the success of the organization and hinders the individual's advancement. Experience suggests many reasons for being reluctant to delegate. Although these reasons are understandable, the results produced are unfortunate. Entry-level engineers and other technical professionals may be reluctant to delegate some of their tasks to others for one or more of the following reasons:

- Delegation, as already noted, often requires an up-front investment of time and energy. The young professional may believe that he or she simply doesn't have the time to provide basic information required to delegate tasks to others. The young professional may even know that delegation is vital, but simply "doesn't have the time" to make the extra effort "now." This procrastination position may become habitual.

- Delegation is sometimes opposed because of excessive pride of authorship or ownership or perhaps even arrogance. That is, the young professional feels that he or she is really the only one who knows how to do the work correctly.

- While delegation may seem attractive to you, you may feel that you have no one to delegate to. As an entry-level professional, you probably do not have supervisory responsibility for full-time personnel in your organization. Recall, however, Figure 4–1 and the related discussion, which argues that even the entry-level individual has opportunities to "manage" some of the efforts of other members of the organization such as clerical staff, technicians, and draftpersons. They are your potential delegatees.

• Delegation might be frustrated by the potential delegator's poor organizational and communication skills. Although you thoroughly understand the tasks that need to be done because you have done them many times, and although you realize the value of delegating some of those tasks to others, you may lack the organizational and communication skills needed to explain what needs to be done and to suggest ways to do it.

• Delegation may be frustrated by fear of losing knowledge-based job security. The young engineer may be hovering around Maslow's second level and believe that his or her job security is linked in part to the exclusive ability to carry out certain tasks.

• Fear of appearing lazy or incompetent may inhibit delegation (Raudsepp, 1978). You will have many opportunities to demonstrate your willingness to carry your share of the work load in a competent manner and thereby gain the respect of your colleagues.

• Finally, you may be hesitant to delegate because you fear advancement or you are concerned that advancement may occur too rapidly as you shed tasks to others and free up your time to take on new responsibilities.

Delegation Isn't Always Down

Delegation is usually downward in the organizational structure, that is, from a supervisor to a supervisee. However, lateral and upward delegation can be an effective way to make optimum utilization of an organization's human resources.

Assume, for example, that the young technical professional has the responsibility to draft a report on a subject that is new to him or her. In order to start with an overall structure that incorporates experience already within the organization, the young professional might ask a senior person to prepare an outline based on that senior person's experience. Similarly, the young professional might delegate the task of preparing minutes of a meeting to someone on his or her level who has talent for preparing effective minutes. In all cases, the delegator, while giving authority to individuals above and next to him or her in the organizational structure, retains responsibility for the delegated tasks.

Recall the decider-director-doer organizational praradigm presented in Chapter 1. Consistent with that enlightened approach to the functioning of an organization, each member of the organization should be receptive to being a delegatee regardless of their position and rank. More specifically, by so doing they demonstrate their ability to carry out the important "doing" or "producing" function.

Delegation Tips

Assume that the delegator is a theory Y person and is about to delegate tasks to another person in the organization. If the potential delegatee is also a theory Y person, he or she will welcome the opportunity to learn more about how the organization does things and to make a higher-level contribution to the organization's objectives.

Provide the potential delegatee with a context of the tasks. Explain how the task is one small, but important part of a major effort. For example, the young engineer working in a civil engineering consulting firm may be delegating field survey work to one of the organization's surveyors. Provide an overview of the project, indicate the expected outcome of the project, and clearly explain how the survey data will be used in project tasks that will culminate in the desired outcome. Explain the desired outcome of the tasks being delegated in appropriate terms such as when it is needed and the quality, size, format, location, and documentation. Identify the resources that are available to carry out the tasks such as budget, personnel, and available data and information.

Resist the temptation to overprescribe "how" the delegated tasks are to be carried out. That is, focus on the desired outcome and the resources available to achieve the outcome. Depending on what you know about the delegatee, delegate the tasks in such a way that you, he or she, the project, and the organization benefit from the "novice" effect.

Consider providing milestones to the delegatee. For example, you might suggest that the delegatee prepare an outline of how he or she will go about doing the delegated tasks and that the outline will be available for the two of you to discuss within two days. Clearly give authority commensurate with what you are asking the delegatee to do. Finally, remember that everybody reports to somebody. All delegators are also delegatees. Delegating is a chain process or, perhaps more properly stated, a lattice process—certainly not a series of isolated events.

When you give an assignment to someone and it is understood and committed to, recognize that there are only three possible outcomes. Expect the first and most desirable outcome to occur most often, but recognize that the second and third outcomes will occasionally occur. The three possible outcomes are

- The work is delivered as needed.
- The work will not be completed as needed, but the delegator is so advised by the delegatee well before the deadline.
- The work is not going to be completed as expected. The delegator learns about the deficiency at or after the time the work was to be completed.

The first result, as already noted, is by far the most common. Although the second outcome is unfortunate, it is much better than the third outcome, because the second provides the delegator with an opportunity to seek an alternative course of action. If a particular act of delegation is not going to be successful, the delegator should be informed immediately.

In reviewing the results of delegated assignments with a delegatee, remember to critique the work, not the person. Avoid "you" messages, especially when the work is deficient. Instead, speak about the work, noting which parts satisfy the requirements and which parts are deficient. Identify results that are of high quality and those that are of low quality. Say "thank you," especially when the work has been carried out in accordance with requirements. Refer to Figure 4–6, 100 Ways to Say "Very Good" for ideas. Consider speaking directly to the person who did the work, calling them, or sending them a handwritten note.

Recognize that the usual consequence of an individual's not adopting a delegation mode of operation within an organization is that he or she are likely to be relegated to the slow track. Through delegation, the young professional demonstrates many important management perspectives and skills, not the least of which is teamwork. Failure to delegate is likely to label the entry-level technical professional as "not a people person" and stymie both managerial and technical advancement.

MANAGING MEETINGS

A committee is a group that keeps minutes and loses hours.

(Milton Berle)

Although Mr. Berle may have found humor in committees and their meetings, many professionals at all levels in organizations find little to even smile about as they sit through long, nonproductive, and sometimes unruly and highly stressful sessions that are often unduly dignified by being referred to as "meetings." For purposes of this chapter, a meeting is defined as three or more people discussing business or professional work. Meetings are often a waste of time and excessively frustrating because of poor planning and execution, which often indicates lack of respect for the time, talent, and feelings of others. The negative aspects of meetings become more frustrating as people progress in their careers because the percentage of time devoted to meetings tends to increase with increased levels of responsibility and with advancement in an organization.

Some meetings are absolutely necessary, because they are the best way to enable groups of people to do certain things. Because so much time is devoted to

1. You're right!
2. Good work!
3. Well done.
4. You did a lot of work today!
5. It's a pleasure to work with you.
6. Now you have it.
7. Fine job!
8. That's right!
9. Neat!
10. Super!
11. Nice going.
12. That's coming along nicely.
13. That's great!
14. You did it that time!
15. Fantastic!
16. Terrific!
17. Good for you!
18. Make it so!
19. That's better.
20. Excellent!
21. Good job (name).
22. Super Fine!
23. That's good.
24. Good going.
25. That's really nice.
26. WOW!
27. Keep up the good work.
28. Outstanding!
29. Fantastic!
30. Good for you.
31. What talent!
32. Good thinking.
33. Exactly right!
34. You make it look easy.
35. YES!
36. Awesome!
37. Way to go.
38. Superb!
39. OK!
40. You're on target.
41. I knew you could do it.
42. Wonderful!
43. You're great.
44. Beautiful work.
45. You've worked hard on this!
46. That's the way!
47. Keep trying.
48. That's it.
49. Let's tell the boss.
50. You're very good at that.
51. You're learning fast.
52. You certainly did well today.
53. I'm glad your approach is working.
54. Keep it up!
55. I'm proud of you.
56. That's the way.
57. You're learning a lot.
58. That's better than ever.
59. Quite nice.
60. You've figured it all out.
61. Perfect!
62. Fine!
63. I'll sign the order.
64. You've got it.
65. You figured that out fast.
66. Very resourceful.
67. You are really improving.
68. Look at you go.
69. You've got that down pat.
70. Tremendous!
71. I like that.
72. I couldn't do it better myself.
73. Now that is what I call a fine job.
74. You did that very well.
75. Impressive!
76. Sharp!
77. Right on!
78. That's wonderful.
79. You mastered that in no time.
80. Very nice!
81. Congratulations.
82. That was first-class work.
83. Sensational.
84. Right!
85. You don't miss a thing.
86. You make my job fun.
87. You must have been practicing it.
88. I'm glad I assigned this to you.
89. You came through again.
90. DYNAMITE!
91. I knew I could count on you.
92. You deserve a raise.
93. How can I help you with this?
94. Go for it!
95. The Best!
96. You have my complete support.
97. MARVELOUS!
98. Clever idea.
99. I am glad you are on our team.
100. Thank you!

Figure 4–6 100 ways to say "very good" (Source: Adapted with permission from Firestien, R. L., "100 Ways to Say 'Very Good'," Williamsville, N.Y.: Center for Studies in Creativity, 1992.)

meetings and because so much important work should be done at meetings, careful meeting management is mandatory. This section of the chapter offers tips for successfully planning and executing meetings and dealing with difficult people and situations at the meetings. Incidently, engineers and other technical professionals occasionally participate in public meetings or hearings. Some of the material presented in this section is applicable to the management of public meetings.

Reasons to Meet

There are only two legitimate types of meetings consistent with the preceding definition of meetings. The first is a working meeting devoted to defining problems, hearing status reports, brainstorming, conceptualizing courses of action, discussing alternatives, and selecting and beginning to implement solutions. The principal feature of the working meeting is informed, positive participation by all attendees.

The second legitimate reason to meet is to provide a briefing on nonroutine, crucial matters such as resignations, promotions, introduction of new staff, acquisition of an organization, and serious financial problems. Although, in contrast with the working meeting, little discussion is expected, the meeting is justified because of the serious implications of the information that is being shared. That is, certain announcements are of major significance and must be fully understood. A memorandum or other form of written communication would not be appropriate.

When Not to Call a Meeting

Interestingly, and perhaps this is the cause of meeting problems, there seem to be more reasons not to call a meeting than to call a meeting. The following reasons not to call a meeting are quoted directly (except for parenthetical comments) from "Time Talk, November 1987" as published in the *Administrator*, January 11, 1988:

- You've made up your mind what to do anyway. Convening others and feigning the seeking of their counsel places your credibility at risk.
- You know what should be done, but don't want to take responsibility. (You call a meeting, indicate what should be done, obtain formal approval or informal acquiescence, and then the "blame" can be "spread" if needed.)
- You don't know what should be done and want somebody else to make the decision. (You are responsible, it is your decision, and therefore you should decide.)

- You're trying to use the meeting to "pull a fast one" on a colleague or your boss. (In other words, he or she is out of town and really responsible, but you call a meeting under guise of "an emergency" to make decisions on his or her project.)
- You are on an ego trip and like the sound of your voice.
- The subject is too important to merit a meeting. (In other words, quick, decisive action is needed and you clearly have the necessary responsibility and authority.)
- You've lost your case elsewhere and you are looking for a life belt.

Tips for Successful Meetings

To the extent you can, try to give attention to the 12 suggestions shown in Figure 4–7. The following discussion of the 12 tips is based, in part, on ideas and information provided by Day Timers (1990), Frank (1989), Hensey (1991), Jones (1987), NIBM (1990), Quartet Manufacturing Company, Tarkenton (1979), Upton (1988), and Wortman (1981).

1. Reason!

2.

3. Agenda

4. Minutes

5.

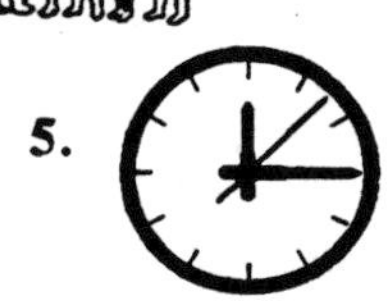

6. Fixed meeting time if you meet often

7. Otherwise, schedule next meeting at current meeting

8. Adequate facilities

9.

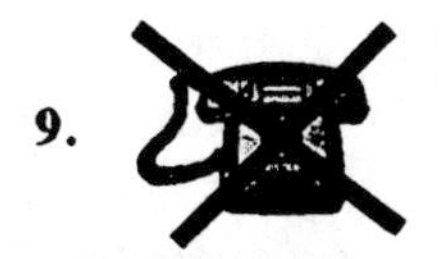

10. All contribute

11. Seek consensus

12. Confidentiality

Figure 4–7 How to manage productive meetings

Tip 1: Meet only when there is a reason. This is one advantage of ad hoc committees and similar groups in that there is less danger of falling into a pattern of periodic but unnecessary meetings. Permanent committees that meet as a matter of practice tend to encourage the creation of agenda items and unnecessary discussion to fill the time available.

Tip 2: Invite only the necessary people. Avoid the "it would be nice to have them there" thinking. Those who need to know the results of the meeting, but do not need to participate in the meeting, can be provided copies of the minutes.

Tip 3: Provide a written agenda prior to the meeting. Refer to Figure 4–8. Note the following features of the example agenda:

- All invitees are identified in the "To" portion of the memorandum. In the example, names are listed in alphabetical order. In a more formal environment, the custom might be to list each person with his or her name followed by position title in intraoffice written communications. In this case, consider listing individuals in order of decreasing rank. Knowing who else will attend a meeting can be useful. For example, this information provides an opportunity to informally discuss other matters with one or more individuals before or after the meeting.

- Both the date and the day of the meeting are indicated. With this practice, invitees are much less likely to miss the meeting by arriving on either the wrong day or the wrong date.

- The meeting starting time and ending time are shown on the agenda. This courtesy enables meeting attendees to plan their day, especially the time after this meeting, because they will know when the meeting is to end.

- The heading "Additional Agenda Items?" appears near the beginning of the agenda. This provides the Chair with the opportunity to add topics of discussion and also facilitates input from attendees, some of whom may not have been involved in preparing the original agenda.

- Individuals having reporting or other responsibilities are identified by name. This practice helps to ensure that they will be prepared to report on their efforts or lead a discussion on the indicated topic.

- Background, support, and other materials are enclosed. This extra effort by the person managing the meeting arrangements provides participants with the opportunity to be prepared so that the meeting time can be used most effectively.

- An action-oriented theme is established by using words and expressions such as "select," "follow-up," and "recommended course of action." In some or-

MEMORANDUM

Date: July 5, 1994

To: Members of the Design Team -- H. O. T. Air, B. Careful, O. U. Kidd, B. Level, and U. R. Liable

From: I. M. Boss, Project Manager

Re: Agenda for Meeting 9, Friday, July 8, 1994, 8:00–9:00 a.m., Conference Room 2.

I. Welcome to new members

II. Additional agenda items?

III. Surveying (see memorandum, included as Attachment A, for alternatives and some pros and cons) (B. Level)

  A. Discuss alternatives.

  B. Select/follow-up.

IV. Design criteria (see pages from State code included as Attachment B) (U. R. Liable)

  A. Recommended course of action.

  B. Decision/follow-up.

V. Alleged design error

  A. Summary of 6/28/94 meeting (see memorandum included as Attachment C) (B. Careful).

  B. Discussion.

  C. Follow-up.

VI. Next meeting

enclosure: Attachments A, B, and C

c: Vice President Quayle (without enclosures)

Figure 4–8 Example agenda

ganizations, such as academic institutions, a meeting might be considered successful if interesting topics are discussed with no decisions made. If the Chair wants to avoid this, the agenda should be structured to encourage and expect action.

• For each agenda item, the invitees are informed via the agenda and the enclosures what the committee is expected or encouraged to do. For example, the person arranging the meeting clearly indicates if the committee is to develop, discuss, and select alternative solutions to a problem (e. g., agenda item III in Figure 4–8) or respond to a recommended course of action developed prior to the meeting by a member of the committee or by others (e. g., agenda item IV in Figure 4–8). That is, the person managing the arrangements for the meeting focuses the energies of the committee members. However, recognize that the committee may elect to broaden its response to the agenda, but is unlikely to do this if trust is established between the Chair and the committee members.

- Consider adding an agenda item, depending on the nature of the meeting, titled "Good News" under which positive happenings are briefly shared. "Bad News" need not be shown on the agenda—it will take care of itself.

Time spent prior to the meeting by the person responsible for meeting arrangements, possibly assisted by a few committee members, can save considerable time for the entire committee. In contrast, a brief agenda, or none at all, and little time spent in preparing for a meeting may result in a long and unproductive session. If you are invited to a meeting and a written agenda is not provided, ask if an agenda will follow. If there is not to be an agenda, insist on knowing what is to be discussed so that you can prepare or arrange to have some other, more appropriate person attend in your place.

Assume there is an important point you, as an invited participant, want to make at a meeting, but there is no directly related agenda item. Either you did not have an opportunity to get your concern on the agenda prior to the meeting or the Chair was not receptive to adding items at the meeting. Carefully prepare your idea or information. Look for opportunities at the meeting to make your point, perhaps in answer to a question posed to you or to someone else.

Tip 4: Provide a written record. Minutes or some other written documentation of the meeting should be distributed soon after the meeting. Action items or follow-up tasks agreed to at the meeting should be clearly indicated, accompanied by the names of the people who have agreed to accomplish the tasks. Consider highlighting the names by using techniques such as underlining, all caps, or bold type. Arrange for someone to serve as secretary, preferably prior to the meeting, and clearly indicate that the secretary's principal responsibility is to produce accurate minutes in a timely fashion.

An undocumented meeting "never happened" or, worse yet, occurred in as many different and conflicting versions as the number of attendees. Without the reality check provided by minutes issued shortly after a meeting, each participant will remember, or think he or she remembers, a scenario that should have occurred or would have been nice to experience. Given the opportunity, especially under pressure, individuals tend to rewrite history—to edit the past so as to serve their present and what they hope to be their future. You, and those you work with and for, can largely avoid this morass by writing and distributing some form of documentation after every meeting.

In a related matter, Doyle and Strauss (1982, pp. 34–54) introduce the concept of group memory. The previously discussed minutes serve the long-term function of the group memory and may be all that is needed for most meetings. The short-term portion of the group memory is the record of that which has happened at the meeting which is in progress and which participants need to recall during the meeting in order to fully participate. Personal, temporary notes taken by participants often serve the short-term memory function. Presumably the notes will not be needed after the minutes are produced.

An alternative, coordinated way to serve the short-term memory function, which may be particularly useful in some sensitive or controversial situations, is to record the key points of the discussion on newsprint, transparencies, or other media readily visible to all participants as the meeting progresses. The Chair, the secretary, a facilitator, or another specially designated person could provide this service. Using this approach, everyone has the same, complete short-term memory, and everyone has an opportunity to question the recorder's interpretation. Better to resolve conflicts at the meeting as they occur than after the meeting when formal minutes are produced.

Tip 5: Start and stop on time. To show respect for the people who have been invited to the meeting, do everything you can to start and stop on time. If a meeting is to be the first of a series of meetings of a group, the pattern established at the first or early meetings will tend to prevail. Consider scheduling meetings late in the morning (e.g., 11:00 A.M.) or late in the afternoon (e.g., 4:00 P.M.) so that impending lunch or dinner encourages focus and brevity. Occasionally schedule a meeting very early in the morning prior to the start of normal office hours, during lunch, during dinner, or in the evening to emphasize urgency or provide variety. Consider setting time limits for each agenda item as a further means of encouraging focus and brevity. Some meeting participants may need to attend only a portion of the meeting. Out of consideration for them, the Chair should schedule their appearance at a specific time and excuse them when they are finished.

Tip 6: Establish a fixed meeting time if the group meets often. A group that meets frequently should establish a fixed meeting time and then cancel a meeting if it is not needed. This practice enables participants to plan their schedules.

Tip 7: Schedule the next meeting at the current meeting. A group that meets occasionally should schedule its next meeting at the current meeting. For this, and other reasons that may arise during the meeting, all attendees should bring their time management systems or their calendars to the meeting. One of the most wasteful practices is to expect a committee member or someone on the support staff to try to schedule a meeting among people with varying responsibilities and already complicated schedules.

Tip 8: Provide adequate facilities. Depending on the nature of the topics discussed and the kinds of presentations to be made, needed facilities, equipment, and supplies may include a blackboard or white board, chalk, flip chart, colored markers, overhead projector, 35-millimeter projector, VCR, refreshments, and access to copy machines, telephone, and rest rooms. Sometimes a very special site may be appropriate such as a retreat setting, a neutral site, or a project location. All of these things should be anticipated and arranged for ahead of time by the person who is managing the meeting. The Chair should arrive early to make sure that everything is in order and arrange to sit in a location where he or she can be easily seen and heard by all participants. The Chair should also be within easy view of the door and the clock.

Tip 9: Limit interruptions. Make arrangements so that participants are not interrupted by telephone calls, disturbed by walk-ins, or distracted by external noises and activities.

Tip 10: Expect everyone to contribute. Presumably everybody in attendance has been invited because useful information is being provided and because he or she has something to contribute. Encourage everyone to participate and consider using techniques like those presented later in this chapter.

Tip 11: Seek consensus. While you must recognize that consensus is not always possible, it is a worthy goal. Try to focus the group on the mission of the organization and the purpose of the project or other activities being undertaken by the group. Individual ideas adopted and committed to implementation by the group should become group ideas. Individuals should detach themselves from their ideas and permit them to be modified as a condition of having their ideas accepted by the group.

Tip 12: Practice confidentiality. The debate and discussion should, in general, end at the meeting and not be carried on in the hallways and offices by subgroups of meeting participants. Participants should recognize that they have an opportunity to make their case at the meeting and should respect the decision of the group.

Dealing with Difficult People and Situations at Meetings

As Chair, you will sometimes encounter intentional or unintentional behavior by members of a committee or by people in attendance at a meeting which, in your opinion, is counterproductive. Some of these behaviors and possible solutions are presented here. Problematic behavior involving one person that continues at a given meeting or is repeated at successive meetings may indicate an interpersonal problem between you and that individual. Your ability to practice empathetic listening, as described in Chapter 3, is very valuable in this situation. Perhaps you and the other individual should privately discuss the matter outside of the meeting environment.

Committee member speaks to one or more individuals out-of-turn in a low, negative, and insulting tone. The Chair could, during the meeting, abruptly ask the person to repeat and explain his or her comment for the benefit of the entire committee. Then invite questions and comments from the committee members to reinforce the idea that participants are to speak to the entire group and show courtesy when others are speaking. Another approach is for the Chair and everyone else to simply stop talking and draw attention to and listen intently to the side conversation. The resulting embarrassment may stifle the negative activity.

Committee member continues to press for course of action even though the committee, by consensus or formal vote, is clearly opposed. The Chair could ask that person to put his or her ideas or arguments in writing for possible future reconsideration or for the record.

Committee member talks too much. Some meeting participants simply talk too much relative to the point they are trying to make or the information being offered and also relative to the speaking time available for other committee members. One of the solutions is for the Chair to make comments such as:

- "Henry, I think you've made your point—let's permit others to make theirs."
- "Heidi, I think we must move on—our agenda is full."
- "Hans, that is interesting, but I think you are moving away from our agenda."

A person does not contribute. This behavior is most likely to occur early in the "life" of a new committee or working group or when a person joins an existing committee or working group. Unlike the preceding four situations, this behavior is not disruptive. Nevertheless, presume that every person has a reason for being at the meeting and has a right and obligation to contribute. A possible solution is to, in a very gentle way and at the right moment, say something like:

- "Nancy, I understand that you have some experience in this area. What is your opinion?"
- "John, we haven't heard yet from the planning department; what are your concerns?"

Individual is opposed to everything. Some individuals are often opposed, in knee-jerk fashion, to new concepts, ideas, and proposals. A person of this persuasion typically takes the position that a suggested new approach will not work or says something like "We tried that ten years ago and it failed." One approach is to determine if the person thinks there is a problem. For example, say: "Emily, do you think we have a problem to be solved?" If the answer is "yes," then ask her to offer a solution at this or the next meeting. If she chooses the latter, indicate that her report will be an agenda item. In summary, if an individual recognizes a problem and takes the position that the solution suggested by others is unworkable, he or she has an obligation to present a feasible solution.

The "know what we should do?" person. Occasionally a committee will include a person who precedes a lot of statements with "You know what we should do?" but does not commit to take action or assume responsibility. In this situation, the Chair should, during the meeting, ask the "idea" person to research one of his or her ideas and present a written or an oral report on it at the next meeting.

Miscellaneous Thoughts

If a manager calls a meeting to "discuss" a topic, the manager gives up the option to act unilaterally on that topic or area. To do otherwise is to risk loss of credibility.

Carefully selected handouts, visual aids, and props can be very effective at meetings. Assume, for example, that you are trying to convince the group that they should utilize a new material in a product being manufactured by your organization. Bring a sample or samples of the material to the meeting so that everybody can see and feel it as you talk about it.

Planned and unplanned supervisor–supervisee interactions and discussions with colleagues constitute special kinds of meetings, certainly informal ones. While such "meetings" usually do not follow a formal agenda that is provided ahead of time, try to bring a list of discussion items to the discussion and indicate to the other person that there are specific topics you would like to discuss. These informal meetings often lead to various kinds of follow-ups. Rather than a single document, such as a set of minutes, the follow-up might be in the form of documents that are related to topics discussed and decisions made at the meeting with your supervisor or supervisee. An example would be a memorandum to a third party with a copy to the person with whom you met.

APPRECIATING AND WORKING WITH SUPPORT PERSONNEL

Support personnel include secretaries, clerical workers, word processing staff, technicians, draftpersons, surveyors, and data entry personnel. Most of the people holding these positions are compensated on an hourly basis. Some support personnel may be salaried and have salaries that are higher than those of entry-level engineers.

One way in which support personnel are different from the professional staff (e.g., engineers, planners, economists) is that most of the work done by the former is at the specific request of the latter. Stated differently, support personnel do not usually unilaterally initiate projects or major tasks or take on portions of projects as do the professional staff.

Another way in which support personnel differ from the professional staff is that the former tend to be more specialized. That is, a support person performs a narrower range of functions than the professional person and typically executes them with a high level of skill and expertise. Furthermore, the areas of expertise within the support personnel tend to be different than the areas of expertise within the professional staff. Thus, support personnel as a group are highly skilled in critical areas such as word processing, particular types of graphics, data entry, surveying, testing, and uses of specialized equipment such as computer-aided drafting workstations and thermal imaging devices.

Essential Members of the Organization

The entry-level engineer or other technical professional is advised to view the support staff as valuable members of the organization because of their special skills and their lower labor cost. One or more support personnel should be brought in for a project whenever and wherever their combination of expertise and cost per unit of production is more favorable than having the tasks performed by professionals. Even if the entry-level person has the necessary skills, the cost per unit of production should govern. Just because the entry-level engineer or other professional could do a task does not mean he or she should do the task, as noted in the earlier section of this chapter on delegating.

The young technical professional should recognize that he or she needs a support staff more than they need the young professional. That is, in well-managed engineering and other technical organizations, everybody, including all the support staff, has much to do. Therefore, if some of the support staff are not assisting a particular technical professional with his or her project, they will be engaged with another project.

In contrast, the young technical professional cannot be productive without significant assistance of support personnel in specialized areas such as word processing, graphics, surveying, and data entry. Although the young professional might be able to do some of these functions, he or she will not be able to do them as well, as measured by productivity (labor and other costs per unit of work completed) and quality, as a support specialist.

Challenges Unique to Working with Support Personnel

Working with the many and varied individuals that populate the engineering and business workplace is a challenge. This is true in working with support personnel, for whom there is a unique set of challenges.

Communication. One challenge is communication—mainly writing, speaking, and listening—between the young technical professional and the support staff. Professionals in highly technical and sophisticated disciplines routinely communicate with each other using terms, concepts, acronyms, abbreviations, and other expressions known only to them. The support staff, because of their narrower and more pragmatic or applied training and interests, will not be as widely conversant. Therefore, there is a communication challenge for the entry-level technical professional if he or she is to be productive, that is, to effectively work with the people and use the resources available in the organization.

Age differential. Most of the support staff will be older than the entry-level technical professional. The 22-year-old engineer will rely on support from the 35-year-old chief draftsperson/CAD operator, the 58-year-old head of the word processing group, the 47-year-old head of the motor pool, and the 32-year-old surveying party chief. Inherent in the age differentials will be a wealth of knowledge in areas such as "how this place really runs" or where you can find or get something. The entry-level engineer or other technical professional should conduct himself or herself so as to benefit from that knowledge. One indication that the confidence and respect of support personnel is being earned is when they make comments to the young person such as "I'm sure you know how you want this done, but if I were you I would consider doing it this way . . . ."

Devaluation. There is an unfortunate tendency among some engineers and other technical professionals to look down on or devaluate support personnel. As the entry-level professional surveys an organization, particularly the larger, diverse public or private entity, he or she is likely to see people in a variety of positions, some of which appear undesirable. For example, from the perspective of the recent college graduate, some jobs may be characterized as simple, boring, dirty, noisy, and hot. Because the entry-level professional would not want to do those jobs, he or she looks down on those who do.

This devaluing perspective may be evident in the young professional's relationships with the support staff. One example is use of the terms "boy" or "girl" in referring to men and women as in "I'll have one of the boys enter the data" or "Let's have one of the girls in the field check this out." Another example is the use of first names without being specifically invited to do so. A final, and particularly disturbing example, is being impolite to and surly with support staff, but polite to and friendly with the professional staff. The overall impact is very negative.

It is presumptuous and arrogant to assume that a person's inherent value is determined by the position he or she holds on an organization chart. While the economic value of people's work varies widely, reflecting their education, experience, and responsibility, their inherent value as individuals is an independent matter. The young engineer or other technical professional is strongly advised to take the position, unless he or she knows otherwise, that everyone in the organization is doing basically what they want to do. Assume that each person has consciously made a series of education and work decisions that led to their current position. People should not be embarrassed by or have to apologize for the position they hold or the work they do, unless the work does not meet the expected quality and quantity for the position.

Tips for the Entry-Level Technical Professional

As already noted, support personnel are functionally different from professional personnel and are essential members of the technically based organization. Furthermore, working effectively with support personnel presents some special challenges. The following suggestions are meant to help the young professional quickly develop an effective working relationship with support personnel.

- Identify the organization's support personnel and learn how they are organized. That is, who are they, where are they, what services do they perform, and to whom do they report? Sources of information include administrative charts, the telephone directory, personnel handbooks, maps or drawings of the physical facilities, and colleagues.

• Introduce yourself to the support personnel or the individuals who direct the work of the support personnel. If these introductions are not provided by others as part of your initial orientation to the organization, take the initiative to do so on your own. Try to complete this process before you begin to request assistance.

• Unless they volunteer, ask people how they want to be addressed or observe how they are addressed by others. Examples are, Ms., Mrs., Mr., Dr., first name, and nickname. This is particularly important when dealing with senior support personnel, who may be offended by excessive familiarity on your part.

• Always be polite. Words and expressions like "please," "excuse me," and "thank you" show respect for individuals and maintain civility in the civilized workplace and encourage it in the uncivilized one.

• Find out how support is requested. Expect procedures to vary widely from organization to organization and even within an organization. The standard operating procedure may be very informal, or there may be a formal, written work request form. Respect the established work request procedure. However, if the requesting procedure is very informal, such as a verbal request, accompany or follow the verbal request with written instructions. There are three reasons for this. First, you are more likely to think through your needs if you write them out. That is, your request will be more clear and complete. Second, there is a much lower probability that support personnel will misunderstand your needs if your request is written rather than only spoken. Finally, when the requested work is completed, your written instructions will remind you how the product is to be used or what is to be done next. Ask that your written instructions be returned to you, preferably with a sign-off by the support personnel who completed the task or tasks.

• Always indicate what you need, that is, the product you want, and when you need it. With respect to the desired product, try to provide an example such as a copy of a previous letter, memorandum, drawing, survey, or data sheet. With respect to the desired completion time, avoid saying things like "as soon as possible," "no hurry," or "when you get around to it" unless there is a clear understanding of what these terms mean.

• Explain, as appropriate, the context or purpose of the work or product you are requesting. Support personnel are much more likely to produce work or products that meet your needs if they understand the context. They may even suggest an alternative approach.

• Minimize your interruptions of support staff. They cannot be making progress on your work or the work of others when they are talking to you or otherwise being interrupted by you. Written work request procedures minimize interruptions. Consider grouping verbal requests to minimize the number of interruptions and consider scheduling a brief working meeting at a mutually agreeable time.

- Do not, through your poor planning or as a means of getting attention, allow too many requests to be in a crisis or panic mode. If you "cry wolf" too many times, you will lose credibility within the support staff. Try to develop a sense of the absolute and elapsed time required to complete certain tasks. The young professional will occasionally encounter bona fide emergencies and, if he or she has credibility with the support staff, the staff will respond favorably. Remember, in keeping with theory Y, that most people want to be helpful.

- Be prepared to prioritize. You may not be able to get everything you want when you want it. If asked by the support personnel, establish priorities and look for alternative ways to accomplish certain tasks.

- Insist that you be told if the product will not be done in the manner requested and/or completed on time as soon as possible. Recall the earlier "three outcomes" discussion under the topic of delegation.

- Provided that you get the necessary product on time, be slow to criticize how support personnel do their work. As you and they develop an effective working relationship, you might ask questions about why they do things the way they do them and perhaps eventually make some suggestions. You will learn much in the process. Ask them how you could be more effective in your joint efforts.

MANAGING YOUR BOSS

The title of this section may sound manipulative or presumptuous, but it is intended to emphasize that you need to proactively approach relationships with personnel all around you in the organizational hierarchy—down, laterally, and up. Gabarro and Kotter (1993) define "managing your boss" as " . . . the process of consciously working with your superior to obtain the best possible results for you, your boss and the company." Use of the word "company" in the definition certainly does not limit the importance of the "boss management" process to business. Clearly managing the boss is applicable to young technical professionals in all employment sectors, including government.

You are encouraged to have high expectations for both you and your immediate supervisor while remembering that you and he or she are in a situation of "mutual dependence between two fallible human beings" (Gabarro and Kotter, 1993). In other words, you need each other, and neither of you is perfect. Refer to Figure 4–9 for useful ideas.

Make sure you understand your boss and his or her context, including:

- **Goals and objectives**
- **Pressures**
- **Strengths, weaknesses, blind spots**
- **Preferred work style**

Assess yourself and your needs, including:

- **Strengths and weaknesses**
- **Personal style**
- **Predisposition toward dependence on authority figures**

Develop and maintain a relationship that:

- **Fits both your needs and styles**
- **Is characterized by mutual expectations**
- **Keeps your boss informed**
- **Is based on dependability and honesty**
- **Selectively uses your boss's time and resources.**

Figure 4–9 Checklist for managing your boss (Source: Adapted with permission of *Harvard Business Review*. Excerpt from Gabarro, J. J. and Kotter, J. P., "Managing Your Boss," *Harvard Business Review,* May–June, pp. 150–157. Copyright by the President and Fellows of Harvard College, all rights reserved.)

CARING ISN'T CODDLING

As you begin to manage your relationships with others, optimally an element of caring will be evident in your actions towards others and in the actions of others towards you. Caring, as used here, does not mean coddling. If caring isn't coddling, what is it?

To answer this, think about those former teachers or professors who you believe really cared about you. They probably demonstrated their concern for you narrowly as a student and broadly as a person through a variety of meaningful interactions. Examples are: delivering well-prepared lectures; making regular and demanding assignments intended to deepen and broaden understanding of the course material; providing opportunities for independent study such as a research paper or laboratory project; encouraging you to participate in cocurricular and extracurricular leadership and service activities; offering an encouraging word at a discouraging time; and praising when nobody else seemed to notice what you had accomplished. The preceding were not offered in a paternalistic, condescending, ostentatious manner; rather these actions were part of a high-expectations–high-support environment intended to stretch without snapping, provide example without expecting cloning, and build confidence without imparting arrogance.

Caring is also exemplified by the parent who said, "If it's worth doing, it's worth doing well" and the colleague who, at a meeting in your firm, has the courage to ask the awkward question or raise the sensitive issue that almost everyone knows must be addressed. Caring is also shown by the manager who says no to the pleading employee who did not strive to meet the established requirements and now wants to avoid the adverse consequences.

You may also recall, with disdain, those teachers, supervisors, colleagues, and others who were generally "nice," but didn't expect all that much of you. Often you delivered in accordance with their expectations. You and they could have done so much more. Perhaps they didn't really care about you, or even themselves.

Caring isn't coddling. Caring is pushing, pulling, admonishing, stretching, demanding, encouraging, urging, challenging, cajoling . . . . Caring is high expectations coupled with high support. Caring helps individuals and organizations meet their goals and realize their full potentials.

REFERENCES

BENTON, D. A., *Lions Don't Need to Roar: Using the Leadership Power of Professional Presence to Stand Out, Fit In and Move Ahead.* New York: Warner Books, 1992.

BREDIN, J., "Confessions of a Management Mole," *Industry Week*, September 19, 1988.

BROWN, T., "Autocracy or Involvement?," Inside Management, *Industry Week*, April 4, 1988, p. 13.

COVEY, S. R., *The 7 Habits of Highly Effective People.* New York: Simon & Schuster, 1990.

Day-timers, Inc., "The Art of Effective Delegation." Allentown, Pa., 1990.

Day-timers, Inc., "The Cost-effective Meeting: How to Convene It, Control It and Benefit from It." Allentown, Pa., 1990.

DOYLE, M., and D. STRAUS, *How to Make Meetings Work.* New York: Jove, 1982.

FIRESTIEN, R. L., "101 Ways to Say 'Very Good'." Buffalo, N.Y.: Center for Studies in Creativity, Buffalo State College, 1988.

FRANK, M. O., *How to Run a Successful Meeting in Half the Time*. New York: Simon & Schuster, 1989.

GABARRO, J. J., and J. P. KOTTER, "Managing Your Boss," *Harvard Business Review*, May–June 1993, pp. 150–157.

GENEEN, H. S., *Managing*. New York: Doubleday and Co., 1984.

HARRIS, S. J., *Winners and Losers*. Allen, Texas: Argus Communications, 1968.

HENSEY, M., "Keys to Better Meetings," *Civil Engineering*, February 1991, pp. 65–66.

JONES, S., ed. "Better Meetings," *Executive Communications*, November 1987.

MARTIN, D. D., "Motivation and the Developing Engineering Manager," *Journal of Management in Engineering–ASCE*, Vol. 2, No. 4 (October 1986), pp. 246–252.

MCGREGOR, D., *The Human Side of Enterprise*. New York: McGraw-Hill, 1960.

MCQUILLEN, J. L., Jr., "Motivating the Civil Engineer," *Journal of Management in Engineering—ASCE*, Vol. 2, No. 2 (April 1986), pp. 101–110.

MOLE, J., *Management Mole: Lessons from Office Life*. London: Bantam Press, 1988.

National Institute of Business Management, "Win At Meetings," *Executive Strategies*, June 5, 1990.

PATTERSON, K. J., "Organizations: The Soft and Gushy Side," *The BENT of Tau Beta Pi*, Fall 1991, pp. 19–21.

Quartet Manufacturing Company, "How to Conduct Interactive Meetings," Skokie, Ill. (no date).

RAUDSEPP, E., "Why Managers Don't Delegate," *Chemical Engineering*, September 25, 1978, pp. 129–132.

SCHERMERHORN, J. R., Jr., "Chapter 12—Leading Through Motivation," *Management for Productivity*. New York: John Wiley and Sons, 1984.

TARKENTON, F., "The Management Huddle: Making Group Decisions Work," *Sky*, July 1979, p. 50f.

UPTON, H., "Failed Meeting? Check the Agenda," *Spirit*, September 1988, p. 23f.

WORTMAN, L. A., *Effective Management for Engineers and Scientists*. New York: Wiley, 1981.

SUPPLEMENTAL REFERENCES

BLANCHARD, K., and S. JOHNSON, *The One Minute Manager*. New York: Berkley Books, 1983.

PARACHIN, V. M., "The Fine Art of Encouragement," *The Rotarian*, February 1992, pp. 18–19.

5

The Organization of Organizations

> *I was to learn later in life that we tend to meet any situation by reorganizing; a wonderful method it can be for creating the illusion of progress while producing confusion, inefficiency and demoralization.*
>
> (Petronius, Roman satirist who died in 66 A.D., quoted by Brown [1992])

Recall the "Playing Field" discussion in Chapter 1, which identified the major types of organizations that employ technical professionals. You might work for a consulting engineering firm, a government agency, a contractor or constructor, or a manufacturing business. Throughout your career, you are very likely to interact with all four organizational types, and you might even be employed by all four. This chapter focuses on organizations, or more precisely how they are organized. The young professional should observe how a potential employer is organized, or, if already on the staff of an organization, should examine the nature of the organization. You are also advised to be aware of how the organizations you interact with are organized.

The organization of organizations is important to you for three reasons. First, as discussed in Chapter 4, you will need to learn how to get things done by working with many and varied people within the organization. Knowing the basic structure of your organization and of organizations you interact with will help you get things done on schedule, within budget, and in accordance with the expectations of others.

The second reason for organizational awareness, particularly your employer's or a potential employer's organization, is that the structure of the organization is likely to determine your potential progress and satisfaction. The way an organization is organized tends to reveal its attitude toward professional and other staff, its desire and ability to expand, and the relative value placed on various functions such as engineering, project management, marketing, finance, and administration. For example, Rosenberg's (1992) thesis is that the morale and, more specifically, the creativity of technical professionals is influenced by the way an organization is organized. Burgower (1990) notes 3M's success in supporting technical professionals by including a "dual ladder" advancement system in its organizational structure. Your values and goals should be consistent with those of the organization where you are employed. For example, if you seek to achieve a high level of technical competence and be recognized and rewarded for it, you are more likely to achieve your goal in a matrix type of organization, as discussed in this chapter, than in a functional type of organization.

The third reason to understand the various ways organizations may be organized is that you will eventually have an opportunity to set up or rearrange or help to set up or rearrange an organization or be part of one. Perhaps it will be your own business. You want to be prepared.

This chapter defines the ingredients of an organization and then introduces the three principal legal forms of business ownership. Four types of organizational structures are discussed—functional, regional, client-oriented, and matrix. A discussion of single- versus dual-ladder advancement systems concludes the chapter.

THE CONCEPT OF AN ORGANIZATION

Schermerhorn (1984, p. 8) defines an organization as a "collection of people working together in a division of labor to achieve a common purpose." Obviously, organizations are formed because of the belief that they are needed. Schermerhorn (1984, p. 8) notes that "Organizations perform tasks that are beyond individual capabilities alone." As clearly suggested by Schermerhorn's definition, a group of people does not necessarily constitute an organization. For example, a 50-member theatrical group preparing to perform is likely to be an organization, whereas 50 theatergoers milling around in the lobby before the performance are not likely to constitute an organization. Even 50 people working in the same company or office may not, in fact, constitute an organization as precisely defined by Schermerhorn. They may not be committed to a common purpose, may not appreciate the importance of their individual roles, or they may have difficulty working together.

Consider further the three ingredients of an organization as set forth in Schermerhorn's definition. The first ingredient is a "common purpose." In the case of a consulting engineering firm or a government unit or agency, the purpose is likely to be a provision of services. In contrast, the purpose of a manufacturer or constructor is to assemble products or erect structures, facilities, and systems. Clear definition and widespread understanding of purpose are essential to the effectiveness of an organization.

"Division of labor" is the second ingredient of an organization. In general, strong organizations are characterized by diversity of ability and talent, with the mix being carefully selected to achieve the organization's purpose. Individuals should understand and appreciate their roles in the organization as determined by their unique sets of talents and skills. The third and last important ingredient of an organization is "people working together." This typically requires some hierarchy of authority which, depending on the factors such as the purpose and size of the organization, might be very simple and basic or extremely complex. Some structure is needed and typically includes reporting relationships and basic administrative process to encourage efficiency and to avoid chaos.

LEGAL FORMS OF BUSINESS OWNERSHIP

Consulting firms, manufacturing organizations, and constructors usually use one of three forms of business ownership from a legal perspective. The three typical options are the individual proprietorship, the partnership, and the corporation. The following discussion of these three forms of business ownership is based on Clough (1986), Martin (1993, pp. 172–174), Morton (1983), Roemer (1989), and Stephens (1993).

Individual Proprietorship

With the individual proprietorship or, as it is sometimes called, the sole proprietorship, an individual owns and operates the business. The advantage of the individual proprietorship is that it is the simplest and least expensive to establish and operate. An example of simplicity is that U.S. income tax reporting consists of completing one tax schedule for inclusion with the proprietor's personal tax return. Furthermore, there is maximum freedom of action in that the individual proprietor is the "boss" and can act unilaterally in all decisions. The most significant negative aspect of the individual proprietorship is that the owner is personally liable, including his or her personal—that is, nonbusiness—assets, for all debts and obligations. In addition, the size of the business and the ability to expand are limited by the individual proprietor's resources or his or her ability to obtain financing.

Partnership

In a partnership, two or more persons own and operate the business, although ownership, decision-making, debts, losses, and profits are not necessarily equally shared. Partnerships are formed and operated under state partnership laws. The laws generally recognize the partnership as an entity separate from the individual partners. For example, partners pay income taxes, not the partnership. Any partner can act on behalf of the partnership, provided the action is in keeping with the scope of the business.

The appeal of a partnership is that it combines the financial assets and talents of two or more individuals who are interested in engaging in the same type of business. On the negative side, particularly when compared to an individual proprietorship, individual partners are restricted in their business actions, but typically have much more latitude than owners of a corporation. For example, a partner cannot sell or mortgage his or her share of the partnership's assets without permission of the other partners. Furthermore, in a fashion similar to the individual proprietorship, each partner is financially responsible for the acts of all the partners to the full extent of his or her personal assets.

Corporation

A corporation is an entity created under state incorporation laws, consisting of one or more individuals, owned by one or more stockholders, and considered to be separate from the employees or the owners. A board of directors, elected by the owners, provides general control. The corporation can buy and sell real estate property, enter into agreements, and sue and be sued. A corporation can be dissolved by surrender or expiration of its state charter with its business obligations settled in accordance with those laws. Owners pay taxes only on dividends received after the corporation pays state and federal taxes. An S corporation is a special U.S. form of corporation available to businesses that have 35 or fewer shareholders and meet other requirements. The advantage of this form of corporation is that it has less tax liability than a standard corporation.

A significant advantage of a corporation, contrasted with an individual proprietorship or partnership, is the limited liability of the owners. Stockholder liability is limited, with a few exceptions, to the amount of their investment in the corporation. Examples of exceptions, that is, situations in which one or more individuals within a corporation risk personal liability, are when fraud is committed or when a corporation is underfunded or underinsured. Other advantages of corporations are the ability to raise large amounts of capital and the corporation's perpetual organizational life, that is, continuation of the organization is not dependent on particular employees or owners, because the corporation is an entity separate from the employees and owners. Another advantage of a corporation is that it provides ease of multiple ownership.

A possible disadvantage of a corporation is that the stockholders are not in any way agents of the corporation. The board of directors controls the corporation, and the officers of the board act as agents for the corporation. Stated differently, although each stockholder owns a part of the organization, he or she cannot act unilaterally on behalf of the organization.

ORGANIZATIONAL STRUCTURES

Recall the three ingredients of an organization according to Schermerhorn (1984, p. 8), that is, purpose, division of labor, and hierarchy of authority. The division of labor and hierarchy of authority ingredients are typically addressed within an organization's structure. Although the purpose or, more broadly, the mission of an organization may not be explicitly identified in the organizational structure, purpose or mission should influence that structure. Schermerhorn (1984, p. 180) defines organizational structure as "the formal system of working relationships that both divide and coordinate the tasks of multiple people in groups to serve a common purpose." The four basic organizational structures presented in this chapter are functional, regional, client, and matrix. Each of the four basic organizational options is described, and positive and negative features are identified. Additional ideas and information on organizational structures are provided by Clough (1986, Chapter 3), Dhillon (1987, Section 2.8), and Schermerhorn (1984, pp. 185–190). Grigg (1988) provides information on the matrix organizational structure in government.

Functional

In the functional organization, the first partition or highest-level "division of labor" is the grouping of individuals with similar expertise. Focus on function typically continues down in lower levels of functional organizations. Universities typically utilize a functional organizational structure from the very top of the organization down through its various administrative levels. As a current or very recent college student, you probably are aware of the functional organizational structure. Refer to Figure 5–1. Note that the highest level or first partitioning of the hypothetical university is one in which individuals with similar interests and expertise are grouped together. Grouping of individuals by function is carried on down through the second and third partitioning of the organization. For example, academic affairs are partitioned into various colleges and schools, which are partitioned into academic departments.

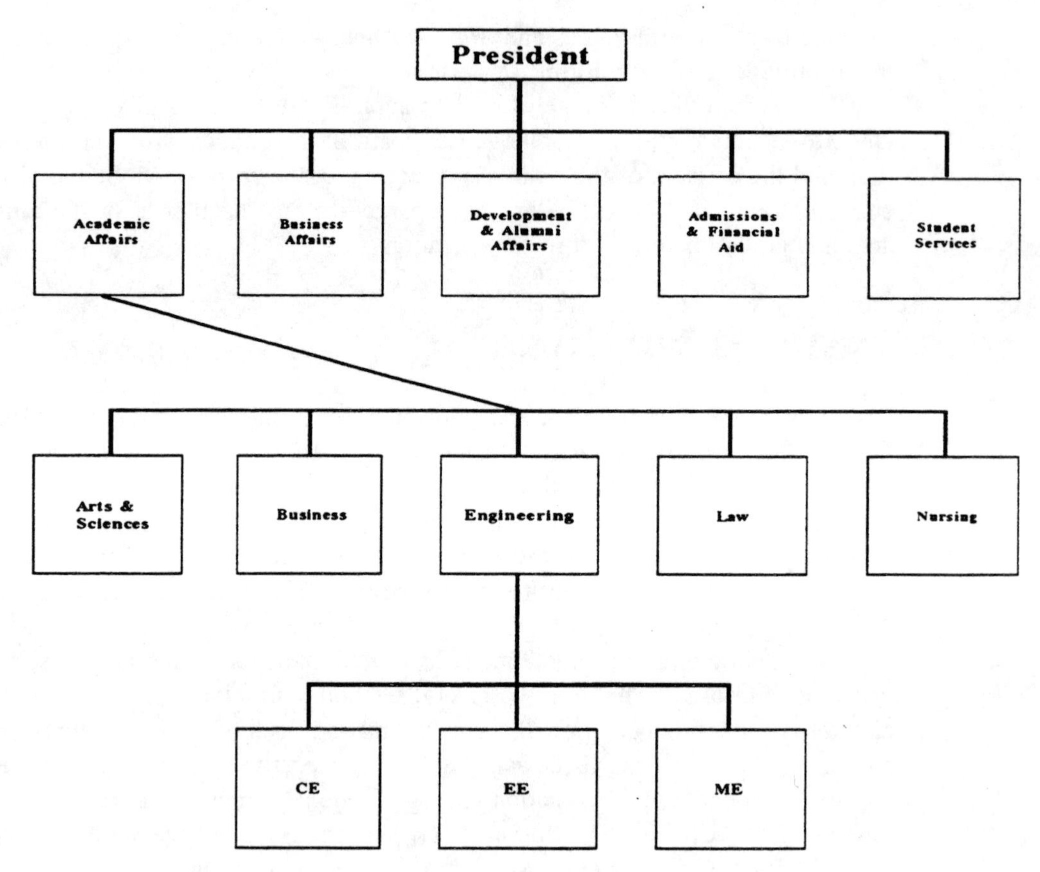

Figure 5–1 Functional organizational structure—hypothetical university

Cities and other governmental units are often organized along functional lines. As illustrated by Figure 5–2, the highest level or first partitioning of the organization focuses on the major functions of planning, engineering, water and wastewater, streets and parks, fire, and police. Within each of these functional areas, further partitioning may be done on the basis of subfunctions.

Many consulting engineering firms and manufacturing organizations are organized functionally. For example, refer to Figure 5–3. Note how similar the highest level or first partitioning of the hypothetical consulting firm is to the hypothetical city. The second level of organizational structure might also be functional, as suggested by the partitioning of the design unit into transportation, environmental, and buildings. Figure 5–4 shows two levels of function-oriented structure. It, like the other three functional organization figures, suggests that the partitioning could go on almost ad infinitum.

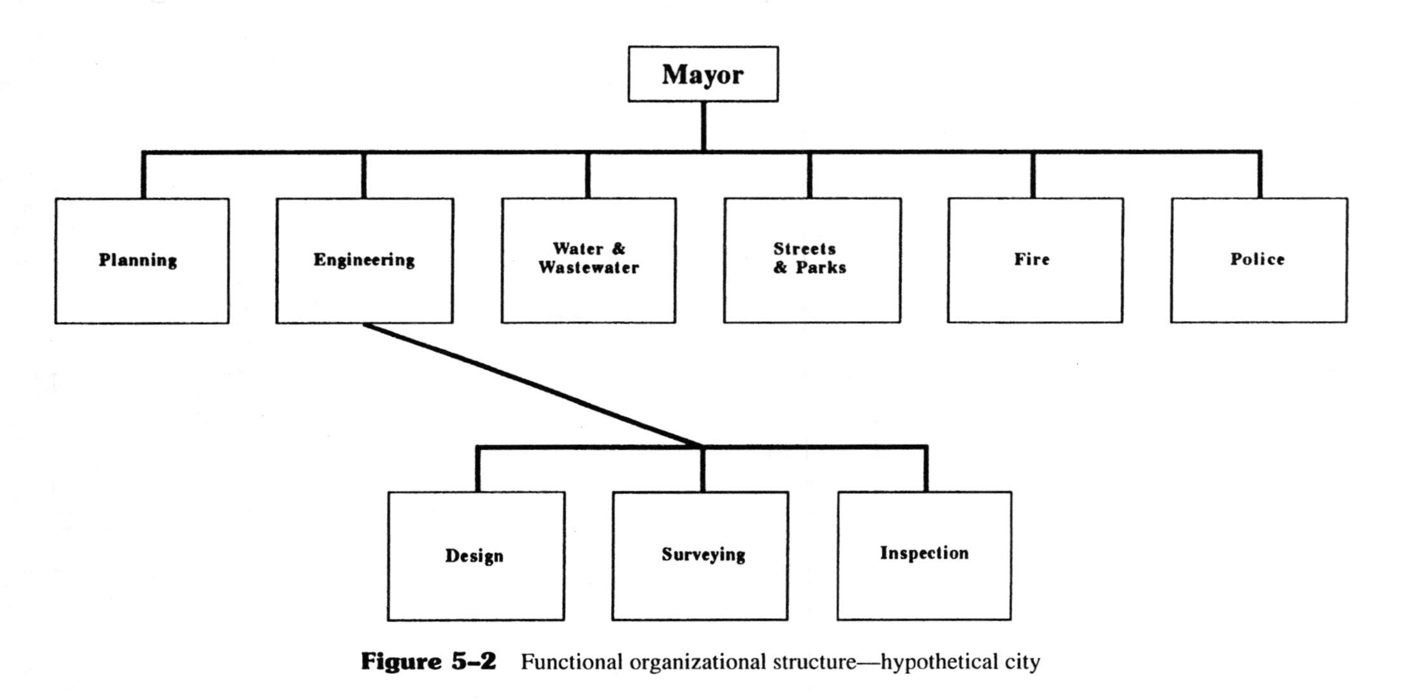

Figure 5-2 Functional organizational structure—hypothetical city

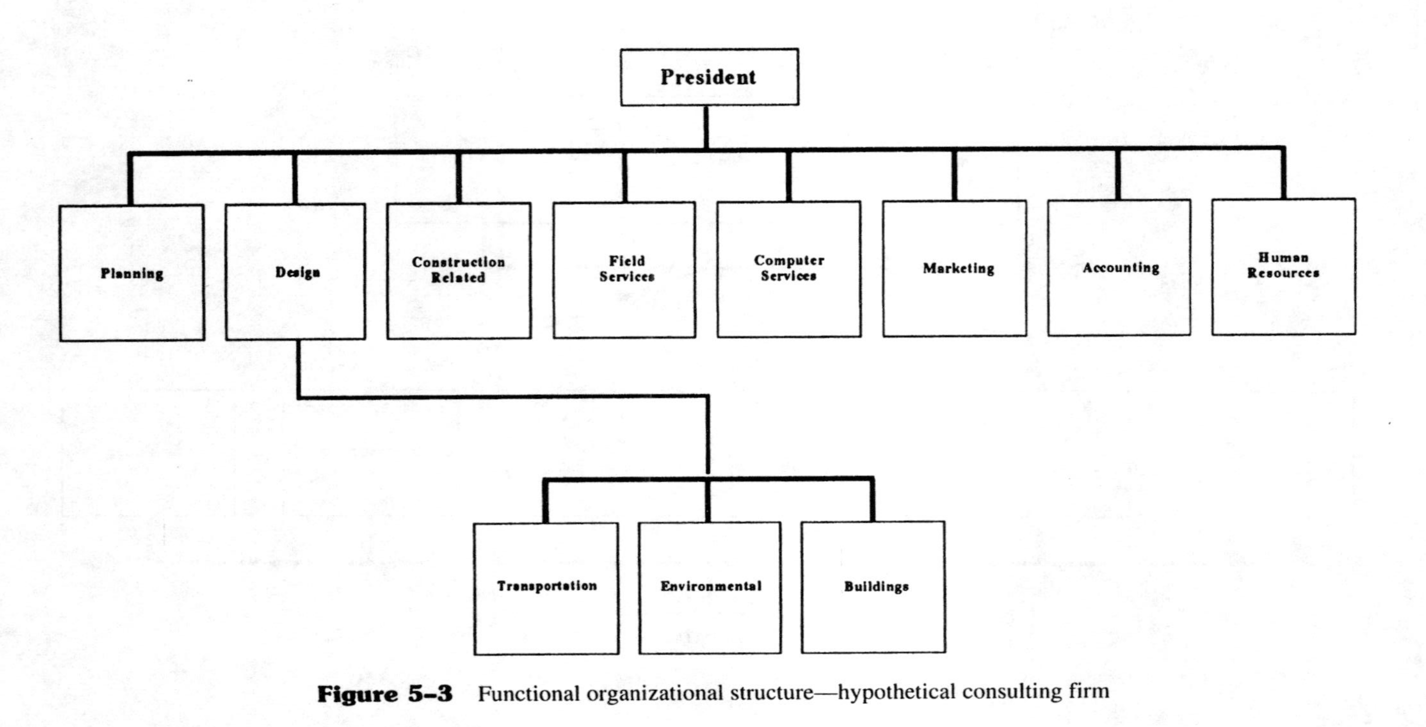

Figure 5–3 Functional organizational structure—hypothetical consulting firm

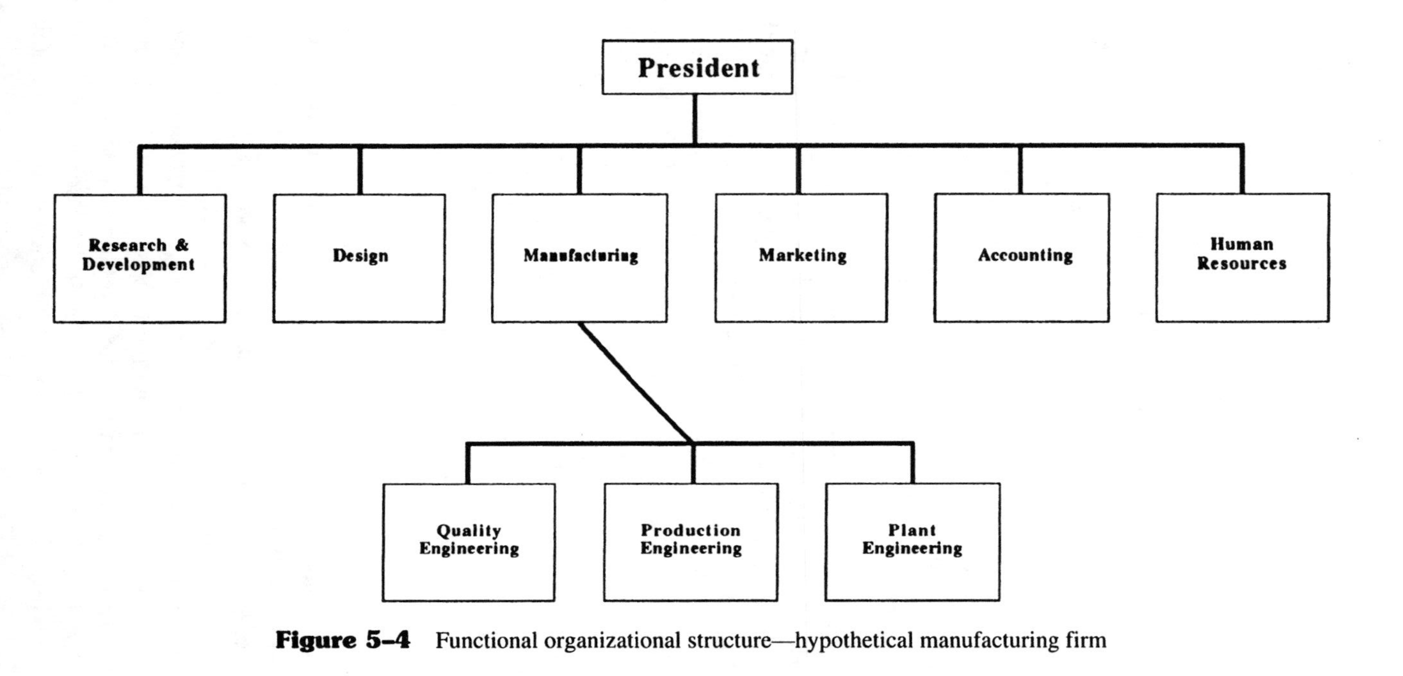

Figure 5–4 Functional organizational structure—hypothetical manufacturing firm

Positive features. One advantage of the functional organizational structure, which is shared with the regional and client-based organizational structure, is that each employee reports to only one individual, that is, has only "one boss." This may seem obvious, but as described later, the fourth and last organizational option discussed in this chapter is the matrix structure, which has the complication of most individuals reporting to "two bosses."

Another advantage of the functional organization is that the grouping of individuals with similar education, experience, and ability tends to facilitate obtaining assistance from others. For example, the entry-level structural engineer assigned to a design unit within a large consulting engineering firm will be surrounded by individuals who, as a group, represent great breadth and depth of design knowledge and experience. A possible third advantage of the functional organization is the homogeneity within each functional group. Some individuals prefer to work primarily with people of similar interests.

Negative features. A negative aspect of the functional organization is that difficulties are often encountered in coordinating activities and projects between functional groups. Consider the university, for example, where the academic affairs functional group makes curricular changes and does not advise the admissions functional group, which leads to miscommunication with potential students. Or perhaps the business affairs functional group changes the tuition and fee structure and does not advise the admissions and financial aid functional group, resulting in the need to rework financial aid packages. Within a manufacturing organization, the product design functional group may lack sufficient communication with the manufacturing functional group, resulting in plans and specifications for products that are difficult to manufacture. Stated differently, designers who are isolated from manufacturing, fabrication, and construction-oriented functional groups may produce plans and specifications that simply cannot be manufactured, fabricated, or constructed without major manufacturing floor or field changes. As discussed in Chapter 10, Legal Framework, failure to adequately consider manufacturability or constructability tends to increase an organization's legal liability.

Over-specialization for specialization's sake is another negative tendency of functional organizations. For example, the computer services unit in a manufacturing organization may purchase and implement new software that they view as being valuable and interesting without consulting with the design, manufacturing, and other functional groups that the computer services unit is supposed to serve. Such myopic actions can lead to purchasing software for which there are no needs, or worse yet, increasing the efficiency of doing something that no longer needs to be done. Functional units risk becoming too self-centered because of the homogeneity of the individuals within the functional group and the homogeneity of the subject matter. They risk losing contact with the purpose of the organization as a whole and the clients or customers it serves. A final disadvantage of functional organizations is that the advancement route tends to be primarily administrative, that is, not technical. If you intend to advance in a functional organization, you will probably have to leave your technical or other specialty and take on primarily administrative functions.

Regional

The regional or territorial organizational structure arranges personnel on the basis of the needs of various regions or territories served. Generally speaking, the regional structure is more client-oriented than is the functional structure. Refer to Figure 5–5. Note how the highest level of the organization is partitioned on the basis of geographic locations. The regional approach is widely used by the U. S.

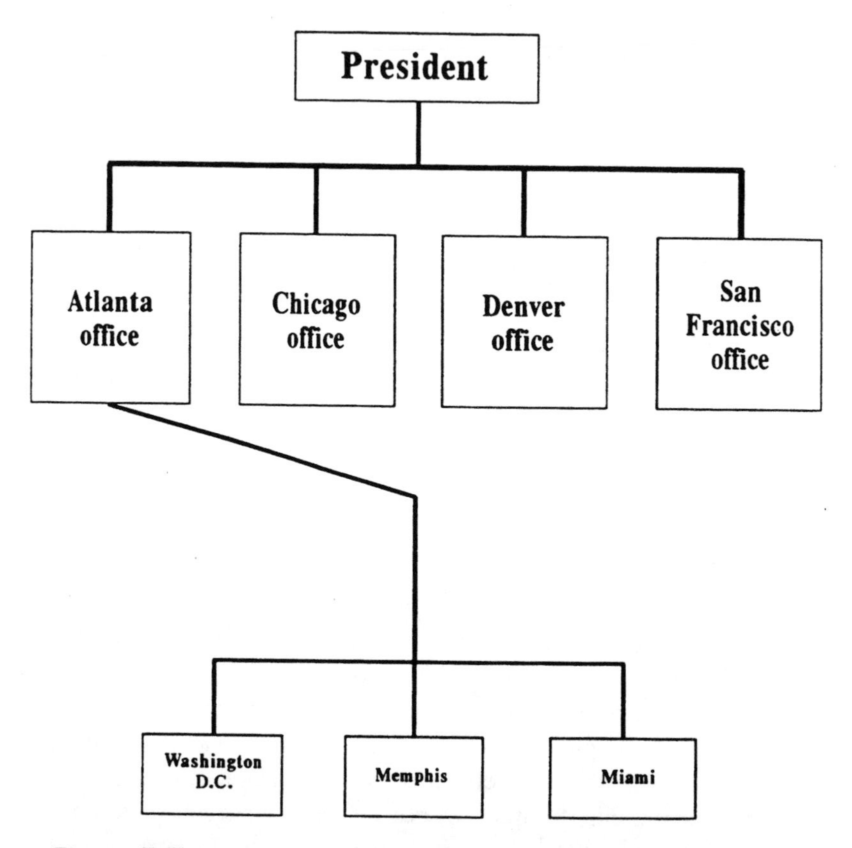

Figure 5–5 Regional organizational structure—hypothetical consulting or construction firm

government. Examples are the Environmental Protection Agency, the Internal Revenue Service, and the U.S. Geological Survey.

Within each major regional office, a refined regional structure may be used, as depicted in Figure 5–5, in which case the principal office serves as a hub with smaller, satellite offices being at the ends of spokes. Another approach is to organize each regional office on a functional basis, thus mixing the regional structure, at the highest level of the organization with the functional structure at the second level of the organization. Regardless of the details of organization, each employee in a regional organization reports to one individual.

Positive features. An advantage of the regional structure is that each employee has one boss. More significantly, the regional structure offers the potential for better service to customers and clients because the resources of the firm tend to be dispersed throughout the geographic area it serves. Equally important, the top focus of the company is not on functions such as planning, design, manufacturing, and construction, which tend to be matters of primarily internal concern, but is instead on location, that is, where the clients are.

Negative features. One negative aspect of the geographic structure in contrast with a single large corporate office is that technical expertise is spread over a wide geographic area and depth may be lacking in some offices. Resolving this may lead to another disadvantage of the geographic structure, that being excessive duplication of some people and equipment, again in contrast with a large central office. Additional management resources may also be required. Finally, the regional structure suffers the same liability as the functional structure in that the principal advancement route is usually administrative, not technical. That is, to advance in the organization, an individual must gradually leave his or her technical specialty and be increasingly involved with budgets, personnel, marketing, and other administrative matters.

Client

As indicated by its name, the client structure is organized so that it is most responsive to client types or categories. This concept is illustrated in Figure 5–6. At its highest level, the organization is partitioned to reflect its various clients, in this case municipal, land development, and industrial clients. The structure of the organization explicitly focuses the professional and support staff on clients. At secondary and lower levels, a function or regional structure may be used. A variation on the client-based structure is establishment of a temporary office in or near a client's office for the purpose of effectively carrying out a major engineering and/or construction project. This strategy may result, during or after comple-

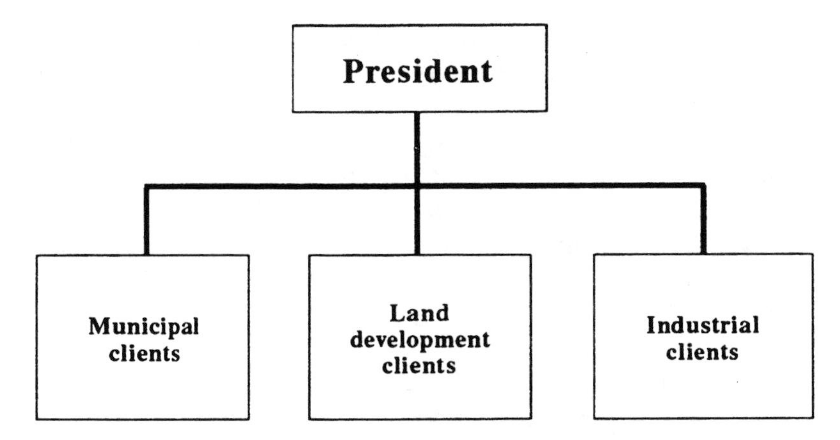

Figure 5–6 Client-based organizational structure—hypothetical consulting or construction firm

tion of the project, in establishment of a new regional office. Another variation on the client-based structure is to organize on the basis of current projects, that is, projects A, B, and C would replace the threes types of clients in Figure 5–6.

The client-oriented organizational structure is clearly focused on clients. However, this does not mean that the other structures presented in this chapter cannot be client-focused while accomplishing other high-priority goals. The matrix structure, for example, has great potential to focus on the entities to be served. The local or closest office of the matrix organization is the point of access for the client, customer, or constituent and the point of delivery of services. The functions or service areas, which cross all of the firm's offices, are the mechanism by which the most appropriate corporate resources are brought to bear on the needs of the client, customer, or constituent. The extent to which individuals and entities are served by an organization is based as much on the corporate culture as on the corporate structure. See *Business Week* (1993) for a representative discussion of the importance of organizing to serve customers.

Positive features. The principal advantage of the client-based organization is the excellent potential it offers for meeting the needs of clients. As with the functional and regional organizational structures, the client-based structure has the advantage of each employee's reporting to only one individual.

Negative features. On the negative side, the client-based organization may spread technical expertise over several administrative units and, therefore, depth may be lacking in some units. As with the regional structure, rectifying this situation may lead to excessive duplication of some people and equipment and, therefore, unnecessary costs. Another disadvantage of the client-based organization is that the professional staff, in their zeal to serve their types of clients, may not be aware of and, therefore, may not utilize resources available in other groups within the organization. For example, a project management tool being effectively used in the industrial group to serve industrial clients might be applicable for use in the municipal client group, but the two groups do not communicate. Finally, the client-based structure, like the regional and functional structures, tends to favor advancement by administrative routes, as opposed to technical specialties.

Matrix

In a matrix organization, the fourth and final type of organizational structure to be discussed, professional and support staff are arranged in two ways. Typically, matrix organizations include a functional dimension and then some other dimension such as regional, a client, or project. Figure 5–7 shows a matrix organizational structure for a hypothetical consulting firm. In this case, functional organization as represented by rows intersects the regional organization as represented by columns. Matrix organizations are rare within technically based organizations. Note that the matrix organization shown in Figure 5–7 was created by combining the functional organization of Figure 5–3 with the regional organization of Figure 5–5. This was done to emphasize the two-dimensional nature of the matrix organization.

A manufacturing firm could also use a function versus location matrix like that shown in Figure 5–7. Another matrix structure suitable for a manufacturing firm is presented in Figure 5–8. The functions, that is, rows, are the same as in Figure 5–7, but products or product lines now form the columns. Another approach, which is a variation on Figures 5–7 and 5–8, is to have projects, rather than offices or products, occupy the columns. Teplitz and Worley (1992) note that having projects as one of the two dimensions of an organization will improve the effectiveness of project management.

A characteristic unique to the matrix organization and not shared by the functional, regional, and client-based organizations, is that each employee reports to two individuals. For example, in the case of the hypothetical firms illustrated in Figures 5–7 and 5–8, each member of the professional staff would report, for some purposes, to the head of his or her functional or technical specialty area and would report, for other purposes, to the person responsible for their regional office or product. Each employee, in effect, has two bosses. You, as a young engineer or other technical professional in the matrix organization shown in Figure 5–7, would confer with your functional head, or his or her designate, on matters such as obtaining assistance on a technical problem, finding a computer program to assist with data analysis, and preparing a technical paper for presentation at a conference. In general, your functional head or his or her designate would not be

President

Locations / Functions	Atlanta office	Chicago office	Denver office	San Francisco office
Planning				
Design				
Construction related				
Field services				
Computer services				
Marketing				
Accounting				
Human resources				

Note: Each interior rectangle contains one or more employees.

Figure 5-7 Matrix organizational structure—hypothetical consulting firm

located in your office and, therefore, most of your communications would be by telephone, voice mail, electronic mail, fax, or conventional mail. In contrast, you would confer with the person in charge of your office, or his or her designate, on matters such as having adequate work space, equipment, and supplies; arranging for vacation time; and representing the company by joining organizations in the local community. The functional head and the office head in a matrix organization usually share responsibility and authority for areas such as hiring staff and interoffice transfers, recognizing and rewarding personnel, quality of products or services, and marketing.

President				
Products / Functions	Cars	Trucks	Vans	After market
Planning				
Design				
Manufacturing				
Field services				
Computer services				
Marketing				
Accounting				
Human resources				

Note: Each interior rectangle contains one or more employees.

Figure 5–8 Matrix organizational structure—hypothetical manufacturing firm

Positive features. The principal advantage of the matrix organization, particularly one that is based on the dimensions of expertise and regional offices as illustrated in Figure 5–7, is that the structure offers the potential for delivering the best the organization has to offer to its clients or customers regardless of where they are. In theory, a client of the firm represented by Figure 5–7 who happens to be in the San Francisco area has access to design capabilities of the firm regardless of where they are. Matrix firms are managed that way, that is, typical office parochialism is minimized and interoffice cooperation within disciplines is optimized. Stated differently, the matrix structure offers excellent opportunity for coordinating the entire organization's resources to the service of its clients. With the matrix structure, an organization can draw globally and deliver locally.

The addition of offices and specialties (i. e., columns and rows) is also readily done with the matrix structure, making this form of organization attractive to the aggressive, growing firm or company. Another advantage of the matrix organization, is that two advancement routes are available to the professional staff. One route is largely administrative/managerial, this being along any of the vertical paths in Figures 5–7 and 5–8. The other route is largely technical/professional, these being illustrated by any of the horizontal paths in Figures 5–7 and 5–8. The dual-ladder feature is discussed further later in this chapter.

Negative features. The major negative aspect of the matrix organizational structure is the complication of each professional and many other personnel having to interact with two individuals, a functional or discipline head and an office (Figure 5–7) or project (Figure 5–8) head. Furthermore, as already noted, the functional head or his or her designate is usually assigned to some other office, further complicating communication. This generates additional administrative and communications efforts in contrast with more conventionally organized firms, such as the need to have carefully coordinated periodic evaluations and the inevitable complication that occurs when individuals from two or more offices are working as a team on the same project. Certain "ills of the matrix" are frankly discussed by Davis and Lawrence (1978) including "tendencies toward anarchy, power struggles, severe groupitis, collapse during economic crunch, excess overhead, sinking to lower levels, uncontrolled layering, navel gazing, and decision strangulation." Each ill is diagnosed, and preventive or remedial measures are suggested. Guterl (1989) argues that the matrix organization is too rigid for multinational companies, whereas others claim the opposite.

Implementing a matrix approach. Given the rarity of fully functioning matrix organizations and their potential significant advantages, how does a more conventionally arranged organization reorganize into the matrix structure and make it work? Poirot (1991) discusses this topic, drawing on 20 years of matrix experience at the consulting firm of CH2M-Hill. As a point of reference, at the time the article was written, CH2M-Hill had almost 4,000 full- and part-time employees, about 700 owners, and 63 offices in the United States. The following tips are quoted directly from the Poirot article:

- Allow time to define responsibilities/authorities.
- Commit senior management time.
- Develop people who want to make the matrix work.

- Make decisions based on what is good for the client and firm.
- Promote open communication with no secrets.
- Eliminate politics at high levels.
- Commit energy to evaluate and compensate on a common basis.
- Use consensus management.
- Hire top people.
- Consolidate net income at corporate level and reward everyone in the firm.

SINGLE- VERSUS DUAL-LADDER ADVANCEMENT SYSTEMS

Most organizations advance personnel along a single ladder like that shown in Figure 5–9. The example ladder, constructed for an engineering organization, is typical of that which usually dominates the previously presented functional, regional, and client-structured organizations. With movement up the rungs of the ladder, titles move away from explicit reference to technology. Obviously, the rungs get shorter with movement up the ladder because of the reduction in the number of positions available at each successively higher level. Professionals

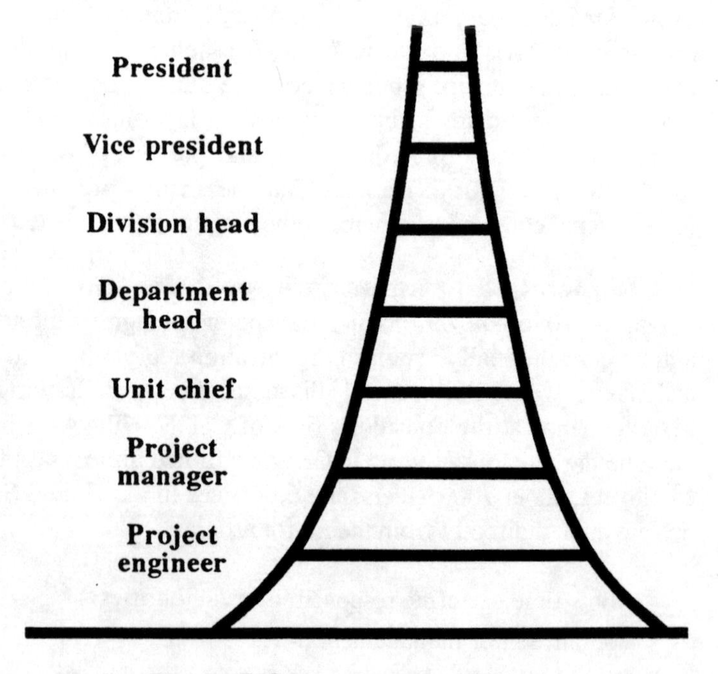

Figure 5–9 The "administrative" ladder

who wish to eventually move away from technical specialization and away from doing primarily technical work will probably be comfortable with organizations that offer the single-ladder "administrative" advancement system.

Some organizations have dual-ladder advancement systems—one for personnel with administrative/management orientation and one for personnel with technical/professional orientation. Figure 5–10 illustrates the second of the two modes of advancement in a dual-ladder system. Note how titles retain explicit reference to technology with progression up the rungs of the ladder.

Toto (1988) lists the "skills, personality traits, and interests" which tend to be associated with professionals desirous of moving up the administrative/management ladder and of professionals whose primary interest is moving up the technical ladder. These skills, traits, and interests are listed in Figure 5–11. You should examine the two lists to determine which list is most likely to define your current interests and goals, recognizing that such interests and goals can change dramatically during your career.

Goldstein (1988) provides a concise history of and prospects for the dual-ladder advancement system. The dual-ladder approach started in the early 1970s in the United States and was motivated by a desire of organizations to retain high-quality professional people whose primary aspirations were not in "management." Dual-ladder systems had problems in the 1970s with low visibility and

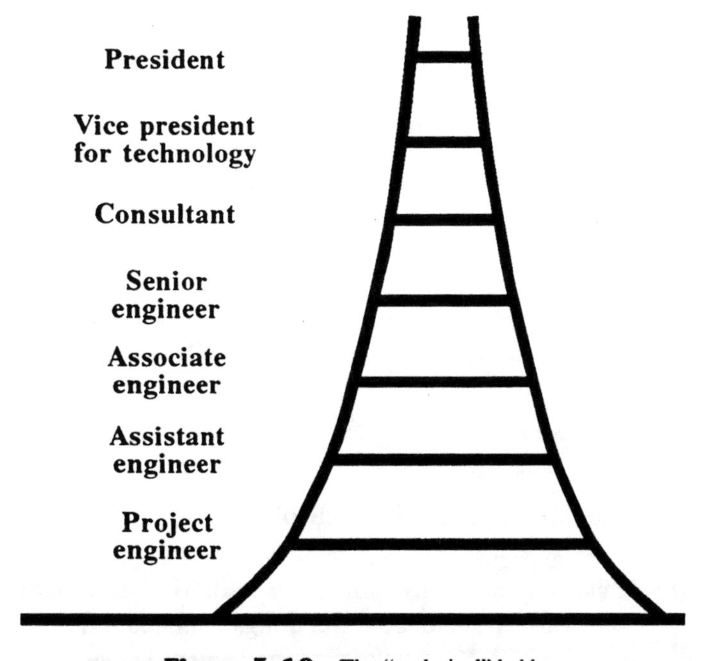

Figure 5–10 The "technical" ladder

Management	Technical
■ Comfort with being removed from the day-to-day work, which means that impact on the work is through others.	■ Knowledge and use of technology being used in the discipline or profession.
■ Decision-making which requires intuition, knowledge, and studied judgments about unique problems and events.	■ Deductive reasoning ability.
■ Resourceful; generally knows where to get resources needed to perform the project.	■ Gratification from one's work.
■ External and internal networking.	■ Satisfaction in dealing with things (data, case histories, materials, etc.)
■ Comfort with complexity.	■ Enjoyment from unique projects, involving creative approaches rather than routine solutions.
■ Skills in planning, organizing, staffing, directing, motivating, leading, and controlling.	■ Personal satisfaction in publishing technical articles.
	■ Recognition by peers.

Figure 5–11 Skills, personality traits, and interests associated with technical professionals in "management" positions vs "technical" positions (Source: Adapted with permission from Toto, J. V., "Consultant Know Thyself," *Journal of Management in Engineering—ASCE*, Vol. 4, No. 2, April 1988, pp. 167–168.)

credibility, but have recently experienced a resurgence. That resurgence is being driven by factors such as:

- Increased importance of technology and, therefore, recruiting and retaining technical and scientific professional people.
- Flatter organizations with fewer traditional administrative positions available. That is, there are fewer rungs available on the administrative ladder, suggesting the wisdom of creating another ladder.

- Concern with inequities—technical and scientific staff ought to share more equally in the fruits of their labor.

Dual ladders are usually associated with a matrix organizational structure, as already noted. However, dual-ladder advancement systems can be superimposed on or housed within organizations that use the more traditional functional, regional, or client-based organizational structures. Regardless of the details of the organizational structure, employers who convert to dual-ladder advancement systems need to provide their technical professionals with a clear understanding of corporate expectations and support. Figure 5–12 suggests activities and achieve-

Figure 5–12 Examples of corporate "technical" ladder expectations

ments that can be legitimately expected of those engineers and other technical professionals who choose to climb the technical ladder. High corporate expectations should be matched with high corporate support. Accordingly, Figure 5–13 shows ways in which an organization can support those professionals who strive to climb the "technical" ladder and reward those who demonstrate achievements. A similar set of expectations and of supports should be articulated for those technical and other professionals who choose to climb the "administrative" ladder.

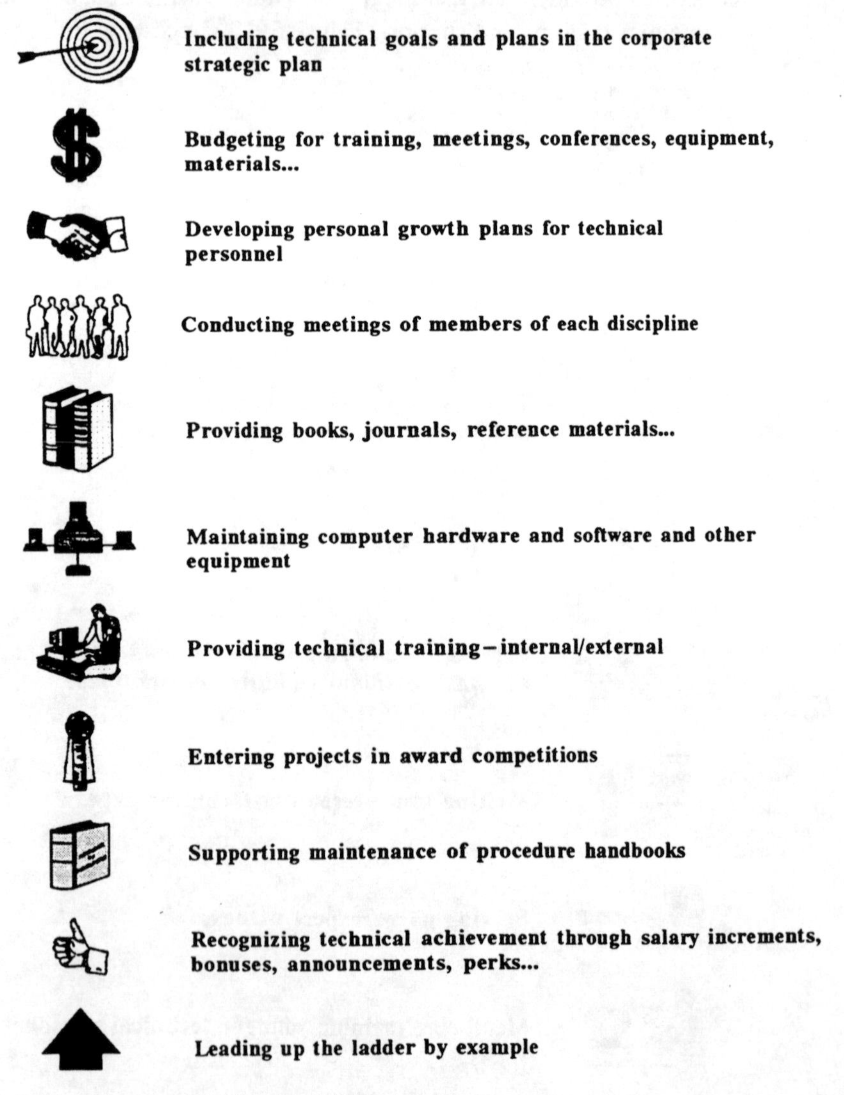

Figure 5–13 Examples of corporate "technical" ladder support

Stated differently, engineering and other technical organizations considering implementation of a dual-ladder system should "build" a technical ladder that:

- Has well-defined and widely spaced rungs—to stretch technically oriented professionals
- Comes with "instructions" on how the user could climb from one rung to another
- Assures climbers that they will receive support *if* they strive and recognition *if* they achieve
- Leans against a worthy wall—the goal of enhancing corporate technical capabilities

Entry-level engineers and other technical professionals who have a strong technical bent, at least for the foreseeable future, or who want to "keep doors open" are advised to seek out matrix or other organizations that offer the dual-ladder system of advancement. For additional ideas and information on dual ladders, refer to Burgower (1990) and Walters (1986).

REFERENCES

BROWN, T., "It's One Thing to Reorganize," *Industry Week*, May 4, 1992, p. 17.

BURGOWER, B., "Sweetening the Lure of the Lab," *Business Month*, August, 1990, pp. 76–77.

Business Week, "The Horizontal Corporation," October 20, 1993, pp. 76–81.

CLOUGH, R. H., *Construction Contracting*, fifth ed., New York: John Wiley & Sons, 1986.

DAVIS, S. M., and P. R. LAWRENCE, "Problems of Matrix Organizations," *Harvard Business Review*, May–June 1978, pp. 131–142.

DHILLON, B. S., *Engineering Management: Concepts, Procedures, and Models*, Lancaster, Pa.: Technomic Publishing Co., 1987.

GOLDSTEIN, M. L., "Dual-Career Ladders—Still Shaky but Getting Better," *Industry Week*, January 4, 1988, pp. 57–60.

GRIGG, N. S., *Urban Water Infrastructure: Planning, Management and Operations*, Chapter 6—Planning and Management Essentials," New York: John Wiley & Sons, 1986.

GUTERL, F. V., "Goodbye, Old Matrix," *Business Month*, February 1989, pp. 32–38.

MARTIN, J. C., *The Successful Engineer: Personal and Professional Skills—A Sourcebook*, New York: McGraw-Hill, 1993.

MORTON, R. J., *Engineering Law, Design Liability and Professional Ethics*, Section I-4, "Business Associations," San Carlos, Cal.: Professional Publications, Inc., 1983.

POIROT, J. W., "Organizing for Quality: Matrix Organization," *Journal of Management in Engineering—ASCE*, Vol. 7, No. 2 (April 1991), pp. 178–186.

ROEMER, C., "Business Ownership," *Michiana Executive Journal*, August 1989, pp. 48–52.

ROSENBERG, D., "Where Does the Passion Go?," *Industry Week*, August 17, 1992, pp. 11–12.

SCHERMERHORN, J. R., Jr., *Management for Productivity*, New York: John Wiley & Sons, 1984.

STEPHENS, S., "Corporations Not Panacea on Personal Liability," *Engineering Times*, November 1993, p. 9.

TEPLITZ, C. J., and C. G. WORLEY, "Project Managers Are Gaining Power within Matrix Organizations," *PM Network*, Project Management Institute, Vol. 6, No. 2 (February 1992), pp. 33–35.

TOTO, J. V., "Consultant Know Thyself," *Journal of Management in Engineering—ASCE*, Vol. 4, No. 2 (April 1988), pp. 165–170.

WALTERS, S., "The Dual Ladder of Advancement," *Mechanical Engineering*, June 1986.

SUPPLEMENTAL REFERENCES

HENSEY, M., "Organizational Design: Some Helpful Notions," *Journal of Management in Engineering—ASCE*, Vol. 6, No. 3 (July 1990), pp. 262–269.

IRCHA, M. C., and J. M. TOLLIVER, "Restructuring Organizations: Alternatives and Costs," *Journal of Management in Engineering—ASCE*, Vol. 5, No. 2 (April 1989), pp. 164–175.

SONAR, D. G., *Getting Started as a Consulting Engineer*, Chapter 3—"Deciding on the Form of Ownership," San Carlos, Cal.: Professional Publications, Inc., 1986.

WHITE, F. D., "Managing Satellite Offices Presents Special Problems," *American Consulting Engineer*, 2nd Quarter, Vol. 4, No. 2 (1993), pp. 24–26.

EXERCISES

5.1 OWNER-ENGINEER-CONTRACTOR RELATIONSHIPS

Purpose

Expand the student's understanding of the terminology and processes involved in the interrelationships among owner, engineer (and architect), and contractor. Most newly graduated engineers and other technical professionals will function within at least one of the points on the owner, engineer-architect, contractor triangle and ought to understand the relationships.

Tasks

1. Interview an engineer or other knowledgeable official (e.g., director of public works, city manager, mayor) employed by a unit of government such as a city,

town, village, or county. In making arrangements for the interview, explain that you are a student in a management course for technical professionals and have an assignment in which you are trying to learn more about relationships between government units (owner), engineer-architects, and contractors. During the course of the interview, which must be a personal face-to-face (not telephone) interview, obtain the answers to all of the following questions:

a. Who does engineering planning and design for the unit—e.g., in-house professional staff, consultants, both?
b. If consultants are used, how are they selected—e.g., qualifications only, qualifications plus proposed fee, other?
c. How are construction contractors selected?
d. How does the unit do construction management—e.g., contractor does it; owner does it?
e. What building code(s) are applicable?
f. What type of organizational structure (functional, matrix, etc.) does the unit use and why?
g. In addition, pose additional questions of your own choosing and obtain answers.

2. Write a memorandum to the instructor reporting the results of your interview. The memorandum must include, but not be limited to, the following:

a. Name, title, address, and telephone number of the interviewee.
b. The date and place of the interview.
c. Answers to the specified questions.
d. Your additional questions and the answers or discussion that followed.

3. Write a brief thank-you letter to the interviewee and provide the instructor with a copy as an attachment to the memorandum. On the letter, indicate that a copy is being sent to your instructor. (*Note:* Thank-you letters or notes are a courteous and thoughtful way to thank others for the assistance they provide us in carrying out our professional work.)

4. This assignment will be weighted as two times a normal assignment.

Timing

Send the thank-you letter to the interviewee within one week of the interview.

5.2 DIAGNOSE ORGANIZATION

Purpose

1. Increase the student's awareness of the basic organizational structure of his or her current or recent past employer. This heightened awareness can enable improved productivity with a current employer or the development of a better "match" with a future employer.

Tasks

1. Provide the following basic information about your current or recent past employer:
   a. Name (OK to omit for reasons of confidentiality).
   b. Type (e.g., manufacturer, consulting firm, government agency, contractor/constructor).
   c. Form of ownership if a private organization (e.g., sole proprietorship, partnership, corporation, other).
   d. Organizational structure (e.g., functional, regional, client, matrix, other).
2. Answer these questions:
   a. Were you fully aware of the form of ownership?
   b. Does or would your awareness of the form of ownership (regardless of when or how you became aware of it) help you be more effective as a member of the organization? Explain.
   c. Were you fully aware of the organizational structure?
   d. Does or would your awareness of the organizational structure (regardless of when or how you became aware of it) help you be more effective as a member of the organization? Explain.
   e. What would you change (assuming you could) in the form of ownership and/or the organizational structure to improve the overall effectiveness of the organization? Explain why, that is, who would benefit and how?
3. Write a memorandum to the instructor regarding tasks 1 and 2.

6

Project Management

> *Why do we never have enough time to do it right but always have enough time to do it over?*
>
> (Anonymous)
>
> *Prior proper planning prevents poor performance.*
>
> (Anonymous)

Chapter 2, which focused on management of self, included a detailed discussion of personal time management. The emphasis was on the importance of managing personal time. Various time-management tips were suggested. Effective time management is also required for projects. Recall that one definition of engineering management presented in Chapter 1 is "the process of deciding what is going to be done, how it is going to be done, who is going to do it, when they are going to do it—and how are we doing?" This chapter focuses on the "when" of project management. Also included are ideas and information on important project management topics: preparation of a project plan, project monitoring and control, and conducting a project postmortem.

The word "project" is used very broadly in this chapter and is not necessarily limited to technical fields. House's (1988, p. 9) definition is very appropriate: "Project can be defined as an interrelated and primarily nonrepetitive set of activities which combine to meet certain objectives." Lock (1992, p. 1) states that a common characteristic shared by all projects is "the projecting of ideas and activities into new endeavors." Inherent in this definition is the idea that familiar, previously used ideas and activities comprise projects; what's new is their relative emphasis and how they are arranged.

Examples of technical projects are preparing a proposal, conducting a data collection program, developing a plan, writing a report, preparing plans and specifications, manufacturing or fabricating a product, constructing a structure or facility, and carrying out a training program. Typically projects involve the cooperative efforts of two or more people. This team effort implies communication and coordination challenges that can be met with the assistance of material presented in this chapter and in Chapters 3 and 4. Furthermore, conduct of even the rare single-person project will be aided with ideas and information included in this chapter.

Although this chapter discusses ways to more effectively manage engineering and other technical projects, clearly many of the ideas and much of the information presented are applicable to other professions and, nonprofessionally, to projects you may lead or participate in for your community, church, synagogue, club, or other organization. For example, Brown (1992) describes the successful application of the critical-path method, which is discussed in this chapter, to a magazine publishing business.

Another way to view this chapter is to recognize, as stated by Covey (1990, p. 99), "the principle that all things are created twice. There is a mental or first creation, and a physical or second creation." This chapter concentrates on the first creation with the conviction that, by so doing, there is a greatly enhanced probability that the second and ultimate creation will be achieved on time and within budget and will meet expectations.

Too many projects are done twice in a wasteful fashion. The first time through they, or major portions of them, are done wrong because of poor or no planning. Then they must be done a second time to get them right. The premise of this chapter is that such waste and the associated frustration; loss of clients and customers, or constituents; and, for businesses, low or no profitability can be avoided by the philosophy "Plan your work; work your plan." This productive way of doing a project twice is illustrated in Figure 6–1.

Although Pirsig (1974, p. 284), in his book *Zen and the Art of Motorcycle Maintenance*, describes a process to ostensibly use in a motorcycle repair project, his advice is applicable to project as defined in this chapter. Before beginning the project, he urges you to:

> *. . . list everything you're going to do on little slips of paper which you then organize into proper sequence. You discover that you organize and then reorganize the sequence again and again as more and more ideas come to you. The time spent that way usually more than pays for itself in time saved on the machine and prevents you from doing fidgety things that create problems later on.*

THE CENTRALITY OF PROJECT MANAGEMENT

Project management is the process by which an organization's sources are marshalled to deliver quality products and services increasingly expected by many

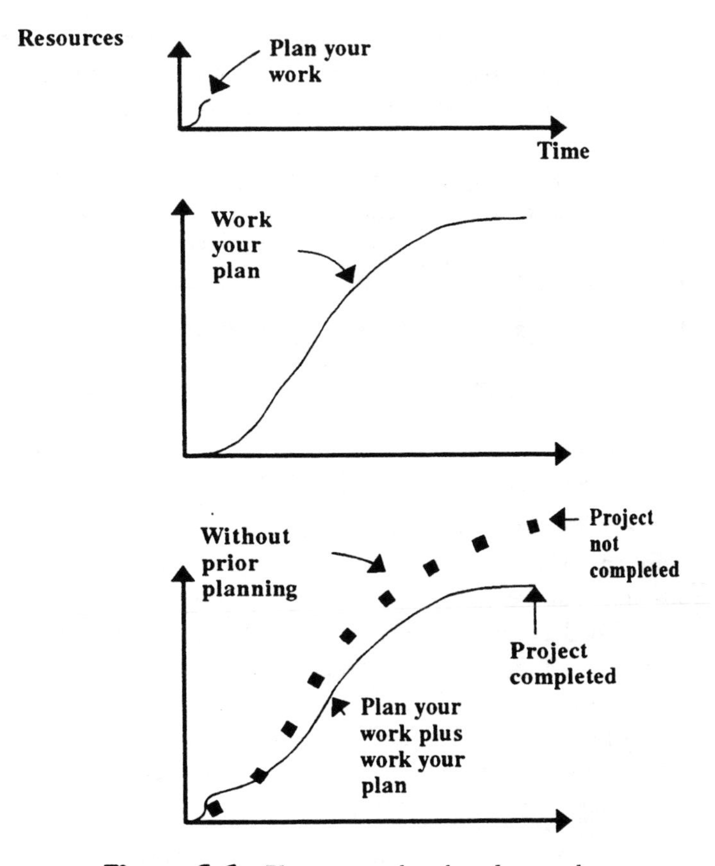

Figure 6–1 Plan your work and work your plan

and varied internal and external clients. Quality, as used in this book, is defined as meeting client expectations and project requirements; it is discussed in detail in Chapter 7.

The manner in which a business organization manages its projects is the key to developing and retaining clients and customers and being a profitable operation. Essentially everyone in an organization, business or otherwise, does projects or is at least indirectly involved in them. Each person can contribute to the successful completion of projects and can derive satisfaction from the team's and organization's achievement.

Some projects are large and some are relatively small, some are very sophisticated and others are very straightforward, but all projects require careful project management.

Successful project management propagates throughout an organization—especially a business firm—in many and varied positive ways, thus significantly contributing to an organization's success. Conversely, mediocre or failed project management has widespread, negative and sometimes devastating impacts upon an organization. Accordingly, project management should be a major, if not the principal, focus of an organization's energies. Effective project management is

- The key to quality, that is, happy, satisfied clients, customers, and constituents. Project management is the activity closest to an organization's clients, customers, and constituents—they immediately and continuously get the results, positive or negative, of the way projects are managed.

- The primary determinant of profitability.

- The focus of efforts to reduce the likelihood of client dissatisfaction and disputes and liability exposure. Refer to the discussion of liability-minimizing strategies and tactics presented in Chapter 10, and note how many occur within or are directly related to the project management process.

- A fantastic on-the-job arena for teaching, mentoring, and learning. There is no "make believe" here. Individual or corporate technical and nontechnical capabilities can be markedly enhanced during the conduct of projects. Experience—good and bad—of senior personnel can be readily shared with junior personnel within the project management forum. In contrast, skepticism and even cynicism will thrive in the project management area, especially among the younger staff, if they are denied the benefit of learning from senior personnel and if they see inconsistencies between what senior personnel say and do.

- The setting in which existing technical and nontechnical methods can be improved and new approaches developed. The idea that "necessity is the mother of invention" is clearly demonstrated in effective project management.

- The forum in which individuals and the organization can identify near-future needs—information, tools, skills, and people. By following through and meeting those needs, an organization's capabilities are enhanced.

- Where new work for existing clients, customers, and constituents lies—because they are pleased with the results of recent and current projects and because project team members learn of additional client plans and needs. Effective project management attracts the trust of the individuals and organizations being served, which is a vital part of the process of earning the privilege of serving them again.

- The source of ideas, experience, and references in the form of satisfied clients, for earning the opportunity to provide similar services or products to new clients, customers, or constituents. Refer to Chapter 14 for a discussion of marketing, noting, in particular, the importance of gaining the trust of individuals and organizations as part of the process of earning the privilege of serving them.

- A source of personal satisfaction through team achievement.

If carefully cultivated and cared for, the project management tree will flourish and bear much fruit for an organization and its members. The continuing harvest will yield fruits like those listed above and illustrated in Figure 6–2. To reiterate, the health of an organization is determined principally by its ability to manage projects, because the doing of projects and the delivery of the results impact so many of an organization's vital interests, functions, and relationships.

Incidentally, project management can be viewed as an application of the engineering method—diagnose situation, define problem, develop alternatives, select a course of action, and implement it. When project management is presented as being an application of the engineering method, it is more likely to engage the interest of younger technical professionals who tend to have a technical and analytic bent.

RELEVANCE OF PROJECT MANAGEMENT TO THE ENTRY-LEVEL TECHNICAL PROFESSIONAL

You may generally concur with the need to place a high priority on project management, but you may not see the relevance to you as an entry-level engineer or other technical professional. After all, you will not be managing a project during your first year of professional work. You are partly correct. If you are in the first year of employment with an automobile maker, you will not manage the design of the next model of the company's sports car. Authority and responsibility for that project will be trusted to a seasoned veteran. But you may be responsible for the design of a component of the vehicle. If you are in your first year on the staff of a consulting engineering firm, you will not manage the design of an airport, but you may be responsible for a component. No matter how small your "piece of the pie" is, it will constitute a project, given the broad definition of project presented at the beginning of this chapter.

You are urged to manage your project well, giving consideration to the ideas and information presented in this chapter. By so doing, you will increase the likelihood that your part of the overall project will be completed on time, within budget, and in accordance with functional and service expectations. Besides short-term, project-specific benefits, your commitment to smart management of your projects will establish you as a person who makes good things happen. That desirable reputation will lead to more challenging project management and other growth opportunities for you. Colleagues noticing your success will inquire about and want to emulate your approach to project management. As a result, your efforts will have a positive ripple effect on the organization. The ability to change the behavior of others by example is one mark of a leader.

Is the preceding, optimistic scenario realistic? Yes, because effective proj-

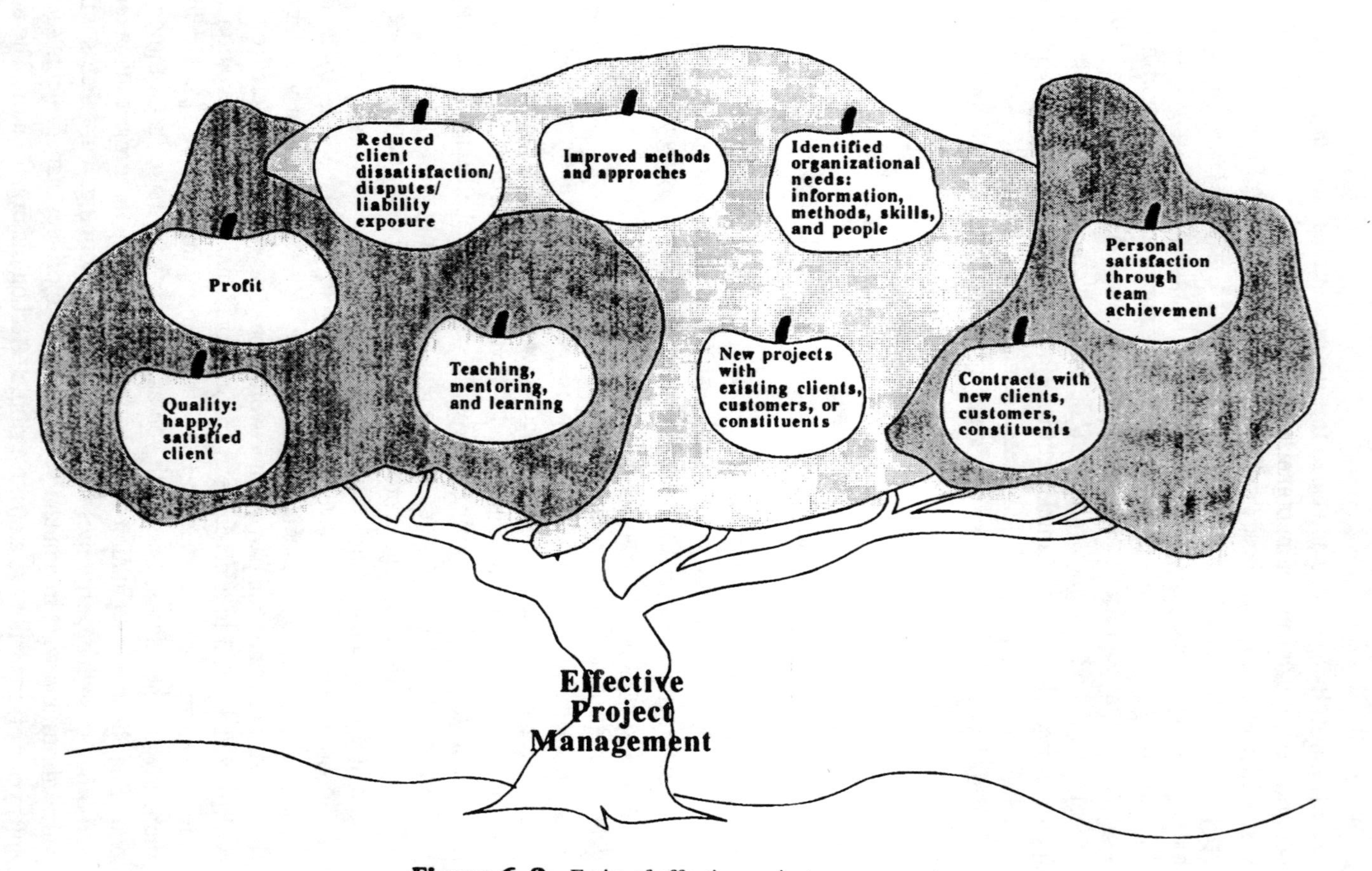

Figure 6–2 Fruits of effective project management

ect management is one of the highest-priority needs in most engineering organizations. You can help to fill that need to your benefit and that of your organization.

PROJECT TIME MANAGEMENT

The "when" or time dimension of projects arises often. As suggested by Figure 6–3, the time dimension of a project is often reflected in the need to answer one or more of the following four questions:

• Question 1. Before a project begins, someone may want to know how long it will take. For example, your supervisor asks you to prepare a conceptual design of a new product to be manufactured by your organization and wants to know how long the design project will take.

• Question 2. During a project, someone may want to know if the project is on schedule. For example, you and the assistant city engineer are doing a transportation plan for your city. The mayor is preparing for a meeting with the city council and wants to know if the transportation planning project is on schedule.

• Question 3. If a problem arises during a project, someone may want to know how much the completion will be delayed. For example, you are helping to manage the construction of a dam that was designed by your design–build firm. A laborer's strike temporarily stops concrete placement operations for six weeks. The client wants to know how much the completion of the project will be delayed, if at all.

• Question 4. After a project is completed, someone may be taking a retrospective look and want to know what can be learned from the project in terms of project time management. For example, your supervisor may ask how much time was actually required to do the various tasks comprising the project.

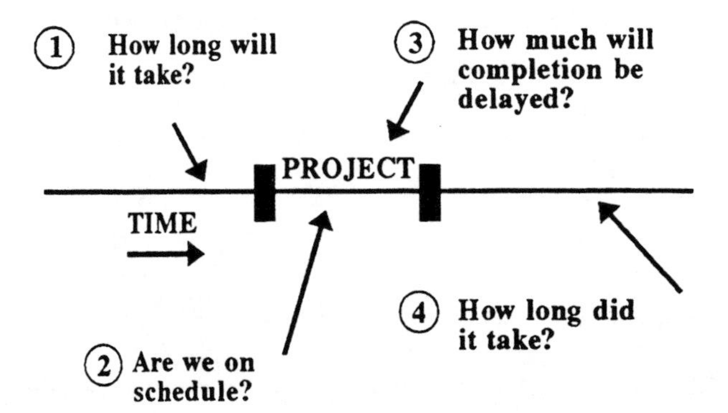

Figure 6–3 Project time-related questions

This section represents three different project time management methods ranging from simple to moderately complex. The idea, of course, is to select and use the technique most appropriate to the project situation. The project time management methods presented are

- Chronological list
- Gantt (bar) chart
- Critical path method (CPM)

The three project time-management methods are presented and illustrated in the listed order.

Chronological List

This two-step method is by far the simplest project time-management technique. It almost seems trivial. The first step is to list tasks that need to be done in approximate chronological order, and the second step is to estimate the elapsed time for each task. Consider, for example, the chronological list method applied to the design of a small dam. Some of the tasks and associated elapsed times are presented in Table 6–1. The most obvious positive aspect of this method is its simplicity. The negative feature of the chronological listing, particularly for projects having many tasks, a large proportion of which overlap each other, is that the interrelations between tasks are not explicitly shown, nor is the extent to which tasks overlap. That is, the chronological list suggests that tasks must be done sequentially. But obviously overlaps arise. For example, presumably task C in Table 6–1, which calls for specification of soil borings, could be done before completion of task A, which involves site reconnaissance and surveying. However, such interrelationships and overlaps are not explicitly shown in the simple chronological listing.

TABLE 6–1 EXAMPLE OF CHRONOLOGICAL LISTING METHOD—DESIGN OF A SMALL DAM

Task	Elapsed time (weeks)
A. Perform site reconnaissance and survey.	2.0
B. Draw site map.	2.0
C. Specify soil borings.	0.5
D. Arrange for and do soil borings.	1.5
E. Submit application for preliminary permit.	2.0
F. etc.	—

Nevertheless, the chronological listing method is often quite adequate and the most appropriate project time-management method. Surprisingly, many projects are undertaken without even identifying and listing tasks that need to be done. Some project managers believe that "activity is progress" and they start the work without planning it. Recall the anonymous admonition at the beginning of this chapter—"Plan your work, work your plan."

Gantt (Bar) Chart

Named after Henry Gantt, the Gantt or bar chart method provides a means of showing task overlap, provided that the overlap is known and understood, and also is graphical in the sense that it provides a "picture" of a project. Four steps are involved, the first two of which are identical to those in the chronological list method, namely, list tasks that comprise the project and estimate the elapsed time for each task. The third step is to estimate the start time for each task, and the fourth and last step is to draw the bar chart.

Figure 6–4 is an example Gantt or bar chart using the first five tasks of the previous example of designing a small dam. An example from a larger and actual planning, design, and construction project is presented as Figure 6–5. The multi-

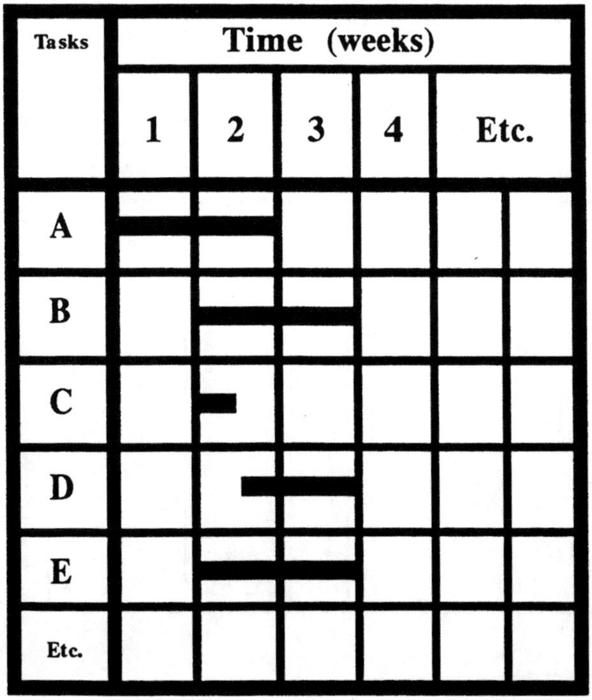

Tasks	Time (weeks)				
	1	2	3	4	Etc.
A					
B					
C					
D					
E					
Etc.					

Figure 6–4 Example of Gantt chart method—design of a small dam

Activity/Event	Year										
	1981	1982	1983	1984	1985	1986	1987	1988	1989	1990	1991
Major flood	•		•								
Preparation of flood control plan			······	······							
Preparation of recreation plan				······							
Preparation of plan and design for bypass highway (a)											
Purchase of fairgrounds site by city					•						
Design of flood control components					······	······					
Design of recreation components						······					
Site clearing, excavation, and rough grading					······	······					
Construction of bypass highway			······	······	······	······	······	······	······		
Construction of diversion, inlet and outlet works and on-site drainage works							······				
Construction of recreation facilities							······				
Design of connection sewers from the west						······	······				
Construction of connection sewers from the west									······	······	······
Flood control works operable								······	······	······	······
Recreation facilities available								······	······	······	······
	YEAR										
Activity/event	1981	1982	1983	1984	1985	1986	1987	1988	1989	1990	1991

a) Initial discussions on the bypass project occurred at least as early as the mid-1940's.

Figure 6–5 Example of Gantt chart—planning, design, and construction of a multi-purpose project

purposed project involved planning, designing, and constructing a by-pass highway, a flood control facility, and a recreation facility. The example Gantt chart was prepared in about 1986 to explain the interrelationship between the project tasks already completed and to depict the way future tasks would lead to the ultimate completion of the project.

One positive aspect of the Gantt or bar chart method is its graphical nature, which seems to aid the understanding of many and varied users. A second advantage is that task overlap can be depicted, provided that such overlaps are known or may be approximated.

A major disadvantage of the Gantt chart method, which is also shared with the chronological listing method, is that actual interrelationships between individual tasks are not shown. For example, does the initiation of tasks B and C in Figure 6–4 require that task A, or at least part of it, be completed before they can start? Is task C dependent only on task B? A second negative aspect of the Gantt or bar chart is that it fails to identify critical tasks, that is, those tasks that will delay the completion of the entire project if they are delayed. For example, in Figure 6–4, should task E, which involves submitting an application for a preliminary dam permit, be started as soon as possible so as not to delay the ultimate completion of the entire project? Or can it wait?

Critical Path Method (CPM)

As noted, although Gantt charts nicely display overlap among project tasks, they do not show all interrelationships, nor do they identify critical tasks. The critical path method (CPM), which is described by Clough (1986), Dhillon (1987, p. 105), Beakley et al. (1986), Schermerhorn (1984), Spinner (1992), and others, does not have these deficiencies.

History. The CPM method was developed in 1956 by the Engineering Control Group—a design and construction unit—of E. I. Du Pont de Nemours and Company. The method was programmed for Du Pont by the Remington Rand Corporation to run on the UNIVAC computer. Construction of a $10,000,000 chemical plant in Louisville, Kentucky in 1957 was the first application of CPM. There were 800 tasks in this project. CPM is now widely used in planning and management of construction projects. However, CPM can also be used for any project as project is broadly defined in this chapter.

A related and somewhat more sophisticated technique or method is the Program Evaluation and Review Technique (PERT), as described by Dhillon (1987, Chapter 7), Beakley et al. (1986), and others. PERT is a method that explicitly accommodates uncertainties associated with completion times of tasks. It is not treated in this book.

CPM steps. As indicated by Figure 6–6, application of CPM can be viewed as consisting of the following five steps:

- Step 1. List tasks that comprise the project. Clough (1986, pp. 321–322) offers several suggestions. First, consider which individual or group will be responsible for completing each task. Try to have each task be the responsibility of one person, one unit within an organization, or one organization. Stated differently, try to avoid multiple-responsibility tasks. Also consider the location at which each task will be carried out. For example, if a dam location feasibility study is being conducted and two sites need to be surveyed, the survey at each site should be a separate task or separate set of tasks. Finally, consider the magnitude of the tasks. Keep tasks small and, when in doubt, partition them further. Tasks can be easily reaggregated later if appropriate. Note that task 1 in the CPM is identical to task 1 in the previously discussed chronological listing and Gantt chart methods.

- Step 2. Estimate elapsed time for each task. Note that this is identical to task 2 in the chronological listing and Gantt chart methods. Typical time units might be work days, calendar days, work weeks, calendar weeks, and calendar months. If work days or similar work units are used, the overall project duration predicted by CPM can be converted from work days to calendar days by multiplying by an appropriate factor such as 7/5.

1. List tasks that comprise the project.

2. Estimate elapsed time for each task.

3. Identify interrelationships among tasks (the logic).

4. Draw the network.

5. Determine critical path and minimum project completion time.

Steps 1 and 2 are the same as in chronological listing and Gantt chart methods.

Figure 6–6 Steps in the critical path method

• Step 3. Identify interrelationships among tasks. This important step is not included in the previously discussed project time-management methods. Assume, for example, that task A is the collection of a specific set of data and task B is analysis of that data. Then, presumably, task B depends on task A, and, even more specifically, task A probably must be completed before task B can be initiated. Task B may also depend on other tasks in the overall project, such as a task that involves selecting the analysis method to be used.

• Step 4. Draw the network. A network is two or more nodes, where nodes represent tasks, connected by ordered (directional) branches. Network fundamentals are discussed in the next subsection of this chapter. The network is a very valuable intermediate or even final product of CPM. That is, to the extent that the network represents all project tasks and their interrelationships, its completion requires a thorough understanding of the project and provides the basis for managing the project. Or, stated differently, the completed network can be viewed as the "Plan your work" part of the "Plan your work, work your plan" advice presented earlier in this chapter. Or, from another perspective, recall Covey's statement (1990, p. 99) ". . . that all things are created twice. There is a mental or first creation, and a physical or second creation." The network that is created as a result of steps 1–4 in CPM can be viewed as the "mental or first creation."

• Step 5. Determine critical path and minimum project completion time. The critical path is that sequence of tasks that cannot be delayed if the project is to be completed in a minimal amount of time. The forward-pass–backward-pass technique used to identify the critical path is explained later in this chapter after the discussion of network fundamentals.

Network fundamentals. As noted, a network is two or more nodes representing interrelated tasks, where the tasks have positive or zero durations, connected by ordered, that is, directional branches. Examples of simple networks are presented in Figure 6–7. This chapter uses the task-on-a-node format. Another option is to use the task-on-an-arrow format (e.g., Clough, 1986; Spinner, 1992). Networks are sometimes called "precedence diagrams" because they establish precedence relationships in that "this task must be completed before that task can be completed" or "this task takes precedence for the time being over that task."

The most important aspect of a network is the connectivity or topology. Relative lengths of branches are irrelevant, as is the overall orientation or relative position of tasks in the network. There are, in effect, an infinite number of ways that any network can be drawn without changing its connectivity. For example, if network 2 in Figure 6–7 were rotated 90 degrees counterclockwise, it would continue to be exactly the same network.

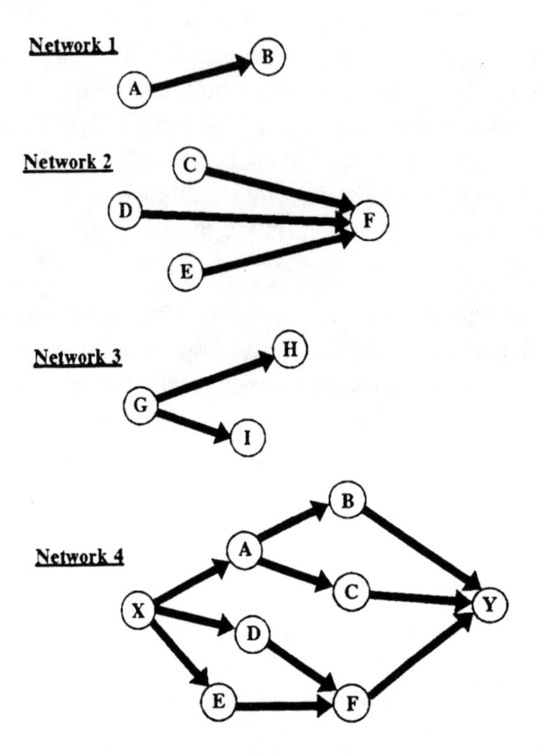

Figure 6–7 Examples of networks

Incidentally, two or more tasks cannot overlap. If they do, they must be broken down into smaller tasks. "Dummy," that is, zero duration tasks are used to tie tasks together. For example, tasks X and Y of network 4 in Figure 6–7 could be, respectively, zero duration "start" and "stop" tasks.

Example application. Assume that steps 1 through 4 in the CPM process as listed in Figure 6–6 are complete and the resulting network is as shown in Figure 6–8. The network consists of ten tasks, two of which are dummies, that is, zero-duration tasks. Each task is identified with either a name (e.g., "Start") or a letter. Task interrelationships are clearly identified through the use of directional branches. Finally, the duration of each task, in days, is indicated on the task symbol below its letter identifier.

As noted, the critical path is defined as the sequence of tasks that cannot be delayed if the project is to be completed in a minimal amount of time. Other important definitions are

- EST—earliest start time, that is, the earliest elapsed time, measured from the start of a project, when a particular task can possibly begin.

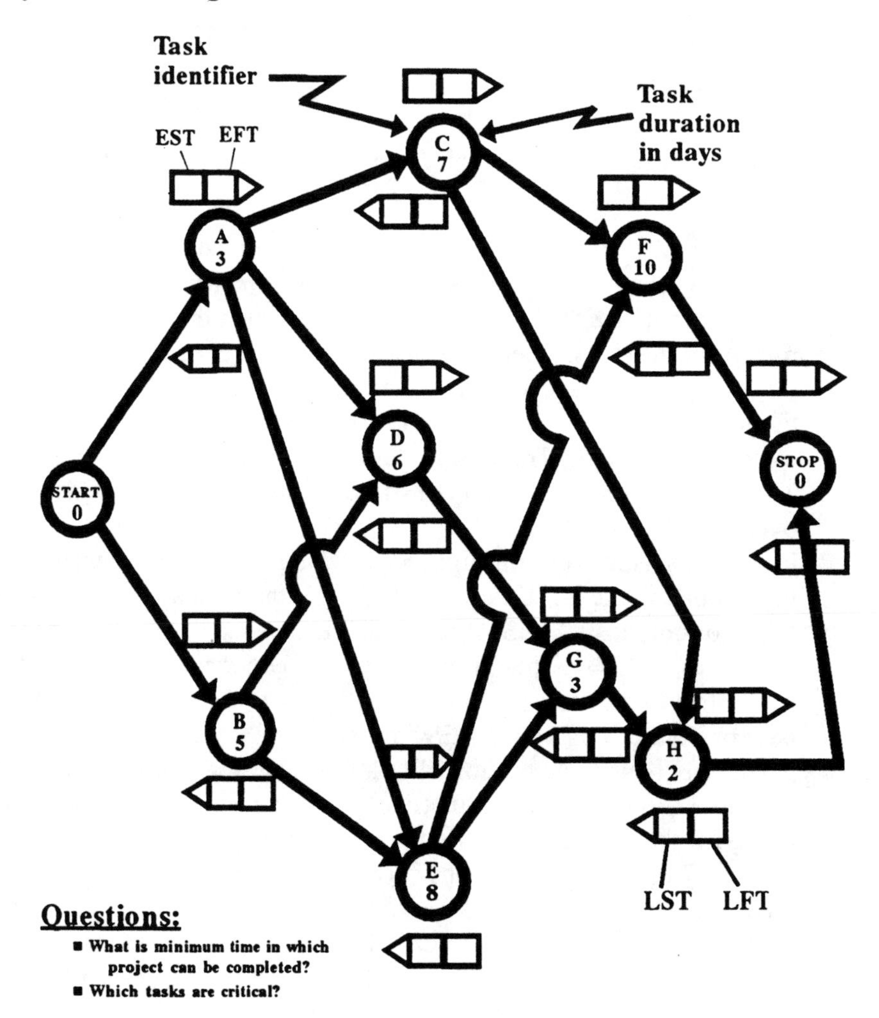

Figure 6–8 Network for CPM example

- EFT—earliest finish time, that is, the earliest elapsed time, measured from the start of a project, when a particular task can possibly be finished.
- EFT = EST + duration
- LFT—latest finish time, that is, the latest elapsed time, measured from the start of a project, when a particular task can be finished without delaying any other task and, of course, the overall completion of the project.
- LST—latest start time, that is, the latest elapsed time, measured from the start of a project, when a particular task can be started without delaying any other task and, of course, the overall completion of the project.

- LST = LFT – duration
- Total float—Amount of time a task can be delayed without delaying the entire project.
- Total float = LST – EST = LFT – EFT

Consider now the determination of the critical path (step 5 in Figure 6–6). The analyst begins by doing a "forward pass" through the network shown in Figure 6–8 from "start" through "stop" during which the ESTs and EFTs are determined for all tasks. Then a "backward pass" is conducted through the network from "stop" to "start" during which the LFTs and LSTs are determined for all tasks. Using the previously defined relationships, the total float is then determined for each task. The critical path is the locus of tasks having zero total float. That is, the critical path contains those tasks the delay of which will delay the overall project.

You should work through the network presented in Figure 6–8 to obtain the minimum completion time and to identify the critical path as shown in Figure 6–9. Consider task A. EST = 0 because task A can be started as soon as the project begins. EFT = 3, that is, EST + the 3-day duration. The EFT of 3 for task A means that its earliest finish time is at the end of three days. Continue the forward pass by considering task C. EST for task C = 3 because initiation of task C depends only on the completion of task A and because the EFT for task A is 3. The EST of 3 for task C means that its earliest start time is at the end of three days, that is, at the beginning of the fourth day. EFT for task C is 10, that is, its EST plus the 7-day duration. As the forward pass continues, task D might be considered next. However, because Task D depends on task B, the analyst must first determine the EST and the EFT for task B.

Having completed the forward pass as shown in Figure 6–9, the analyst will find that the minimum completion time for the project is 23 days. The "backward pass" process then commences, beginning with the task labeled "stop" and proceeding backward through the network. The total float for each task is determined. The critical path, that is, the locus of tasks having zero total float is determined to consist of the following sequence of tasks as is evident in Figure 6–9: Start, task B, task E, task F, and Stop.

Some observations about CPM. The generally horizontal format is standard, at least in the construction industry (Clough, 1986, p. 325), but does not necessarily have to be followed. Assuming that an overall directional format, such as horizontal, is selected, arrows may be omitted from the interconnecting branches. Recall again that line lengths have no significance, and neither does the orientation of nodes. Only connectivity is crucial to the accuracy of a network.

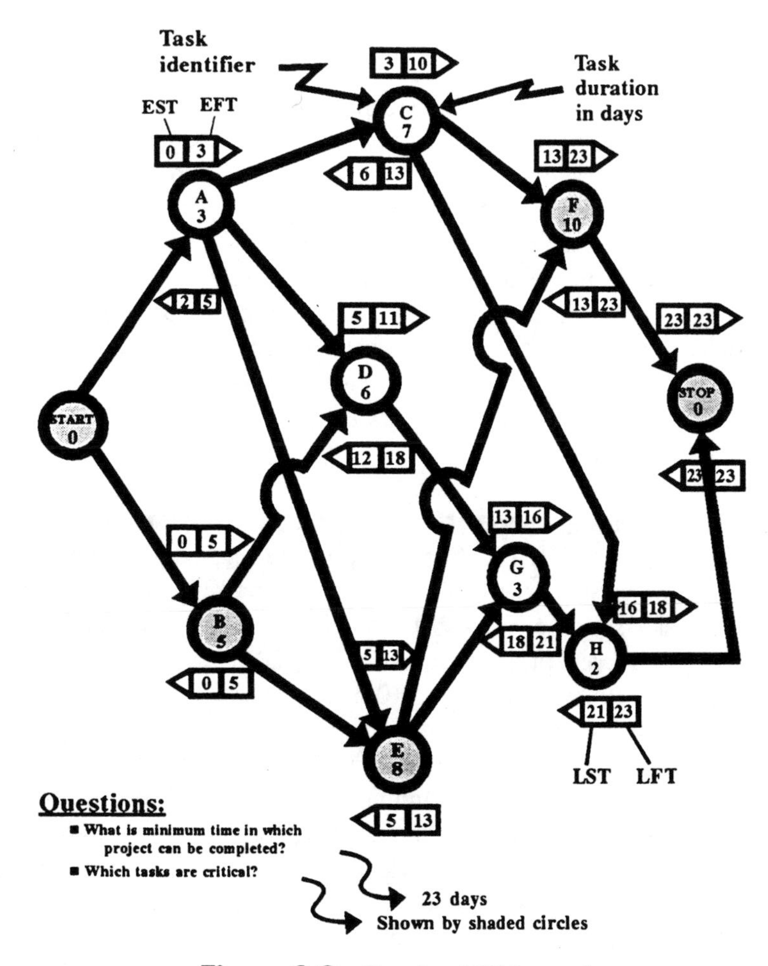

Figure 6–9 Completed CPM example

All paths must be traversed during the project. CPM is not an exercise in taking the longest path between two points, although the critical path could be found in that fashion.

Crossings of directed branches are acceptable, and, in fact, usually impossible to avoid in actual networks containing many interconnected tasks. When crossings occur, they should be clearly indicated by using symbols such as the half circles appearing in Figures 6–8 and 6–9.

There must always be at least one critical path. However, there could be two or more critical path segments "in parallel" through all or some of the network. Any delay in a critical activity automatically delays project completion the same amount unless some compensatory action is taken. The critical path is the longest path in terms of time. Time contingencies can be added to each activity or to the overall project.

The CPM example presented in Figures 6–8 and 6–9 is very small for illustration purposes and to facilitate rapid manual calculations. Because actual CPM applications usually involve many more tasks and much larger networks, which, as explained later, must be frequently updated, manual manipulations are usually not feasible. Commercial computer programs incorporating the basic algorithms described in this chapter are available for production application of the CPM. Lock (1992, Chapter 8) provides a comprehensive discussion of use of the computer in applying the CPM and related scheduling techniques. Included is a detailed checklist to help identify features needed in software. Vandersluis (1994) provides a synopsis of the history of project management software and a glimpse into the future.

Creating a Gantt chart from a network and critical path. As noted earlier in this chapter, a positive feature of a Gantt chart is the simplicity of the graphics particularly for project participants who are focused on one or few tasks. A completed CPM analysis, like that shown in Figure 6–9, can be easily used to construct the corresponding Gantt chart in the usual left-to-right format as illustrated in Figure 6–10. For example, the horizontal bar for each task is started (its left end) at a time corresponding to its EST. The length of the horizontal bar is depicted as the duration of a task. The float for a noncritical task is shown as a dashed line beginning at the right end of each bar, that is, beginning at the task's EFT and extending to its LFT. The solid line plus the dashed line for each task thus depicts the "window" within which each noncritical task would need to be started and completed. Each critical task does not have a dashed line segment indicating that the float is zero.

Updating a critical path analysis. In practice, networks and their critical paths are updated frequently during the course of a project, perhaps on a weekly or monthly basis. Updates are required for a variety of reasons including unexpected delays in starting or completing tasks, faulty estimates of task durations, missing or unnecessary tasks, and flawed logic, that is, incorrect connectivity. The updating process is simply a matter of showing the new information on the network and then doing a new "forward pass" and "backward pass." As a result of the update, the absolute completion time and the location of the critical path may change.

Example updating of a critical path analysis. Refer to Figure 6–8, which is the network used to illustrate CPM. Assume that the project is underway and, at the end of the tenth day, the status of each task is as follows:

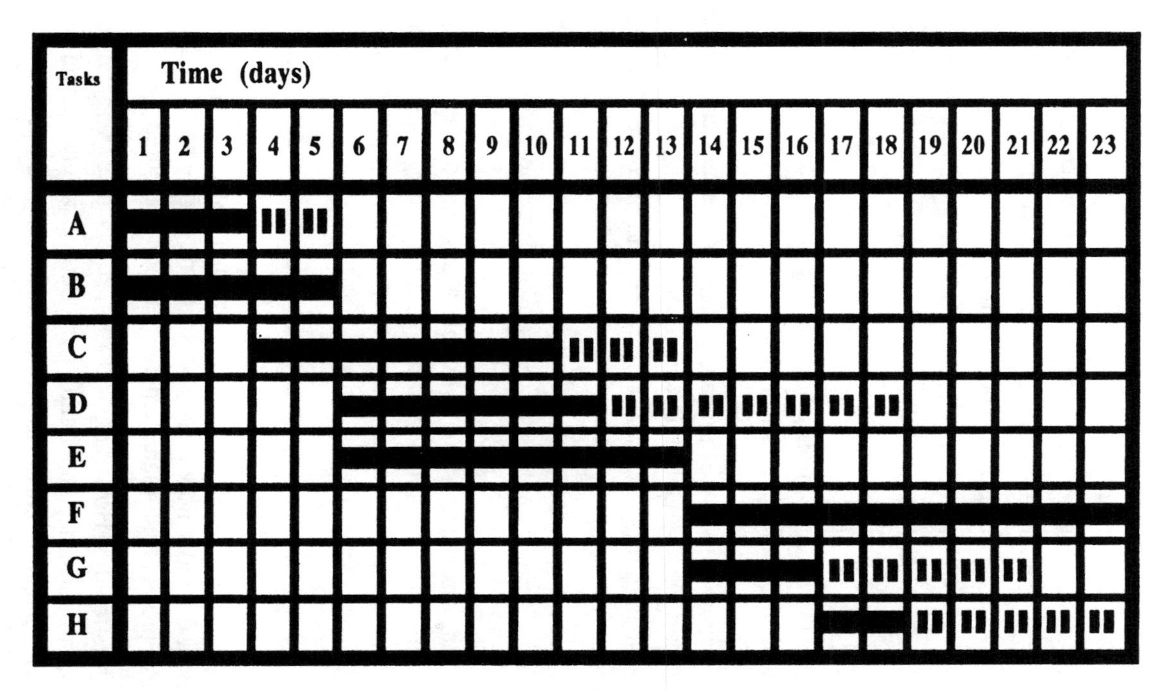

Figure 6–10 Gantt chart constructed from CPM solution

Task	Status
A	Done
B	Done
C	Underway with 5 days of work to be completed
D	Underway with 1 day of work to be completed
E	Underway with 2 days of work to be completed
F	Not started; duration estimate is increased to 12 days
G	Not started; no change in estimated duration
H	Not started; no change in estimated duration
I	A new task that depends only on H and has a 2-day duration

Using the given information, apply the forward- and backward-pass processes and determine the following:

- The critical path—does it change? (Yes—the new critical path is defined by Start, task A, task C, task F, and Stop.)
- The time required, measured from the end of the tenth day, to complete the project. (17 days)
- The projected absolute completion time, (i.e., referenced back to the original start time) compared to that absolute completion time projected at the beginning of a project. (27 days)
- With the information now available at the end of the tenth day, can the original projected completion time still be met? (No—unless some internal changes are made such as one or more of those mentioned under Question 3 in the next section of this chapter.)

Review of Earlier Questions

Recall the typical project time-management questions posed at the beginning of this section. Consider how these questions might be answered with one or more of the three project time-management methods presented.

- Question 1. How long will the project take? CPM clearly provides the answer to this question. The Gantt chart method would answer the question if activity overlaps can be adequately estimated. If the Gantt charts were constructed from the CPM, then clearly the Gantt chart would provide a solid estimate of project duration. The chronological list will not answer this question unless the project is very simple, that is, has few tasks and has relatively little task overlap.

- Question 2. Is the project on schedule? CPM provides a clear answer to this question. For example, the user could look at the status of each activity relative to its LFT. Or the user could update the critical path analysis, the usual practice, and compare the revised absolute completion date to the original absolute completion date. Similarly, if the Gantt chart is constructed from the CPM results, the Gantt chart will effectively display those tasks that are not on schedule. However, if CPM were not used and a Gantt chart or chronological list were used, the nature of the project schedule would be difficult to determine.

• Question 3. One or more activities have been delayed—how much will the project completion be delayed? CPM answers this question very well when the CPM is updated, thus determining if a delay in one or more tasks will delay the absolute completion date of the project. The Gantt chart, assuming that CPM was not applied, and the chronological list would be of essentially no value in determining delay in project completion.

Incidentally, if the answer to question 3 revealed that the absolute project completion date would be delayed an unacceptable amount, the project manager could consider corrective actions. For example, he or she might examine the tasks on the critical path and try to reduce the duration of some of them by increasing the number of personnel on those tasks or by working extra hours per day and/or extra days per week. In considering this strategy, however, the project manager must update the entire network, because reducing the duration of one or more tasks on the critical path may have the effect of moving some or all of the critical path to other tasks. The Construction Industry Institute (1988) offers an exhaustive list, with discussion of 94 techniques that might be used during the design, construction, manufacturing, or fabrication phases of a project to compress the schedule. Furthermore, the likely effectiveness of each method during various phases of a project is indicated.

• Question 4. How much time was actually required? Assume that a significant time underrun or overrun occurred in a project and the project is now being looked at in retrospect. If CPM was used, a retrospective examination may reveal flaws in connectivity, task identification, task duration, and assignment of personnel. If the series of updated CPMs were retained, either in hard copy or on a computer system, the type, frequency, seriousness, and types of errors which occurred could be determined. A postmortem analysis using the CPM can be a very valuable exercise, particularly if similar projects are to be undertaken in the future. Lessons learned on the just-completed project can benefit subsequent similar projects.

Key Ideas

The principal value of all three project time-management methods presented in this chapter is that they require that the project be done on paper before it is done in reality. Project time-management methods steer project managers away from thinking that activity is progress. The smart project manager works out the project "map" and then begins the project "journey." The not-so-smart project manager just starts the "journey." Finally, the most powerful of the three project management methods presented in this chapter is CPM. And much of the value of applying CPM can be realized by carrying it through step 4, drawing the network.

PROJECT PLAN

In keeping with the admonition to "Plan your work, work your plan," a formal project plan, that is, a written document shared with project team members, should be prepared by the project manager. The project plan should be drafted, discussed, and refined before any other significant work is done or resources are expended on the project.

Project plans should be created even for actual or deceptively simple projects, because as the project manager drafts the project plan, he or she is likely to discover previously unnoticed complications. Furthermore, a project that may appear simple to the project manager, because he or she has completed similar projects, may appear complex to team members who are new to this type of project. They and the project will benefit from a written project plan. The goal is to optimize communication about the project among project team members at the beginning of and during the project.

Project Team Kickoff Meeting

After the project manager completes the draft project plan, he or she should invite all or key members of the project team to a kickoff meeting. This working meeting should focus on understanding, refining, and committing to the project plan. Shortly after the kickoff meeting, each team member should receive the revised project plan plus other documentation of the meeting as may be appropriate. As the project proceeds, the project manager should update the plan and keep the team informed.

As an entry-level engineer, presumably you will be on the receiving end of project plans. You may receive copies of project plans and attend kickoff meetings. If so, you are fortunate to be in an organization that expects and supports sound project management, including the use of project plans. As noted earlier, you should apply sound project management methods to your part of the major project. A good place to start is to be certain that you have a written project plan, that is, you know at least the what? why? how? and when? of your miniproject.

Work Plan Format

A what?–why?–how?–when?–who?–where? format might be used by the project manager in developing the project plan (Cori, 1989). Obviously, this is not the only structure for a project plan, but it provides a disciplined way to address the necessary breadth and depth of topics. For example:

• **What?** Describe "what we are going to do" in written (e.g., narrative, list) or graphic form (e.g., Gantt chart, network). Identify new or unique aspects of the project. List the products or deliverables that will be provided to the client, customer, or constituent as a result of the project.

• **Why?** Explain the purpose of the project from the perspective of the client, customer, or constituent. Describe the expected results. If a proposal was submitted or an agreement was developed as part of the process of securing the project (sometimes both will exist), it should be reviewed and salient features, such as the scope of work, shared with members of the project team.

• **How?** To the extent that they are already known, identify the approaches, the techniques, and the tools that will be used on the project. Indicate if new methodologies will be needed and suggest how they might be obtained or developed.

• **When?** Develop, at least in a preliminary form, the time schedule for the project, using one or more of the project time-management methods presented earlier in this chapter. If a detailed time schedule is not feasible this early in the life of the project, identify milestones by name and date.

• **Who?** Prepare an organization chart showing each member of the project team and how they relate to others. Describe what each person will be working on and indicate their responsibility and authority. Assign labor and expense budgets to tasks or individuals.

• **Where?** Indicate where some or all of the project work is likely to be done. Possible sites include various offices of multioffice organizations, offices of other entities, including the client or customer, and field locations.

Work Plan Avoidance Syndrome

Cori (1989) discusses reasons that some project managers offer for not preparing a work plan. They include:

- Solving problems as they arise is more satisfying.
- Cannot find time or environment for the necessary concentrated thinking.

- Required labor time is not in the budget.
- Written plans provide a means to monitor and measure the effectiveness of the project manager, and some project mangers fear such an assessment.

Absent an organizational directive to prepare project plans, each project manager needs to decide for himself or herself if a formal project plan is a prudent investment of time resources. As an entry-level engineer, you need to make similar decisions for your miniprojects. Once again, the admonition "Plan your work, work your plan" seems compelling.

PROJECT MONITORING AND CONTROL

Assume that the project plan is essentially complete and the project is underway. Recall the project definition presented near the beginning of this chapter: " . . . an interrelated and primarily nonrepetitive set of activities which combine to meet certain objectives" (House, 1988, p. 9). The project manager has prime responsibility for monitoring and controlling the ". . . interrelated and primarily nonrepetitive . . . activities . . ." so that the objectives are met. Those objectives typically include staying within the budget, meeting or exceeding the time schedule, and fulfilling the needs of the client, customer, or constituent. Typically, the project manager in a business organization carries out the following monitoring and control tasks, most of which are adaptable to nonbusiness organizations:

- Tracks the project budget and subbudgets and corresponding work progress. Notes accidental or other illegitimate labor or expense charges. If illegitimate charges have occurred or if the costs incurred are moving ahead of the products produced, corrective action is needed.

- Compares tasks completed and milestones achieved to the project schedule, perhaps using periodic updates of a critical path analysis as described earlier in this chapter. If critical tasks are or soon will be behind schedule, takes action as discussed earlier in this chapter.

- Remains alert to changes in scope requested by or attributable to the client or customer, especially those that will increase the cost of doing the project. The preferred remedy to client-driven scope creep is to seek additional compensation commensurate with the additional services. PSMJ (1980) notes that the likelihood of receiving additional compensation will be increased if the agreement under which the services are being provided contains a well-written scope-of-services section, if scope increases are discussed with and documented to the client or customer as soon as they appear or begin, if the service organization is knowledgeable about the client or customer's source of funding, and if the client or customer is provided with detailed documentation supporting a compensation increase. When additional funds are not available, alternatives include reducing the scope of services in other parts of the project and extending the duration of the project so that it extends into the client or customer's next budget cycle.

• Guards against internally driven increases in scope. Well-intentioned members of the project team may expand the breadth or increase the detail of portions of the project beyond that set forth in the agreement or contract, expected by the client, or in any other way required by the circumstances. They may cause scope creep because they believe more is better. Any task can be executed better, but, as discussed in the next chapter, significantly exceeding requirements does not constitute a quality project. Significantly exceeding requirements tends to jeopardize schedules, budgets, and client or customer relations.

• Communicates with the key client or customer representative with emphasis on empathetic listening. The client's or customer's perception of project progress should be determined and, if not consistent with reality, corrected. The project manager responds to questions and addresses concerns in a timely fashion. If the project is to be performed by a private organization having one or more marketing personnel who regularly call on clients or customers, the project manager should ask a marketer to supplement the client contact by asking the client to share his or her views on the project's progress.

• Makes sure that all aspects of the project, including meetings and internal and external communications, are being adequately documented. The project manager should firmly take the position that the project is not done until documentation is complete.

• Determines the adequacy of internal support services such as word processing, drafting, computer services, and surveying. As appropriate, expresses appreciation for responsive assistance and takes action regarding deficiencies.

• Stays in touch with subconsultants and remains informed about their contributions to confirm that the subconsultants, as part of the project team, are meeting their schedule and deliverable obligations.

• Updates the project plan as needed and distributes it to members of the project team.

• Bills the client, generally in proportion to work completed and in accordance with provisions in the agreement or contract.

• Choreographs the project postmortem as described in the next section.

In carrying out his or her project monitoring and control functions, the project manager is like a juggler who successfully knows the location of and controls many balls. To do this, the project manager must be very organized. He or she should also be assertive and positive—not passive or negative.

PROJECT POSTMORTEM

The road to success is not doing one thing 100 percent better but doing 100 things one percent better.

(Anonymous)

Realize, and prove to your own satisfaction, that every adversity, failure, defeat, sorrow and unpleasant circumstance, whether of your own making or otherwise, carries with it the seed of an equivalent benefit which may be transmuted into a blessing of great proportions.

(Napoleon Hill)

With the benefit of hindsight, even a successful project could have been done better. Mediocre and failed projects contain the seeds of major future improvements in project management. Accordingly, each project should be immediately followed by a postmortem analysis to determine what can be learned for the benefit of near-future projects. Two important components of this postproject review are input from the client, customer, or a constituent and a meeting of the project team. The client or customer interaction should precede the team meeting, so that the team has the benefit of the client or customer input.

Client Input

Various approaches are possible for obtaining input from the client, customer, or constituent depending on circumstances. At the informal end of the spectrum, postproject input might be obtained through a private and casual one-on-one conversation between a representative of the organization that received services and the project manager or another liaison person. On the other extreme, a formal questionnaire, like those presented in Bates (1991) and Hensey (1989) might be sent to the client or customer. The client-input method must fit the situation and be based on a sincere desire to view the just-completed project or, more specifically, the services delivered from the client or customer's perspective, and use what is learned to improve the management of subsequent projects.

As an entry-level engineer, you are not likely to conduct the postproject review with your organization's client. But, as a member of a project team, you served internal "clients," that is, members of your organization. Make an effort to meet one-on-one with them. Ask for a frank evaluation of the "services" you provided with the goal of doing an even better job the next time. Many individuals are reluctant to volunteer encouragement or criticism, particularly the latter, but will comment on your efforts if you ask.

Team Meeting

At the initiative of the person who managed the just-completed project, all or key members of the project team should meet. As noted by Bonar (1994), the purpose of the meeting "is not to say who was right or wrong, but is to seek out what was right or wrong." Project successes such as new approaches developed, schedules and budgets met, positive client comments received, and extra efforts contributed should be celebrated at the meeting. That is, the group should resist the natural tendency to dwell on or discuss only negatives. Possible agenda items include schedule, budget, documentation, internal and external communication, and quality. Problems should be identified and analyzed with the idea of avoiding them in future projects. A succinct, proactive memorandum should document the post-mortem meeting focusing, again, on "what," not "who."

REFERENCES

BATES, G. D., ed., "Management Forum," *Journal of Management in Engineering—ASCE*, Vol. 7, No. 1 (January 1991), pp. 5–20.

Bonar and Associates, Inc., *Project Management Handbook*, April 1994.

BROWN, J. W., "The Week of Chaos: Critical Path Techniques Applied to Magazine Publishing," *PM Network*, Project Management Institute, Vol. 6, No. 2 (February 1992), pp. 36–38.

BEAKLEY, G. C., D. L. EVANS, and J. B. KEATS, *Engineering: Introduction to a Creative Profession*, fifth ed., Chapter 14—"Engineering Design Phases," New York: Macmillan Publishing Company, 1986.

CLOUGH, R. H., *Construction Contracting*, Chapter 11—"Project Time Management," fifth ed, New York: John Wiley & Sons, 1986.

Construction Industry Institute, "Concepts and Methods of Schedule Compression," University of Texas at Austin, Publication 6–7, November 1988.

CORI, K. A., "Project Work Plan Development," Presented at the Project Management Institute and Symposium, Atlanta, Ga., October 1989.

COVEY, S. R., *The 7 Habits of Highly Effective People*, New York: Simon & Schuster, 1990.

DHILLON, B. S., *Engineering Management: Concepts, Procedures and Models*, Lancaster, Pa.: Technomic Publishing, 1987.

HENSEY, M. ed., "Management Forum," *Journal of Management in Engineering—ASCE*, Vol. 5, No. 3 (July 1989), pp. 209–219.

HOUSE, R. S., *The Human Side of Project Management,* Reading, Mass.: Addison-Wesley, 1988.

LOCK, D., *Project Management*, fifth ed, Hants, England: Gower Publishing, 1992.

PIRSIG, R. M., *Zen and the Art of Motorcycle Maintenance,* New York: Bantam Books, 1974.

PSMJ, "Getting More Money," *Professional Services Management Journal*, Vol. 7, No. 6 (June 1980), pp. 1–3.

SCHERMERHORN, J. R., Jr., *Management for Productivity*, Chapter 16—"Production and Operations Control," New York: John Wiley & Sons, 1984.

SPINNER, M. P., *Elements of Project Management: Plan, Schedule and Control*, Englewood Cliffs, N.J.: Prentice Hall, 1992.

VANDERSLUIS, C., "Project Management Computer Software Systems," *The Project Manager*, Association for Project Managers, Winter 1994, pp. 37–39.

SUPPLEMENTAL REFERENCES

HARTLEY, K. O., "How to Make Project Schedules Really Work for You," *Journal of Management in Engineering—ASCE*, Vol. 9, No. 2 (April 1993), pp. 167–173.

HRIBAR, J. P., and G. E. ASBURY, "Elements of Cost and Schedule Management," *Journal of Management in Engineering—ASCE*, Vol. 1, No. 3 (July 1985), pp. 138–148.

KERKES, D. J., "Precepts of Project Management," *Civil Engineerring—ASCE,* (September 1994), pp. 70–72.

EXERCISES

6.1 CONSTRUCT A NETWORK

Purpose

Improve the student's understanding of networks and network logic.

Given

Operational logic between all the activities in a project.

1. A is the initial task.
2. E and F can be performed simultaneously and cannot be started before B is completed.

3. I depends on F, G, and H.
4. K can begin only after E and I are finished.
5. L must follow J and K.
6. J cannot start until E is completed.
7. C must be completed before G can begin.
8. B, C, and D cannot begin until A is completed and can be performed simultaneously.
9. D must be completed before H can begin.
10. L is the final task.

Tasks

1. Construct the network.
2. Use the letters A–L to identify tasks.
3. Show arrows on the lines that connect nodes.

6.2 APPLY THE CRITICAL PATH METHOD

Purpose

Improve the student's understanding of identifying the critical path and related indicators such as project duration and total float.

Given

The network developed in Exercise 6.1 and the following task durations:

TASK	DURATION (DAYS)
A	0
B	7
C	13
D	19
E	6
F	23
G	25
H	10
I	6
J	14
K	21
L	0

Tasks

1. Determine minimum project duration.
2. Identify the critical path by highlighting the path on the network.
3. Determine the total float for each task, that is, list each task with its total float.

6.3 CONSTRUCT A GANTT CHART

Given

Results of Exercises 6.1 and 6.2.

Tasks

1. Develop a Gantt chart for the project.
2. Position the "bar" for each task as beginning with its EST and ending with its LFT.
3. Show task duration with a solid line and activity float with a dashed line. *Note:* The result is, in effect, a more meaningful Gantt chart.

6.4 APPLY CPM AND CONSTRUCT A GANTT CHART

Purpose

Apply CPM to an actual project.

Given

"Real" project submitted earlier as Exercise 1.2

Tasks

1. Estimate the duration of each task.
2. Perform the complete CPM process, including constructing the network, applying the CPM, highlighting the critical path and developing the Gantt chart with total float shown (that is, do as in Exercises 6.2 and 6.3).

6.5 UPDATING A CRITICAL PATH ANALYSIS

Purpose

Show how a previous critical path analysis can be updated once a project is underway to provide new project time-management information.

Tasks

1. Refer to the solution for Exercise 6.2.
2. Assume that the project is underway and, at the end of 20 days, the status of each task is as follows:

TASK	STATUS
A	Done
B	Done
C	Done
D	Underway with 8 days of work to be completed.
E	Done
F	Underway with 2 days of work to be completed.
G	Underway and on schedule (i.e., 18 days of work to be completed).
H	Not started; duration estimate is increased to a total of 11 days.
I	Not started.
J	Underway with 4 days of work to be completed.
K	Not started.

Note: There are no changes in the "job logic," that is, in the number of tasks and their interrelationships. However, such changes could very easily have arisen at this stage in the project and would have to be reflected in the updated critical path analysis.

3. Determine the following:

a. The new critical path.

b. The time required, starting now, to complete the project.

c. The projected absolute completion time compared to that projected at the beginning of the project. That is, plus or minus how many days?

7

Total Quality Management

> *Quality is free. It's not a gift, but it is free. What costs money are the unquality things—all the actions that involve not doing jobs right the first time.*
>
> (Crosby, 1979, p. 1)
>
> *. . . care and quality are internal and external aspects of the same thing. A person who sees quality and feels it as he works is a person who cares. A person who cares about what he sees and does is a person who's bound to have some characteristics of quality.*
>
> (Pirsig, 1974, p. 247)

Total quality management (TQM) has been increasingly discussed by all kinds of organizations during the past decade and has been diligently applied by a few of them. Chronologically, interest in TQM in the United States began in the manufacturing sector and then moved through government, consulting engineering firms, professional societies, and academia. Numerous questions, such as the following, are suggested by this recent TQM movement:

- Is TQM another passing management fad?
- Is TQM really new—or just a "gussied up" re-introduction of tried and true managerial and business concepts, ideas, and practices?
- What are the origins of TQM? Who started it? Who are its principal advocates?
- Why is TQM, or at least intensive conversation about it, so popular now?
- Because "quality" is at the center, literally and figuratively, of TQM, what is "quality" in the context of TQM?

- Assuming that TQM has value, how does an organization implement it? What are the costs? What are the benefits? How long would it take to reap benefits? What obstacles are likely to be encountered?
- Are there any success stories?

One purpose of this chapter is to answer, at least in part, the preceding questions. More importantly, this chapter is intended to be a primer for you, the entry-level engineer or other technical professional. Regardless of whether or not you or the educational institution in which you are currently studying or recently studied or the organization in which you are currently employed embraces TQM, in whole or part, all young professionals should be familiar with TQM terminology and basic ideas, because TQM is part of today's business and professional culture.

The chapter begins by reviewing several meanings of quality and focusing on the precise definition of quality used in TQM. An elaboration of that definition follows in terms of the stakeholder idea, and principles of TQM as set forth by one of the founders of the TQM movement. Common tools and techniques typically used in the practice of TQM are described. The chapter concludes with examples of successes said to be based on the application of TQM and with an introduction to the Malcolm Baldrige National Quality Award.

QUALITY DEFINED

Quality as Opulence

The word *quality* in the context of professional work or in a general context, may suggest opulence, luxury, "gold-plating," and overdesign. Examples of products or results consistent with this opulence concept of quality might be Mercedes Benz automobiles, Rolex watches, cashmere sweaters, and safety factors of 3.0. That is, such products and results generally go well beyond what is needed for functional purposes, but not necessarily beyond what may be desired by a few individuals or organizations.

In an individual or organizational environment of unlimited or at least great resources, the opulence definition or understanding of quality might be acceptable. The opulent approach to quality, however, is not useful in the vast majority of engineering and business situations.

Quality as Excellence or Superiority

Another approach to quality is the concept of excellence or superiority as suggested by the dictionary (Neufeldt, 1994) definition of quality, which includes

". . . degree of excellence . . ." and "superiority." Offering a superior, standard-setting product or service is certainly admirable, but is not likely to be practical for most engineering and business situations. Although clients may value "excellence" and "superiority," they may not want to pay for it. Furthermore, while notions of excellence and superiority may engender positive reactions, they may be too vague and expensive.

Nevertheless, some technical professionals will argue for a superiority approach to quality. As an example of this perspective, Huntington (1989) says, "Quality, however, will not come from automation, but from obsession—a craftsman's obsession with making a thing as good as it can be made." While TQM advocates would agree that quality will not necessarily come from automation, they would tend to take great issue with the "as good as it can be made" understanding of quality.

Crosby (1979, p. 14), says "The first erroneous assumption is that quality means goodness, or luxury, or shininess, or weight." In the context of TQM, quality must mean something other than opulence or superiority in kind to the enlightened, progressive, but practical technical professional and/or business person. This leads to a third definition of quality, one that is widely accepted as being fundamental to the TQM movement.

Quality as Meeting Requirements

In TQM, quality is conformance to requirements. Crosby (1979, p. 15) says, "We must define quality as conformance to requirements." Lewis elaborates by saying, "Quality is conformance to requirements. It is achieved by conforming to properly developed criteria which make the services meet all of the customer's needs and expectations." Snyder (1993, p. 53) offers this definition of quality:

> *Quality in engineering is a measure of how well engineering services meet the client's needs and conform to governing criteria and current practice standards.*

This definition is useful because it elaborates somewhat on the concept of "meeting requirements." Certainly, the client, owner, or customer has a major role in defining requirements. But the total definition of requirements may need to go beyond what the client, owner, or customer needs or believes he or she needs. For example, the engineer must strive to satisfy government regulations and codes of ethics, and to be consistent with the standard of care of the profession as defined in Chapter 10 of this book. In the final analysis, a potential client, owner, or customer's requirements may fall short of what the engineer is willing or able to do. In such situations, personal and corporate ethical standards may require termination of the relationship.

Quality definitions used by professional organizations embody the same concepts. For example, the Quality Management Task Force of the Construction Industry Institute says "Quality is conformance to established requirements" (Davis and Ledbetter, 1988). The "quality manual" of the American Society of Civil Engineers (ASCE, 1990, pp. 1–2), which focuses on achieving quality in the constructed project, states:

> *. . . quality is defined as meeting established requirements. Quality in the constructed project is achieved if the completed project conforms to the stated requirements of the principal participants (owner, design professional, constructor) while conforming to applicable codes, safety requirements, and regulations.*

Consider some hypothetical examples of quality as it is typically used in TQM. If a Yugo meets a person's defined transportation needs, then, for that person, a Yugo is a better-quality car than a Mercedes Benz. On the other hand, if a person's defined automobile transportation needs include leather upholstery, headlight wipers, and other special features typically found on the Mercedes Benz or other similar luxury automobiles, then the Yugo will clearly not meet the defined needs and will not, in the context of TQM, be a quality product. If a hand calculation of peak discharge from an industrial development using the rational method meets the design needs for sizing a length of storm sewer, then, for the client, the rational method is a better quality technique than a sophisticated hydrologic-hydraulic computer model. Or, to use a construction-related example quoted from the aforementioned *ASCE Quality Manual* (ASCE, 1990, p. 2):

> *Thus, a temporary, sheet-metal enhoused, pump station, with low capital cost, high operating costs, short expected life, and aesthetic deficiencies, may well be a quality project if it meets the expectations and requirements of the three principal participants. Conversely, a Taj Mahal with all of its beauty and durable materials may not qualify as a quality project if its construction results in costs or overruns, litigation, environmental controversy, or negative impact on public health and safety.*

Quality as meeting requirements, as opposed to quality as opulence or superiority, is widely used in the TQM literature and is standard operating procedure in TQM practice. You, as an entry-level engineer or other technical professional, are advised to use this definition when discussing or applying TQM.

Frankly, the young professional may have difficulty accepting the idea of quality as meeting, but not significantly exceeding, the established requirements. Almost any technical activity or project, such as field investigations, laboratory tests, a planning study, and a design culminating in plans and specifications for a manufactured product or constructed facility, can be done better than expected.

After all, technical professionals tend to be very bright and usually have access to many and varied sophisticated tools and techniques. Furthermore, the young professional's education may have encouraged him or her to go well beyond what was needed. But, in the world of practice and business, going well beyond what is needed tends to increase labor and other costs and cause delays, both of which are ultimately disruptive to a profitable business and to a satisfied customer or client. You need to weigh your personal desire to produce a superior or even opulent product or service against the best interests of your employer and those individuals or organizations served by your employer.

STAKEHOLDERS

If quality as used in TQM means "meeting requirements," the next logical question is "whose requirements?" This gives rise to the notion of stakeholders. ASEE (1993) offers this broadly applicable definition of stakeholder:

> *A stakeholder is an individual or organization having a significant interest in the results of another individual's or organization's actions and activities. Stakeholders might be suppliers of goods, services, or personnel to an organization or they might receive the products of an organization in terms of goods, services, or personnel.*

The ASEE report, which was written for engineering colleges, may be paraphrased to give examples of stakeholders relevant to technically based organizations. Within the context of consulting firms, manufacturing companies, and other engineering organizations, the appropriate set of stakeholders is likely to vary widely depending on the organization's mission, goals, strategies, and tactics. From the perspective of those who benefit from the services provided by and the products created by technically based organizations, stakeholders would include clients, customers, and the public at large. The community within which the organization is located is a stakeholder because it derives benefits such as employment and property taxes. There are also many stakeholders within the technically based organization. Various offices, divisions, departments, and other units are interdependent. Each employee is a stakeholder in that he or she relies on the organization for income and for growth opportunities. Surely each person holding stock in a technical organization is also a stakeholder. Figure 7–1 illustrates a technical organization's typical stakeholders.

The functioning and overall satisfaction of many and varied individuals, suborganizations, and organizations—stakeholders—are inextricably tied to the typical engineering organization. As stated by ASEE (1993), the organization's

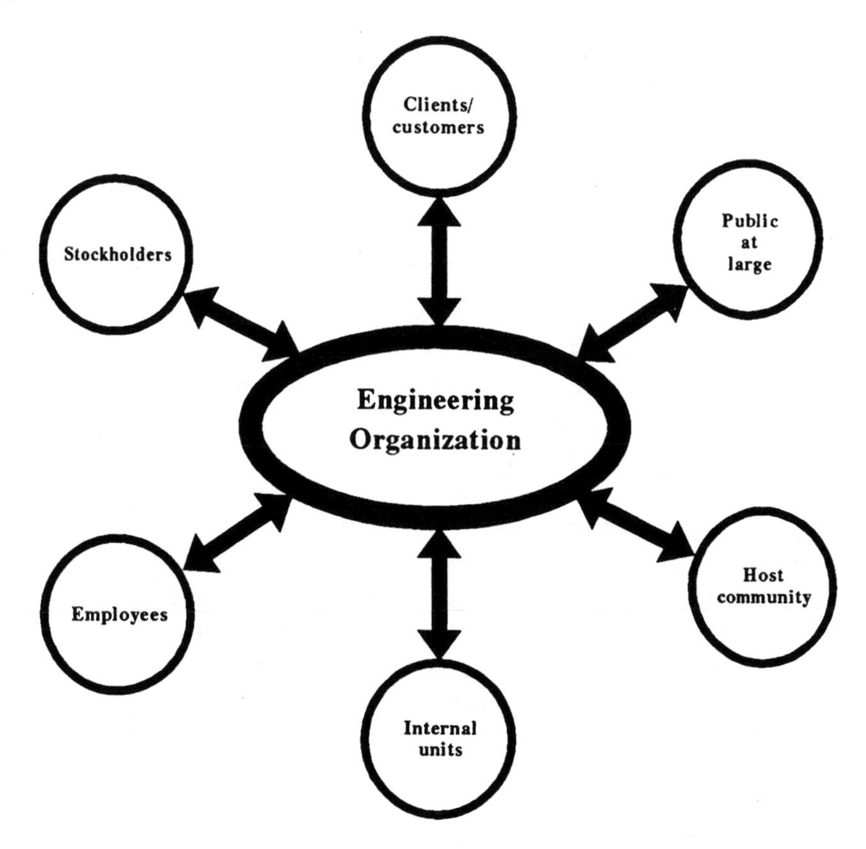

Figure 7–1 A technical organization's typical stakeholders

" . . . planning process should include identification of its stakeholders and focus on developing mutually beneficial relationships with those stakeholders that are crucial to its mission."

TQM DEFINED

If quality is defined as the degree of conformance to established requirements, what is TQM? TQM may be defined as the combination of philosophy, principles, and methods that enable people working together to meet requirements. An elaboration of this definition of TQM appears in the Department of Defense *TQM Management Guide* (Simon):

> *Total Quality Management (TQM) is both a philosophy and a set of guiding principles that represent the foundation of a continuously improving organization. TQM is the application of quantitative methods and human resources to improve the ma-*

terials and services applied to an organization, all the processes within an organization, and the degree to which the needs of the customer are met, now and in the future. TQM integrates fundamental management techniques, existing improvement efforts, and technical tools under a disciplined approach focused on continuous improvement.

C. Jackson Grayson defines TQM by elaborating on the three words. He says (Cook, 1991):

- *Total, meaning all people, all functions, customers, and suppliers.*
- *Quality, meaning not just products, but processes, reliability, and quality of work life.*
- *Management, meaning senior management strategy, goal-setting, organizational structure, compensation, and profits.*

The preceding definitions of TQM support the idea that TQM is the management system needed to achieve quality as defined earlier. To this point, the discussion of TQM has consisted of providing a special definition or understanding of quality, introducing the stakeholder idea, and suggesting the need for a management system to achieve that quality. That is not specific enough to be practiced or to be operational. This leads to a discussion of the principles of TQM.

PRINCIPLES OF TQM

The principles of TQM were first articulated by Dr. W. Edwards Deming, whose background is worthy of note. Dr. Deming, who was born in Sioux City, Iowa, earned a B.S. in electrical engineering and a master's and a Ph.D. in mathematics and physics. In 1925, he began his career by working as a statistician at Western Electric in the Chicago area and later served on the faculty of New York University's Graduate School of Business. While serving as a U. S. Census Bureau statistician and a consultant to the U.S. War Department, he began lecturing and conducting seminars in Japan. He told the shocked Japanese not to emulate and copy the approach of American corporations, but instead to view production as a system and seek constant improvements throughout the system. He urged the Japanese, unlike the Americans, to be responsive to customers and view vendors as partners.

Largely as a result of applying Deming's principles, Japan went from a post-World War II producer of very poor products to a producer of quality products. As an expression of appreciation, the Japanese established the Deming Prize in 1951, which is the highest business award in Japan, second only to a personal citation from the emperor.

Dr. Deming, who died in 1993, authored eight books, over 170 technical papers, and many musical compositions. Into his early 90s, he was still leading seminars and serving as a consultant throughout the United States. His advice was finally being valued by business and professional leaders in his own country (Modic, 1988; Yates, 1992).

The principles of TQM, cast in the form of Deming's 14 Points for Management, are summarized in Figure 7–2 and quoted with permission in their entirety (Modic, 1988) as follows:

*1. **Create constancy of purpose for the improvement of product or service.** Declare to the world—customers, suppliers, and employees—your intention to stay in business by providing a product and/or service that will help people live better and that will*

1. Create constancy of purpose for the improvement of product or service.

2. Adopt that new philosophy.

3. Cease dependence on mass inspection.

4. End the practice of awarding business on the basis of the price tag alone; instead, minimize total cost by working with a single supplier.

5. Improve constantly and forever every process for planning, production, and service.

6. Institute training on the job.

7. Adopt and institute leadership.

8. Drive out fear.

9. Break down barriers between staff areas.

10. Eliminate slogans, exhortations, and targets for the workforce.

11. Eliminate numerical quotas for the workforce and numerical goals for management.

12. Remove barriers that rob people of pride of workmanship, including the annual rating or merit system.

13. Institute a vigorous program of education and self-improvement for everyone.

14. Put everybody in the company to work on accomplishing the desired transformation.

Figure 7–2 Deming's 14 points for management (Source: Modic, 1988, p. 91.)

have a market. Establishment of constancy of purpose requires top management's unshakable commitment to quality and productivity. It also requires a commitment to innovate in new products and service, materials, methods of production, job skills, training, and marketing; to invest in research and education; and to constantly improve the design of products and service.

2. Adopt that new philosophy. *We are in a new economic age created by Japan. Western management must rise to the challenge, must learn its responsibilities, and must take on leadership for change.*

3. Cease dependence on mass inspection. *Routine 100% inspection to improve quality is equivalent to planning for defects. Inspection to improve quality is too late, ineffective, and costly. Quality comes not from inspection, but from improvement of the production process.*

4. End the practice of awarding business on the basis of the price tag alone; instead, minimize total cost by working with a single supplier. *Price has no meaning without the measure of the quality being purchased. American industry and government are being rooked by rules that award business to the lowest bidder. A long-term relationship between purchaser and supplier is best. How can a supplier be innovative and develop economy in his production processes when he can look forward only to short-term business from a purchaser?*

5. Improve constantly and forever every process for planning, production, and service. *Quality must be built in at the design stage. Teamwork in design is fundamental. There must be continual improvement in test methods and an ever-better understanding of the customer's needs and of the way he uses (and misuses) the product. Putting out fires does not improve the process; it merely puts the process back to where it should have been in the first place.*

6. Institute training on the job. *The greatest waste in America is failure to use the abilities of people. Training must be totally reconstructed. Japanese managers start their careers with a year's-long internship on the factory floor. Likewise, Western management must learn all about the company—from incoming material to customer needs. A major problem in training and leadership arises from a flexible standard of what is acceptable work; the standard too often depends on whether the foreman is pushing to meet his daily quota.*

7. Adopt and institute leadership. *The job of management is not supervision; rather, it is leadership. There was a time when foremen selected, trained, and worked with their people. Now, most foremen were never on the job that they're supervising. Most supervisors do not gain the confidence of their people because they are concerned only with numbers; they are incapable of helping the production worker improve his performance.*

8. Drive out fear. *No one can perform at his best unless he feels secure. Fear takes on many faces. One is a widespread resistance to knowledge. Improvement of the kind needed in Western industry requires knowledge, yet people are afraid of knowledge.*

Pride may play a part in that resistance; new knowledge brought into a company might disclose some of our failings.

9. Break down barriers between staff areas. *Teamwork is sorely needed throughout the company. Teams made up of people in design, engineering, production, and sales could contribute to designs of the future while improving product, service, and quality today.*

10. Eliminate slogans, exhortations, and targets for the workforce. *Eliminate slogans and posters urging the workforce to increase productivity. These things are directed at the wrong people. They stem from management's supposition that the production workers can, if motivated to work harder, accomplish zero defects, improve quality and productivity, and all else that is desirable. The charts and posters take no account of the fact that most of the trouble comes from the system, not the individual workers. Exhortations and posters generate frustration and resentment. They advertise to the production worker that management is unaware of the barriers to pride of workmanship. Management has to learn that the main responsibility to improve the system is theirs.*

11. Eliminate numerical quotas for the workforce and numerical goals for management. *A quota is a fortress against the improvement of quality and productivity. I have yet to see a quota that includes any trace of a system that can help anyone do a better job. Work standards, rates, incentive pay, and piecework are manifestations of an inability to understand and provide appropriate supervision. Management by numerical goal is an attempt to manage without knowledge of what to do, and is usually management by fear.*

12. Remove barriers that rob people of pride of workmanship, including the annual rating or merit system. *Such barriers and handicaps rob the hourly worker of the right to be proud of his work, the right to do a good job.*

13. Institute a vigorous program of education and self-improvement for everyone. *An organization needs not just good people, but people who are also improving themselves with education. People require—even more than money—ever-broadening opportunities to add something to society, materially and otherwise.*

14. Put everybody in the company to work on accomplishing the desired transformation. *Agree on the direction you want to take; have the courage to break with tradition; and explain to employees via seminars and other means why the change is necessary and the fact that it will involve everyone. Enough people in the company must understand the obstacles and the solutions. Otherwise, management is helpless.*

Given Deming's education and the positions he held early in his career, his approach to achieving quality might be assumed to be narrow and very focused on quantitative means including statistics. But, although Deming advocated the use of statistics, his approach as suggested in the 14 points is, in effect, a broad and deep management philosophy (e.g., Modic, 1988). Note how the theory Y concept, discussed in Chapter 4, underlies and is woven through Deming's 14 points of management.

In a useful attempt to simplify matters, Deming's 14 points may be condensed into the following nine points (quoted from Simon):

1. *Continuous process improvement.*
2. *Total employee involvement.*
3. *Concentration on prevention, not correction.*
4. *Supplier–customer relationship.*
5. *Customer satisfaction.*
6. *Problem-solving teams.*
7. *Tools and techniques.*
8. *Measurement.*
9. *Recognition.*

So what is so new or innovative or revolutionary about the 14 points or Simon's condensation of them? Don't they all seem like various combinations of common sense? Don't they describe an environment within which intelligent, sensitive people wanting to contribute would thrive? Is the issue the difference between knowing what to do and doing it; saying what you believe versus acting on those beliefs? You, as the entry-level technical professional, will have to decide based on what you hear in your organization, what you see in your organization, and what your organization achieves. If you are already practicing your profession, perhaps you are in a TQM environment, although it may not be called that. If, on the other hand, you see considerable variance between what your organization is doing and what TQM prescribes, perhaps based on what you learn in this chapter you can be one of the agents of change.

COMMENTS ON SOME OF DEMING'S 14 POINTS FOR MANAGEMENT

Deming's 14 points raise essentially limitless discussion possibilities. A few of the points, more specifically points 3, 5, 6, and 13, are discussed here. Point 3 (stop end-of-line mass inspection) and point 5 (constantly improve every step in a process) are extremely powerful. The traditional American way is to operate on the premise that everybody is doing his or her best and we will "check" the final product (e.g., coffee pot, report, automobile, plans, and specifications) to make sure they are or it is OK. This has been referred to as the "downstream" approach (e.g., Hayden, 1990). With the downstream approach, rejects are sent back for rework, or fixed on the spot, or are discarded. The reason for the rejects is not a major concern—they just happen and are inevitable.

In contrast, the Deming way is to identify, chart (e.g., prepare a network diagram as discussed in Chapter 6) all steps in a production process and seek continuous improvement of every step. This is the "upstream" (e.g., Hayden, 1990) or source approach. Rejects still occur, but there will be fewer and fewer of them because the occurrence of a reject focuses attention on its origin(s) in the process. The discovery of a probable origin leads to improvements of the step or steps that cause the rejects. The focus is on "error prevention," not on "error catching" (Hayden, 1990).

While the Deming way of eliminating end-of-line mass inspection and seeking to constantly improve every step in the process may seem very logical, there are many obstacles in traditional American practice. For example, opposition to change is likely to be reflected in comments such as the following:

- "We don't have an understanding of the process; things just happen."
- "Developing a real understanding of what we do and why would be a lot of work."
- "Even if we understood the process, there are so many people involved and some are even in other administrative units, so change will be very difficult. Besides, we've always done it this way. Let's just try harder."

Next consider Deming's point 6 (institute on-the-job training) and point 13 (expect and support self-improvement for everyone). The traditional American positions on training and self-improvement are reflected by the idea that organizations hire only those people who are already adequately trained and educated. Furthermore, self-improvement is a personal matter. Finally, in difficult economic times expenses are being reduced because business is "down" and, therefore, professional development budgets are among the first to be cut.

In contrast, the Deming way would include formal orientation and training for all entry-level people. Furthermore, the Deming way expects that everyone in the organization, from top to bottom, is continuously involved in self-improvement. The organizational culture supports and expects continuous self-improvement of everyone.

TOOLS AND TECHNIQUES

Many procedures are available for implementing a TQM program, that is, for doing an even better job of meeting requirements. Hensey (1993) provides a useful summary. Examples of tools and techniques are presented here with the recognition that the methods selected and the manner in which they are used must be tailored to particular organizational situations.

Metrics

What gets measured, gets done.

(Anonymous)

Metrics is the art and science of measuring. Metrics helps managers understand how things are going in an absolute sense and in terms of historic trends. The term *operating statistics* might also be used. According to Paramax (1992), "The term metric implies a thorough understanding of such factors as the limitations of the measuring tool and the confidence you have in the measure . . . in short what the measure really says about quality." Quantifying, to the extent feasible, various steps in any process and studying the numbers is an important part of TQM. "If you can't measure it . . . you can't manage it" and "If you don't measure it . . . you probably won't be able to improve it" (Paramax, 1992).

Consider for example, a consulting engineering business. The following metrics might be used on a quarterly or perhaps annual basis:

- Percentage of projects completed early, on time, and late.
- Percentage of projects completed under budget, on budget, and over budget.
- Number and percentage of new clients served compared to total clients served where "clients served" might be defined as opening a set of project numbers for a new project.
- Number and percentage of contracts signed by state or country of origin of client. Monetary size of contract could also be used.
- Number and/or dollar amount of liability claims and statistics on how they were resolved.
- Direct or chargeable hours as a percentage of total hours worked, expense ratio, and multiplier, as discussed in Chapter 9.

A manufacturing organization might employ metrics such as:

- Time required to respond to a request for a price quote.
- Time required to design and test a new product or a major modification to an existing product.
- Number of products manufactured per shift.
- Percentage of each component rejected per shift.
- Cost per manufactured product.
- Warranty claims on each product per year.
- Market share.

Similarly, a wide variety of metrics are available to municipalities, perhaps most appropriately on an annual basis. Examples include number of flooding complaints; number and percentage of construction projects completed early, on time, and late; and number and percentage of employees who receive formal training. Metrics or operating statistics are crucial to any management effort. Consider, for example, colleges and universities, which might track metrics such as alumni with advanced degrees, graduates placed in employment or graduate school, Scholastic Aptitude Test scores of incoming first-year students, teaching load credits per faculty member, and funds raised from external sources.

Written Procedures

The simple, but rarely used, TQM tool of written procedures is applicable to all tasks or series of tasks that are likely to be repeated and are not documented elsewhere. Written procedures can be prepared for a wide range of functions in an organization. For example, within the typical technically oriented organization, written procedures can be used for technical tasks done on projects, project management tasks, various administrative and clerical tasks, inspection and maintenance tasks, and marketing tasks. Written procedures should be drafted and continuously updated primarily by the individuals who apply them.

As a result of creating, maintaining, and using written procedures, an organization and its members will derive many long-term benefits. However, realizing these long-term benefits requires an initial investment of effort to create or update the procedures. Accordingly, an initial set of procedures may be expected to evolve over a period of months or years. Furthermore, when the first set of procedures is completed, each procedure will always be subject to continuous improvement.

Eight benefits resulting from written procedures are

• **Eliminate valueless or marginal activities.** Unnecessary, redundant, outdated, and other marginal or valueless tasks and steps are likely to be identified as a result of creating or maintaining procedures. The process of thinking about, discussing, and then describing, in writing, steps to be taken or tasks to be accomplished to achieve an objective or produce a product inevitably identifies valueless or marginal activities. Finding and eliminating tasks and steps—sometimes entire procedures or processes—that do not add value reduces expenses and frees up staff for more productive tasks.

• **Increase efficiency.** As a result of thinking through steps comprising a procedure, more efficient approaches are typically discovered. For example, some steps previously done in series may subsequently be done in parallel, thereby reducing elapsed time. Tasks formerly done by personnel with high hourly rates may subsequently be accomplished just as effectively by personnel with lower hourly rates, thus reducing costs. Delegation of tasks and steps is simplified, because ready reference can be made to written material. Steps previously done manually may, as a result of the insight gained by analyzing a process, subsequently be done with computer programs, thereby reducing costs and saving time.

- **Avoid reinventing the wheel.** Knowledge acquired and experience gained by an organization's personnel in doing their work, assuming that it is reflected in the written procedures, is in a form to be readily shared with other personnel. As a result of the use of procedures that integrate knowledge and experience gained on earlier projects, current projects are more likely to be completed on time and within budget and to meet requirements.

- **Facilitate interdiscipline and interoffice projects.** The multidiscipline–multioffice organization should be structured and operated to provide its clients, customers, and constituents with the optimum mix of personnel and other resources regardless of their physical and organizational "homes" in the company. Written procedures that are used across discipline and office lines facilitate the desired corporate team approach. The lack of procedures frustrates interdiscipline and interoffice cooperation even when personnel want to work as a team.

- **Train new or transferred personnel.** New personnel and personnel transferred from one office, division, or unit to another can be provided with the appropriate set of written procedures as part of their on-the-job training. This training method requires less supervisor time and is more specific.

- **Reduce liability.** Negligence, the principal cause of liability claims in the consulting engineering business, is reduced. Errors and omissions are less likely to occur when work is guided by tested, written procedures.

- **Reduce negative impact of personnel turnover.** Some personnel turnover is inevitable even in the best managed and led organization. Contributions that departed personnel made to an organization are more likely to remain with it if some of those contributions were captured and documented in the form of procedures prepared by the now departed personnel.

- **Support marketing.** External and internal clients, customers, and constituents are increasingly concerned about the quality of the services and products they receive. The test is: Do or will these services or products meet my needs? The existence and use of written procedures is one way of demonstrating a unit or organization's commitment to quality, especially a desire to do things right the first time.

Flow Charting

Flow charting uses the networking technique described in Chapter 6 of this book. The idea is to create a network for those processes within an organization that are yielding substandard products or services to internal or external clients. Hypothetical examples of such processes and substandard results are

PROCESS	SUBSTANDARD RESULTS
Manufacturing a product	Excessive reject rate
Water quality sampling	Too many samples contaminated
Staff meetings	Accomplish too little
Writing reports for clients	Too many spelling and grammatical errors.

Once the network has been constructed by a group of individuals who collectively understand the entire process and there is general agreement that it represents the current state of affairs, the network should be closely examined for illogical sequencing, redundant steps, missing steps, and problematic steps. Corrected actions should be proposed, shown on the network, and tested to see if substandard results are diminished.

Fishbone Diagrams

Fishbone diagrams, also referred to as cause and effect diagrams, are another effective graphical technique. Like flow charting, the creation and analysis of fishbone diagrams can be and, in fact, probably should be done as a group project. This diagnostic tool starts with a problem or deficiency (the effect) and works backward to identify influencing factors (the causes) with the goal of pinpointing the cause or causes most likely to solve the problem or reduce the deficiency. Figure 7–3 uses a common situation to illustrate the concept. Of course, considering potential causes of problems or deficiencies is common in technical disciplines. However, the fishbone or cause and effect diagram systematizes that process and portrays it graphically, thereby facilitating a group effort and increasing the likelihood of a successful diagnosis.

Pareto Analysis

Recall the "vital few–trivial many" rule, the 20/80 rule, or Pareto's law discussed in Chapter 2 under the topic of time management. The approach introduced here is called Pareto analysis after Vilfredo Pareto, an Italian sociologist and economist, who is credited as the source (Paramax, 1992).

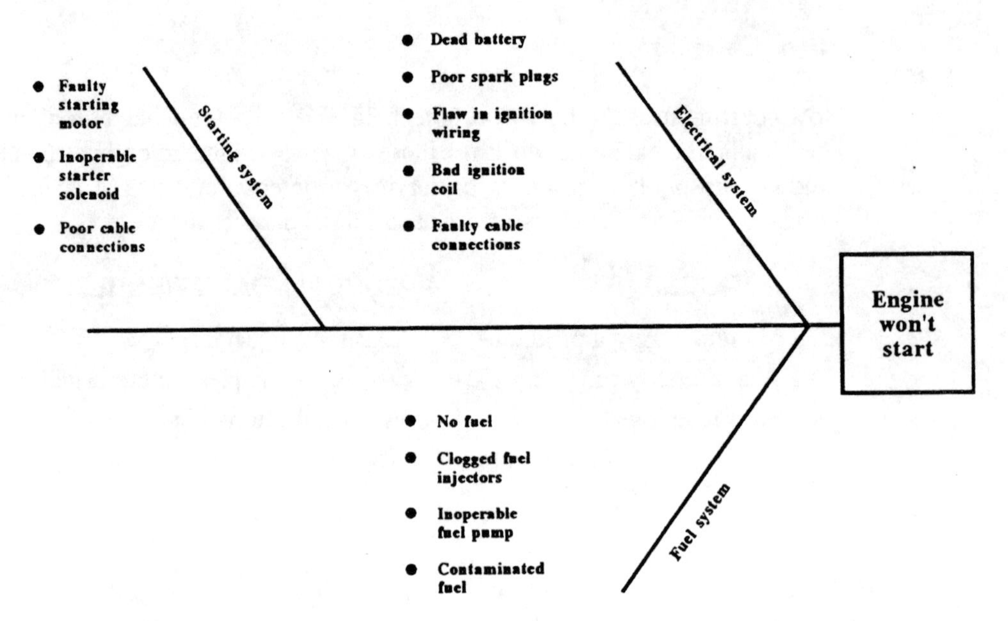

Figure 7–3 Fishbone diagram analysis applied to an automobile engine (Source: Adapted with permission from Paramax Systems Corporation, "Total Quality Management," 1992, p. 25.)

When agreed-upon quality is not being achieved, Pareto analysis can be used to identify the most influential causes of substandard performance so that they can be addressed first. An example of Pareto analysis is presented in Figure 7–4, which shows the relative importance of major causes of defects in circuit card assemblies and clearly points to the most logical solutions. This example suggests many and varied possible applications of Pareto analysis.

Brainstorming Sessions

Don't stand there and jeer
And throw up your handza;
That stupid idea
Might start a bonanza!

(Harrisberger, 1986, p. 77)

Having selected a particular problem area in terms of a substandard product or service for internal or external clients, all involved individuals, or probably more practically, representatives of all involved individuals, should be invited to a non-threatening environment. Invitees are encouraged to participate in a fast-paced

Circuit Card Assembly Defect Analysis

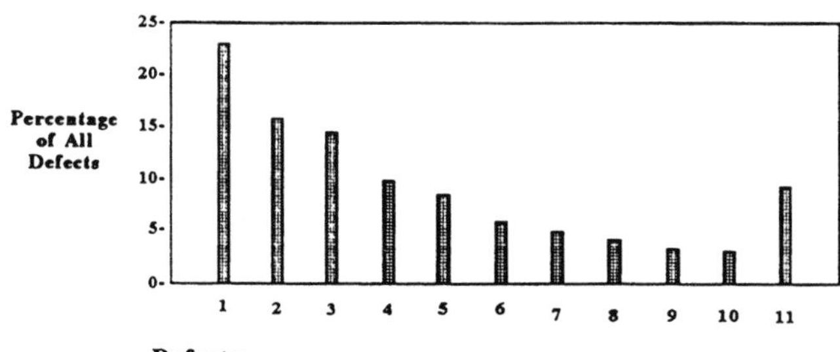

Defects

1. **Component in solder**
2. **Poor wetting**
3. **Excess solder**
4. **Solder bridges**
5. **Improper cleaning**
6. **Lead not visible**
7. **Blowholes**
8. **Insufficient solder**
9. **Reversed polarity**
10. **Raised component**
11. **Other defects**

Figure 7–4 Example of Pareto analysis (Source: Adapted with permission from Unisys Corporation, "Total Quality Management," 1990.)

idea-generation exercise during which judgments are deferred. Results are reviewed later and appropriate corrected actions are identified (Paramax, 1992).

Benchmarking

"Benchmarking is the continuous process of measuring products, services, and practices against the toughest competitors or those companies recognized as industry leaders" (Camp, 1989, p. 10). Although this definition is taken from "industry," it could easily be tailored to apply to other engineering-related areas, such as consulting, government, professional organizations, and academia. For example, replace "companies" in the definition with "organizations" and drop the word "industry." A shorter definition is "Benchmarking is the search for industry's best practices that lead to superior performance" (Camp, 1989, p. 10).

The term *benchmarking* is borrowed from the surveying field, where a benchmark is a recognized reference point or, more broadly, an acknowledged standard. Some key features of benchmarking, as explicitly indicated in the first definition, is that it is continuous, it involves measurement, and it applies to products and services and to the processes that lead to products and services.

Figure 7–5 presents benchmarking as a ten-step process in which the steps are organized in the four categories of planning, analysis, integration, and action. Consider the following comments on some of the steps:

- Step 1. Identify the product, service or practice that needs improvement.

- Step 2. Look first for internal benchmarks. Then look externally at direct competitors and at leaders in particular functions. Assume, for example, that

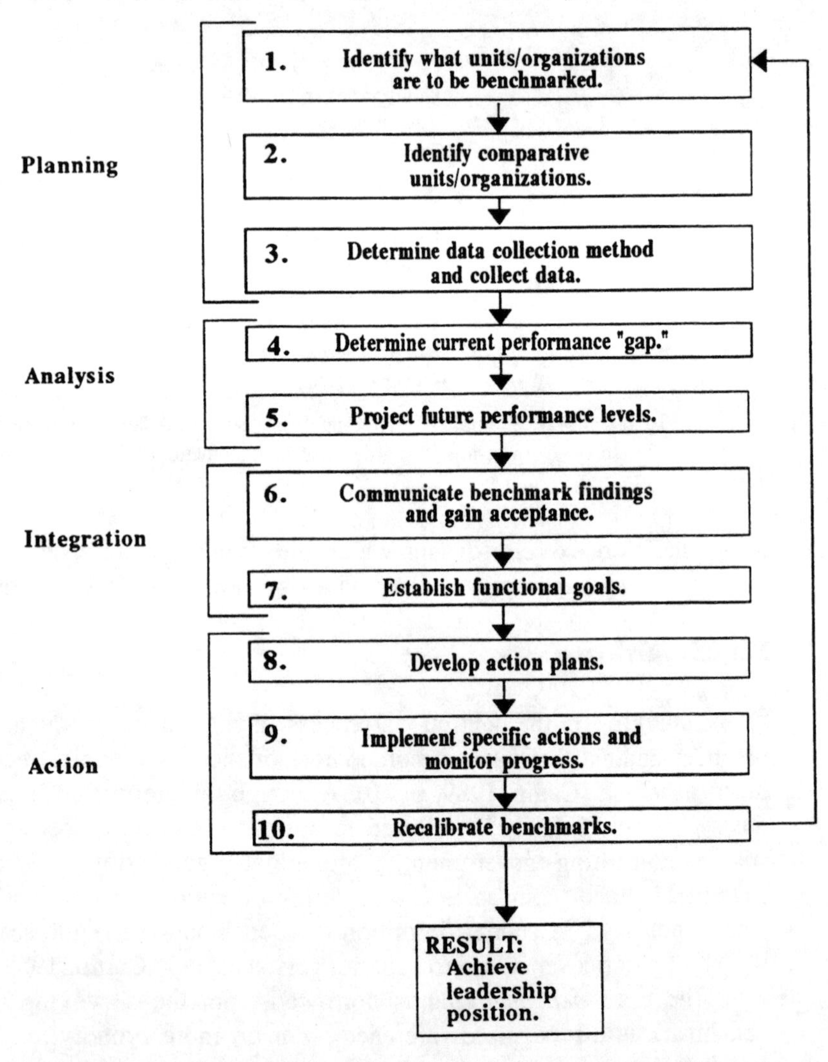

Figure 7–5 Benchmarking process (Source: Adapted with permission from F. C. Camp, *Benchmarking: The Search for Industry Best Practices that Lead to Superior Performance,* Milwaukee, WI: ASQC Quality Press, 1989, p. 17.

an engineering consulting firm's top management is dissatisfied with its marketing group's ability to bring new clients to the firm. One approach is to study a competing consulting firm's techniques. An alternative is to study the new client prospecting techniques of the marketing operation in a different kind of business on the premise that there are sufficient similarities between the two businesses to provide new and valuable insights to the organization doing the benchmarking (Camp, 1989, pp. 60–65). As a specific example of this transferability process, in 1982 Xerox studied and learned from the product distribution processes used by the L. L. Bean Company (Swanson, 1992).

- Step 3. Possible approaches to data acquisition include reading reports published by the organization or the "industry," asking the organization for data and information, and/or retaining a consultant. Incidently, Swanson (1992) says that the data acquisition step in benchmarking always involves the active cooperation of one or more other organizations. This view is not, however, widely held.

- Steps 4 and 5. Is the organization being used as a benchmark really doing better with respect to the product, service, or practice under study? How much better and why and what changes should be made in the organization doing the benchmarking and what will they accomplish?

- Steps 6 and 7. Convince others within the benchmarking organization that there is a serious "gap" and that it can be closed.

- Steps 8, 9, and 10. Implement changes, monitor results, and in keeping with the continuous improvement theme of TQM, iterate; that is, start over with the benefit of new experiences and information.

Partnering

Recall the emphasis on teamwork in the "Principles of TQM" section of this chapter. Partnering is a recent U.S. movement which provides a highly structured process for the formation and functioning of multidiscipline teams. Williams (June, 1994) indicates that the mission of partnering is to:

> *. . . develop a proactive effort and spirit of respect, trust and cooperation among all key players in a contractual relationship or between cross-functional interactive divisions within an organization. It utilizes a structured systematic methodology for developing a spirit of teamwork and cooperation through shared goals, open communication, problem identification and resolution, conflict escalation procedures and the monitoring of team performance.*

Harback et al. (1994) indicates that partnering involves " . . . profound changes in the way business is approached and conducted." They go on to say:

> *Partnering allows us to move from adversarial relations to cooperative teamwork, a win–lose strategy to a win–win plan, a stressful project to a satisfying one, a litigation focus to solutions and accomplishments, and finger pointing to a handshake mind-set; it also lets bureaucratic inertia dissolve and risk-taking be endorsed.*

In the manufacturing sector, a partnering team might be composed of the owner, consulting engineer, suppliers, and contractors. In the municipal market typically served by consulting engineers, the partnering team could consist of the owner, the consulting engineering firm, and the contractor. In each of the preceding cases, the owner would have multirepresentation to include such diverse functions as procurement, operations, and maintenance.

One of the principles of partnering is very early involvement by individuals and entities who will eventually be involved anyway in the project or its results. Another way of stating this is that instead of having planners doing only planning, designers doing only designing, contractors doing only constructing, operators doing only operating, and maintainers doing only maintaining, everybody is involved in planning and design, the two project phases that have the heaviest influence on life cycle costs of the product, structure, facility, or system.

Once team members are identified and have been provided with contract documents (or at least a scope of work) and other background materials, a team workshop is conducted. Typically, the workshop is of one to two days' duration and held at a location at which there will be minimal interruptions. Workshops are usually conducted with the assistance of a facilitator, who guides team members through a process but does not make decisions or comments for them. A partnering workshop might include the following six elements (quoted from Williams, June 1994):

- *Learn to know one another—develop relationships.*
- *Develop mission statement, goals, and objectives.*
- *Identify problems, issues, or opportunities for the project.*
- *Develop problem resolution/escalation process.*
- *Develop evaluation process.*
- *Sign charter.*

On the basis of the trust and process foundation laid at the workshop, team members are enabled to resolve problems and seize opportunities as soon as they appear and build a mutually beneficial project. Examples of significant dramatic improvements in quality attributed to partnering are provided by CII (1993, pp. 22–34), Harback et al. (1994) and Williams (April 1994). Cited accomplishments include improved safety during construction, reduced costs to owner, downsizing of owner's staff, increased profitability for consultants and contractors, shortened design and construction schedules, and reductions in disputes and claims. Partnering costs, such as the up-front workshop, are usually greatly offset by monetary savings during the project.

Stakeholder Input

As noted earlier in this chapter, you or your organization's stakeholders are those internal or external individuals or organizations that have a significant interest in what you and your organization do. Your actions or inactions affect them. The best way to find out how well you are doing is to go to the source—ask your stakeholders. Your organization would probably use formal surveys to do this, as discussed in Chapter 6 in the section titled "Client Input." As an entry-level professional, you can informally query your internal and external stakeholders, especially the former, to determine how you are doing and how you can improve.

RESULTS OF TQM

Assume that you, as an individual, or your organization, view TQM as something new that may warrant implementation. A logical question is "What is the evidence that TQM yields results, particularly in light of the apparent major effort, including profound behavior change, that would be required to implement it?" The best that one can do is be aware of examples of successes that are said to be attributed to TQM and determine if some might be related to your operations. Some examples of successful application of TQM are

- Using TQM, as discussed in this chapter, the proposal staff of Paramax Electronics, Inc. of Montreal, Quebec, increased the average number of proposals produced per month from 2.2 to 3.8. How? They reduced duplication of paperwork, the number of steps in the process, and the number of feedback loops (Unisys, 1990).

- Xerox copy machine sales dropped to 10 percent of the market in 1984. Then Xerox moved into TQM. "Xerox eventually re-organized its entire work force of 1,000 employees into quality teams, spending some $1,300 per person for training. Since then, the company's share of the copier market jumped to nearly 15%; its defect rate dropped to 93% of pre-TQM levels. In 1988, quality teams saved Xerox a reported $116 million by reducing scrap, tightening production schedules, and other measures" (Gill, 1990).

• Using TQM, Oregon State University reduced the average duration of campus remodeling jobs by 23 percent and reduced grant and contract process time by 10 percent (Bemowski, 1991).

• Cadillac, as a result of TQM, reduced warranty costs by 30 percent from 1986 to about 1991 (Perry, 1991, p. 3).

• As a result of TQM, IBM reduced the duration of its software production cycle from four or five years to two or three years (Perry, 1991).

Incidentally, Lewis notes that, because TQM seeks participation by everyone, small organizations probably have an advantage over large organizations in implementing TQM.

MALCOLM BALDRIGE NATIONAL QUALITY AWARD

The Malcolm Baldrige National Quality Award is ". . . an annual award to recognize U.S. companies which excel in quality achievement and quality management." Award criteria are leadership, information and analysis, strategic quality planning, human resources development and management, management of process quality, quality and operational results, and customer focus and satisfaction. Awardees have been in small business, manufacturing, and service categories (U. S. Department of Commerce, 1993). Recipients of this highly valued award, which was first presented in 1988, are

- 1988: Globe Metallurgical, Motorola, and Westinghouse Commercial Nuclear Fuel Division.
- 1989: Milliken & Company and Xerox Business Products and Systems.
- 1990: Cadillac Motor Car Company, IBM Rochester, Federal Express Corporation, and Wallace Company.
- 1991: Marlow Industries, Solectron Corporation, and Zytec Corporation.
- 1992: AT&T Network Systems Group Transmission Systems Business Unit, AT&T Universal Card Services, Granite Rock Company, Ritz-Carlton Hotel Company, and Texas Instruments Defense Systems and Electronics Group.
- 1993: Ames Rubber Corporation and Eastman Chemical Company.

CLOSING THOUGHTS

Whether you view TQM as a new management movement, at least in the United States and some other countries, or as a new name for old, tried and true management principles and practices, you are urged to embody TQM's substantive features as you begin your professional career. These are

- Always striving to understand and meet requirements. Listen, ask, and seek to serve.
- Being aware of and responsive to all the individuals and organizations—internal and external stakeholders—having an interest in the products or services you produce.
- Constantly seeking ways to understand and improve the processes used by you, and those you interact with, as you do your work. Apply appropriate monitoring and diagnostic tools and techniques.
- Expecting and enabling everyone to contribute to the organization's efforts.

REFERENCES

American Society of Civil Engineers, *Quality in the Constructed Project—A Guideline for Owners, Designers and Constructors—Volume I*, 1990.

American Society for Engineering Education, Task Force on Quality Improvement in Engineering Education, *Report of Task Force*, May 1993.

BEMOWSKI, K., "Restoring the Pillars of Higher Education," *Quality Progress*, October 1991, pp. 37–42.

CAMP, R. C., *Benchmarking: The Search for Industry Best Practices that Lead to Superior Performance*, Milwaukee, Wis.: ASQC Quality Press, 1989.

Construction Industry Institute, "Team Building: Improving Project Performance," University of Texas at Austin, Publication 37–1, July 1993.

COOK, B. M., "Quality: The Pioneers Survey the Landscape," *Industry Week*, October 21, 1991, pp. 68–73.

CROSBY, P. B., *Quality Is Free: The Art of Making Quality Certain*, New York: Mentor Books, 1979.

DAVIS, K., and W. B. LEDBETTER, "What Is It and How Do You Get It?," Forum, *Civil Engineering*, July 1988.

GILL, M. S., "Stalking Six Sigma," *Business Month*, January 1990, pp. 42–46.

HARBACK, H. F., D. L. BASHAM, and R. E. BUHTS, "Partnering Paradigm," *Journal of Management in Engineering—ASCE*, Vol. 10, No. 1 (January/February 1994), pp. 23–27.

HARRISBERGER, L., *Engineersmanship . . . The Doing of Engineering Design*, second ed, Belmont, Cal.: Brooks/Cole Engineering Division, Wadsworth, Inc., 1982.

HAYDEN W. M. JR., "Reducing Quality Control Costs," Management Forum, *Journal of Management in Engineering—ASCE*, Vol. 6, No. 2 (April 1990), pp. 139–140.

HENSEY, M., "Essential Tools of Total Quality Management," *Journal of Management in Engineering—ASCE*, Vol. 9, No. 4 (October 1993), pp. 329–339.

HUNTINGTON, C. G., "A Craftsman's Obsession," Forum, *Civil Engineering*, February 1989, p. 6.

LEWIS, W. D., "Quest for Quality Begins By Asking, 'What Is It?'," *American Consulting Engineer*, pp. 12–13.

MODIC, S. J., "What Makes Deming Run?," *Industry Week*, June 20, 1988, pp. 84–91.

NEUFELDT, V., ed., *Webster's New World Dictionary of American English*, third college ed., Englewood Cliffs, N.J.: Prentice Hall, 1994.

Paramax Systems Corporation, "Total Quality Management," 1992.

PERRY, W. E., ed., *National Quality Award Newsletter*, Vol. 1, No. 1, 1991.

PIRSIG, R. M., *Zen and the Art of Motorcycle Maintenance*, New York: Bantam Books, 1974.

SIMON, R. C., "Total Quality Management: A Formula for Success," *American Consulting Engineer* (Date unknown), pp. 15–21.

SNYDER, J., *Marketing Strategies for Engineers*. New York: American Society of Civil Engineers, 1993.

SWANSON, R. C., "Benchmarking: Search for 'Best' Practices," *The Journal of Applied Manufacturing Systems*, Winter 1992, pp. 37–43.

U. S. Department of Commerce, "Malcolm Baldrige National Quality Award—1994 Award Criteria," 1993.

Unisys Corporation, "Total Quality Management," 1990.

WILLIAMS, R. C., "The Partnering Process," prepared for the Northwest Indiana Business Roundtable, June 16, 1994.

WILLIAMS, R. C., "Partnering Successes in Arizona's Transportation Industry," Construction Group, Arizona Department of Highway, April 1994.

YATES, R., "Game Plan—On the Road With the Messiah of Management," *Chicago Tribune Magazine*, February 16, 1992, pp. 14–22.

SUPPLEMENTAL REFERENCES

BREITENBERG, M., "ISO 9000—Questions and Answers on Quality, the ISO 9000 Standard Series, Quality System Registration, and Related Issues," NISTIR 4721, U. S. Department of Commerce, July 1992.

CULP, G., A. SMITH, and J. ABBOT, "Implementing TQM in Consulting Engineering

Firm," *Journal of Management in Engineering—ASCE*, Vol. 9, No. 4 (October 1993), pp. 340–356.

DEFFENBAUGH, R. L., "Total Quality Management at Construction Jobsites," *Journal of Management in Engineering—ASCE*, Vol. 9, No. 4 (October 1993), pp. 382–389.

HAYDEN, W. M., JR., "Management's Fatal Flaw: TQM Obstacle," *Journal of Management in Engineering—ASCE*, Vol. 8, No. 2 (April 1992), pp. 122–129.

HIAM, A., "Does Quality Work? A Review of Relevant Studies," The Conference Board, Report No. 1043, New York, 1993.

UNDERHILL, B. N., "Total Quality Management—Another Fad?" *The Journal of Applied Manufacturing Systems*, University of St. Thomas, St. Paul, Minn., Vol. 6, No. 1 (Fall 1993), pp. 40–41.

EXERCISE

APPLY TQM TOOLS AND TECHNIQUES

Purpose

Provide an opportunity to find a solution to a real problem using TQM tools and techniques.

Tasks

1. Select a real problem faced by you or an organization you work for or belong to. The problem does not have to be an engineering or technical problem, and it does not have to be major. It does have to be real and worth solving.
2. Apply one or more of the TQM tools and techniques described in this chapter or that you know of or learn about from other sources and identify one or more solution steps.
3. If feasible, implement, or start to implement, the solution. However, this task is not necessary for the successful completion of this exercise.
4. Prepare a memorandum that presents the problem selected under task 1 and describes the method(s) used and results obtained in task 2. If task 3 is applicable, include a summary in the memorandum.

8

Decision Economics

> *... mere parsimony [stinginess] is not economy ... Expense, and great expense, may be an essential part in true economy ... Economy is a distributing virtue, and consists, not in saving, but in selection. Parsimony requires no providence, no sagacity, no powers of combination, no comparison, no judgement. Mere instinct, and that not an instinct of the noblest kind, may produce this false economy in perfection. The other economy has larger views. It demands a discriminating judgement, and a firm, sagacious mind.*
>
> (Burke, 1795, pp. 564–565)

Engineers and other technical professionals help others—other professionals, clients, and customers—make choices and decisions. Private and public resources are limited and should be invested prudently. The technical work is not an end in itself, but provides input, in combination with the technical professional's experience, to the choosing and decision-making process.

Figure 8–1 shows the various decision steps that might be required for evaluating alternative highway alignments, a new integrated work station system for an engineering office, or improvements to a manufacturing process. The fourth step in the process is determination of economic feasibility. The analytic tools typically used to complete this step are engineering economics, also called decision economics. Decision or engineering economics is the application of economic criteria to help select the best of a group of technically feasible alternatives. Decision economics permits economic comparisons. It facilitates bringing costs and benefits occurring at various points in time to a common basis. The use of decision economics was led by Arthur M. Wellington, who used it to compare alternative railroad routes and published a book on the subject in 1877 (James

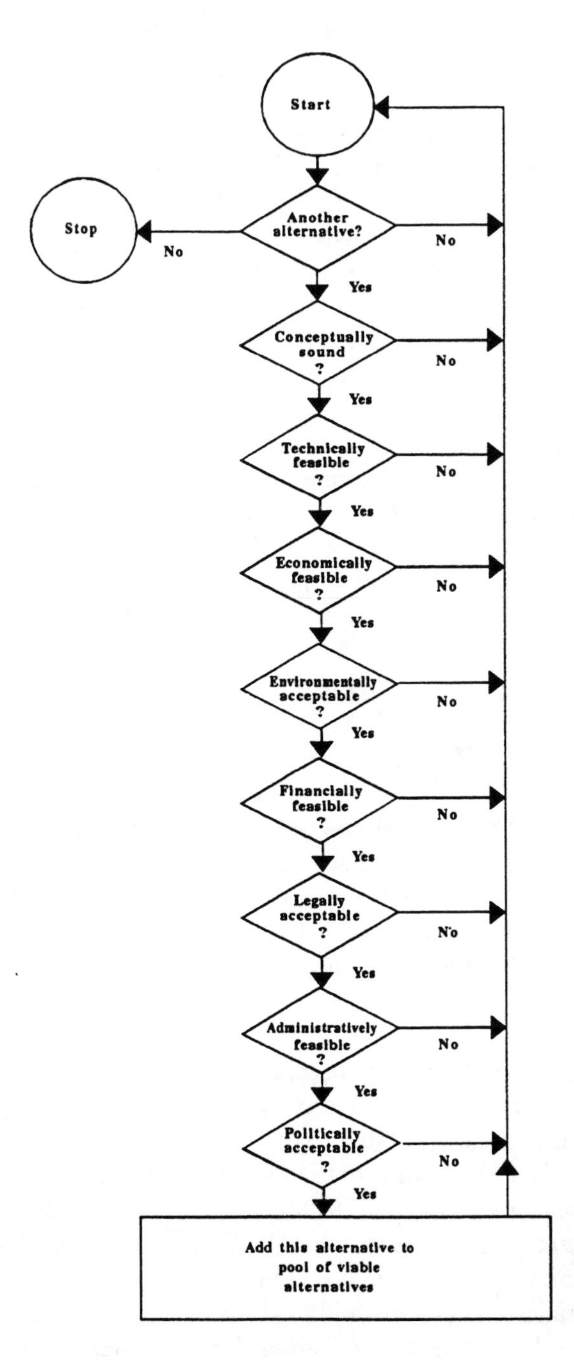

Figure 8–1 Decision process (Source: Adapted with permission from S. G. Walesh, *Urban Surface Water Management*, New York: John Wiley, 1989, p. 474.)

and Lee, 1971, p. 1). While decision economics is an important step in the decision-making process, it is, as illustrated by Figure 8–1, only one of many important steps.

You, as an entry-level technical professional, should understand that a primary function of your work is to help others make choices and decisions. Furthermore, you should know how to use decision economics—a highly quantitative and important step in the decision-making process. Finally, these economic tools will also help you manage your personal financial affairs.

The broad, that is, professional and personal, applications of decision-economics principles are introduced in this chapter followed by a discussion of the distinction between economic analysis and financial analysis. A six-step model of decision-economics analysis is presented followed by a presentation of discounting factors. Benefit-cost analysis is discussed and illustrated with examples. Sensitivity of benefits and costs to interest rate and to period of analysis is presented. The importance of determining the sensitivity of cost of products, structures, facilities, or systems to load, capacity, or other measures of service is emphasized. A discussion of rate of return or return on investment concludes the chapter.

BROAD APPLICABILITY OF DECISION-ECONOMICS TOOLS

Besides being a powerful tool in professional technical work, decision economics uses fundamental principles that are applicable to many facets of nonprofessional and personal life. For example, the young engineer may be arranging financing for a new automobile and want to understand and confirm the accuracy of the agreed-upon monthly payment. Or the young architect may be trying to determine which of two or more alternative price, interest rate, and finance period combinations is most advantageous. Several years into his or her career, the young computer professional may be considering the pros and cons of prepaying an automobile loan or evaluating the consequences of refinancing a mortgage on a residence because of a dip in interest rates.

Decision economics is a powerful tool to use in the area of personal finance. Stated differently, the concepts and equations used in decision economics are the same as those used in banking and other areas of business and personal monetary affairs. Decision economics can give the young professional the tools needed to make prudent personal finance decisions and to verify finance terms determined by others. Never assume that finance calculations done by others are correct—use the tools presented in this chapter to check those calculations.

DISTINCTION BETWEEN ECONOMIC ANALYSIS AND FINANCIAL ANALYSIS

The decision process presented as Figure 8–1 includes a determination, as already noted, of economic feasibility and also includes a determination of financial feasibility. Economic analysis and financial analysis are related but different.

Economic feasibility means that benefits, in whatever quantitative or qualitative way they are determined, exceed costs. Financial feasibility means that individuals or organizations are willing and able to expend funds. Economic feasibility does not necessarily mean financial feasibility. For example, the benefits of an engineering alternative may exceed the costs, but there may be a very small likelihood of being able to finance the alternative because of other factors. Or, at the personal level, assume that you are provided an opportunity to invest in a bona fide venture that promises a high return, but that you must raise $10,000 by the end of the business day. The venture would be economically feasible, but probably not financially feasible.

Closer to the technical professions, a manufacturing firm may be examining a new computer hardware and software system. The economic analysis may be clearly positive in that it shows that benefits will probably exceed the cost over the life of the new system. However, decision makers in the organization may not be willing to risk financing the proposal—they want to be 100% sure. Therefore, the proposal is not financially feasible, at least as perceived by them.

A large, regional flood control facility planned to serve several communities may be conceptually sound, technically and economically feasible and environmentally acceptable. However, the proposal may fail because of financial feasibility driven by the low likelihood of communities being able to work together to raise the capital needed to construct the facility and to then commit to joint efforts in operating and maintaining the facility.

Many proposed products, structures, facilities, and systems that are clearly economically feasible often fail to be implemented because of financial infeasibility. As suggested by Figure 8–1, economically feasible proposals may also fail because they are unacceptable for environmental, legal, administrative, or political reasons.

STEPS IN A DECISION-ECONOMICS ANALYSIS

Refer again to Figure 8–1, noting the position of the determination of economic feasibility in the overall decision process. This section of the chapter presents a six-step process for determining economic feasibility. Consider a hypothetical situation in which three fundamentally different types of structures are being considered for a bridge across a river. More specifically, note the concrete, steel, and timber alternatives illustrated in Figure 8–2. Assume that all three alternatives are conceptually sound and technically feasible as determined by the preliminary design process. Assume further that construction costs and ongoing operation and maintenance expenses have been estimated. Therefore, the six-step process of determining economic feasibility can begin.

Determine Physical and Economic Lives of Project Components

The physical life of a structure or any of its components is defined as the time over which the component could perform at least some of its intended functions, assuming reasonable, but not extreme care. Examples of physical lives of structures or structure components encountered primarily in civil engineering are presented in Table 8–1. Economic life, which is the primary concern in economic analysis, is the period of time during which incremental benefits of use are likely to exceed incremental costs. Or stated differently, the economic life of a component ends when the incremental benefits of use become less than the incremental costs.

Clearly, determinations of physical life and economic life are judgments. Generally, the economic life of an engineered product, structure, facility, or system

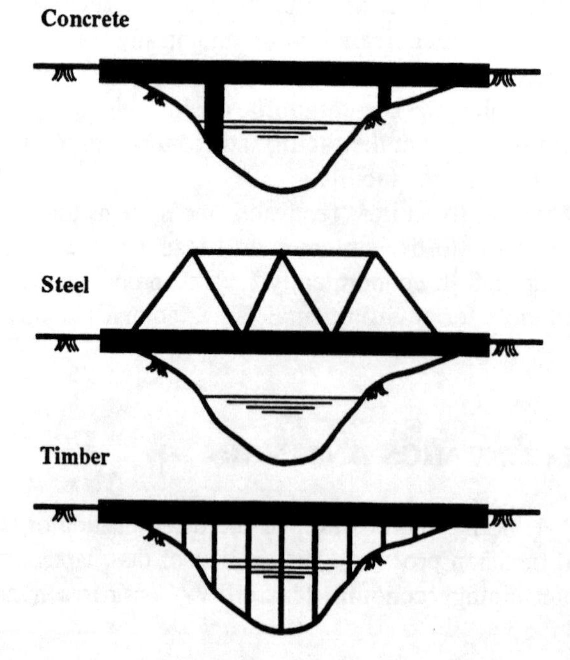

Figure 8–2 Alternative bridges

TABLE 8–1 EXAMPLES OF PHYSICAL LIVES

Item	Years
Canals and ditches	75
Coagulation basins	50
Construction equipment	5
Flumes	
concrete	75
steel	50
wood	25
Nuclear power plants	20
Tanks	
concrete	50
steel	40
wood	20
Wells	45

(SOURCE: Adapted with permission from Linsley, R.K. and Franzini, J.B., *Water Resources Engineering,* Third Edition, New York: McGraw-Hill, 1972. p. 381).

is equal to or less than its physical life. For example, tanks, pipes, and other components in a wastewater treatment plant may have physical lives of up to 50 years. But given the rate of change of treatment technology, significant improvements are likely to occur in the time span of much less than 50 years. Therefore, owners will want to or will be required to implement those improvements. Accordingly, prudence suggests an economic life for such components of much less than 50 years.

One reason that physical life is important in the determination of economic feasibility is that salvage value can be a significant aspect of economic analysis, as illustrated later in this chapter. Salvage value is that monetary worth assigned to a component at the end of its economic life, but before the conclusion of its physical life.

Diagram Revenue and Construction, Manufacturing, Replacement, and Operation and Maintenance Expenditures

Figure 8–3 illustrates a cash flow diagram like that typically used in determination of economic feasibility. If a potential project is revenue-producing, then projected revenue is included in the diagram. This concept is illustrated for a hypothetical hydroelectric, fossil fuel, or nuclear power plant in Figure 8–4. Note the convention used in determining the direction of arrows. Expenditures or outflow is shown by arrows pointing down, whereas income, revenue, or inflow is shown

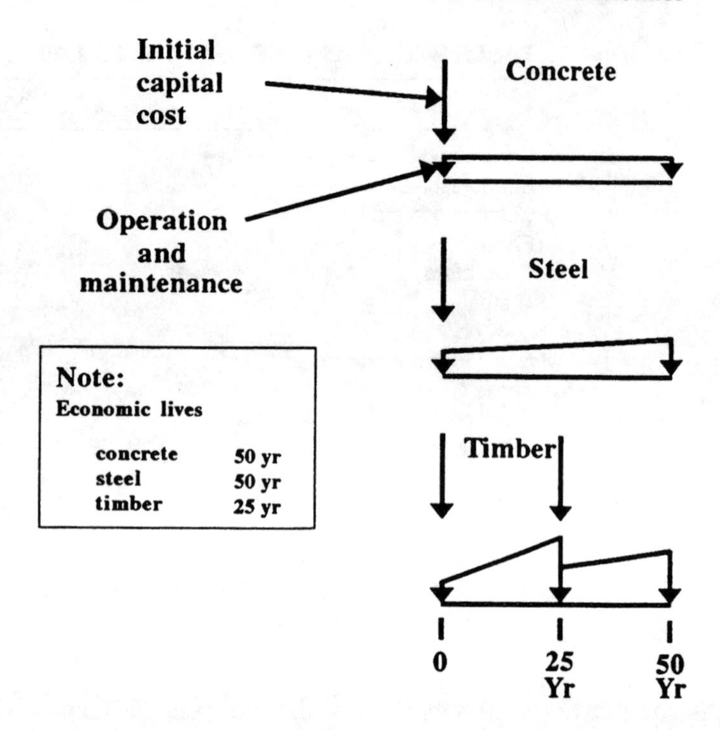

Figure 8–3 Diagram of projected capital and operational/maintenance costs for alternative bridges

by arrows pointing up. While a formal diagram may not always be needed or always appear to be needed, its use is highly recommended so that the situation is clearly understood and so that the discount factors (discussed later in this chapter) are correctly applied. Experience indicates that misrepresentation of cash flow is a major cause of errors in decision-economics analyses.

The total period of analysis is selected so that it is equal to or full multiples of economic lives of all alternatives. Although the timber bridge alternative has an economic life of 25 years compared to the 50-year economic lives of the concrete and steel bridge alternatives, the timber bridge cash flow diagram is extended to 50 years so that the period of analysis for all three alternatives is identical, that is, 50 years. Physically, the timber bridge would be built and used and then rebuilt and used during the 50-year analysis period, whereas the concrete and steel bridges would each be constructed once. If, for example, the economic lives of the concrete, steel, and timber bridge alternatives were, respectively, 40 years, 30 years, and 20 years, then the total period of analysis would, strictly speaking, have to be 120 years because 120 years is the smallest period of analysis that contains full multiples of the economic lives of all three alternatives.

The diagrams presented in Figures 8–3 and 8–4 suggest a comparison prob-

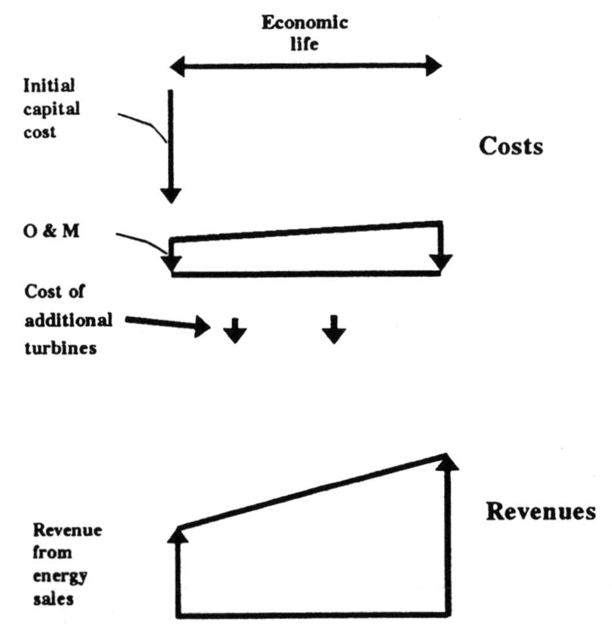

Net = ?

Figure 8–4 Diagram of projected costs and revenues for a power plant

lem in that a dollar spent or received today, that is, at time zero, does not have the same value as a dollar spent or received at some point in the future. The comparison problem is illustrated in Figure 8–5. Assume that you win the state lottery and the prize is $100,000. What is better—to receive $100,000 now or $100,000 two years from now, as illustrated in Figure 8–5. Obviously, you would select the option of taking the $100,000 now, because you could invest or otherwise work

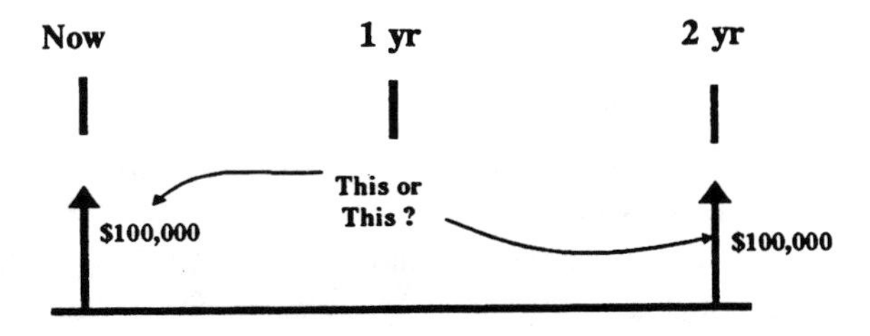

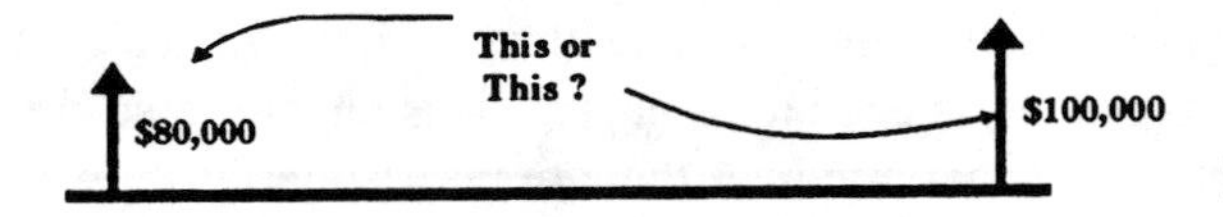

Figure 8–5 Time value of money

with the funds so that two years from now the value could be significantly greater than $100,000. But what if the choice were to accept $80,000 now or $100,000 two years from now, as also illustrated in Figure 8–5? Now the best decision is not obvious—it depends on how effective you would be in investing the $80,000 received today.

Similarly, the cost of an engineered product, structure, facility, or system relative to alternatives depends on when the cost is incurred—not just the magnitude of the cost. Likewise, the monetary benefits of alternatives depend on when the benefits occur—not just their magnitude. Generally, the value of a monetary transaction is determined by its absolute value and when it occurs. Cost and benefits have absolute and time value. Discounting factors, discussed in the next section of this chapter, account for the time value of monetary transactions.

Clearly an accurate analysis of the total cost of an engineered product, structure, facility, or system requires inclusion of more than the initial cost. Nevertheless, there is a tendency to focus too much or even exclusively on initial cost. Wolochuk (1988) refers to this as the "first cost syndrome" and argues, as does this chapter, for applying "lifetime cost considerations in comparing engineering alternatives." Novick (1991) also advocates lifetime economics analysis, noting that maintenance and rehabilitation costs within the public works sector can be large compared to initial costs. Similarly, when benefits can be quantified in monetary terms, a lifetime approach should be taken.

Select Interest Rate

Interest rate is used in accounting as an index of the time value of costs and benefits. Interest rate is defined as the amount (i, stated in percent) earned or charged per unit of time (one year unless stated otherwise) with n indicating the number of time units per unit of principal (P, amount spent, invested, earned, borrowed, etc.). Again, always assume that a given interest rate is for a one-year period unless explicitly stated otherwise. This convention is used within this chapter, in the exercises at the end of this chapter, and in engineering and business practice.

Selecting the exact interest rate to be used in a decision-economics analysis is a matter of judgment. However, that judgment should be guided by common practice. Consider using a rate that approximates the owner's cost of borrowing capital. James and Lee (1971, p. 126) suggest the preceding plus consideration of the current interest rate for risk-free investments such as government bonds. For a private entity's project, James and Lee suggest using an interest rate no smaller than the organization's marginal internal rate of return.

The consequences of selecting interest rates that are unrealistically high or low are discussed and illustrated later in this chapter. This chapter also discusses the consequences of selecting economic lives that are unreasonably short or long.

Conclusion: Within reasonable interest rate ranges and economic life ranges, the results of decision-economics analyses are much more sensitive to interest rate than to economic life. Therefore, give much more attention to interest rate determination than to establishing economic life.

Inflation is typically not explicitly accounted for in a decision-economics analysis. James and Lee (1971, p. 209) state: "Trends and general price levels should never be incorporated into economic analysis." They go on (p. 512) to note that differential inflation in the value of particular goods and services relative to general price level should be reflected if it is known. One reason for not trying to account for, or more specifically quantify, inflation is that federal policy is intended to control inflation. Second, and more importantly, inflation is very difficult to predict. Furthermore, Howe (1971, p. 81) argues that inflation, if used, affects benefits about the same as costs and, therefore, there is no significant net effect.

Put Costs and Benefits on a Comparable Basis and Calculate Benefit–Cost Ratio or at Least Cost

All costs and, as appropriate, benefits are placed on a comparable basis by taking into account the time value of money. Discounting factors, discussed in the next section of this chapter, are used to take into account the time value of money. Further discussion of putting costs and benefits on a comparable basis is deferred until after the discount factors are introduced. Once costs and benefits are on a comparable basis, then the economically most feasible alternatives can be identified by examining their costs or their benefit-to-cost ratios.

Consider Intangible Benefits and Costs

As used here, intangible means that a benefit or cost is important but not quantifiable in monetary terms. For example, the appearance of a manufactured product is clearly part of its intangible benefit. Many individuals would assign a higher aesthetic quality to a timber bridge crossing a stream in a natural setting than to a steel bridge in the same situation. Another example of an intangible benefit would be the preservation of a historic structure such as an old mill as part of a flood control project.

The difficulty presented by absences of quantification is not an excuse for ignoring intangible factors. Lest there be any misunderstanding, although the conduct of a decision-economics analysis is often called for in a technical project, the results of an analysis are rarely the only determinate of the course of action. Figure 8–1 clearly illustrates many other factors in the decision process. As an entry-level engineer or other technical professional who has just completed a technical and highly quantitative undergraduate education experience, you may be predisposed to focus on the more quantitative steps in the decision process illustrated in Figure 8–1. Prudence and experience suggest a balanced approach.

Recommend Best Alternative

Informed by the results of the decision economics analysis and by consideration of the many other factors in the process, the technical professional in consultation with others recommends a course of action to the client or owner. The client or owner's receptivity to the recommendations will depend, in part, on the perceived thoroughness of the economic and other analyses. But other forces will be at work, not the least of which will be the lead professional's personal reputation, which, as suggested by the discussion of this subject in Chapter 2, helps determine the client or owner's receptivity to the professional's recommendation. That receptivity will also be influenced by the technical professional's ability to communicate his or her recommendations and the basis for them. Development of effective communication skills is discussed in Chapter 3.

DISCOUNTING FACTORS

The fourth step in the previously presented five-step decision-economics analysis calls for placing costs on a comparable basis. As noted earlier, discounting factors facilitate this process. More specifically, discounting factors are used to convert a set of discrete and continuous costs and revenues to a common point in time or to a common period. Discounting factors account for the time value of money and enable the professional to "compare apples to apples." Eight discounting factors are derived, or at least presented, in this section of the chapter and at least one example is provided for each factor.

To better understand the need for and power of discounting factors, refer again to Figure 8–4 which shows various discrete and continuous costs and revenues projected for a power plant. In order to determine the relative magnitude of costs and benefits and to answer the fundamental question "Will revenues exceed costs?" the costs and revenues could each be brought to time zero by using discounting factors and then compared. Another approach would be to use discounting factors to convert all costs and revenues to equivalent annual uniform costs and revenues and compare them. A third approach would be to use discounting factors to carry all costs and revenues out to the end of the economic life of the project and compare them. The number of ways in which discounting factors could be used to put costs and revenues on a comparable basis is endless. Interestingly, regardless of which way is used, the ratio of revenue to cost will be unchanged.

The preceding observation about the constant revenue to cost ratio suggests a convenient method for checking the mathematics in decision-economics calculations. The approach is to do each calculation twice. For example, first do the calculations on a present-worth basis and then do the calculations on an annualized basis. The ratios of revenues to costs should be identical. If not, a mathematical or logic error occurred.

Single-Payment Simple-Interest Factor

Refer to Figure 8–6 for the derivation of the equation $F = P(1 + ni)$. The expression $(1 + ni)$ is the single-payment simple-interest factor.

The cash flow diagram presented in Figure 8–6 is explicitly linked to the resulting derived equation. If, for example, the present sum had been depicted as

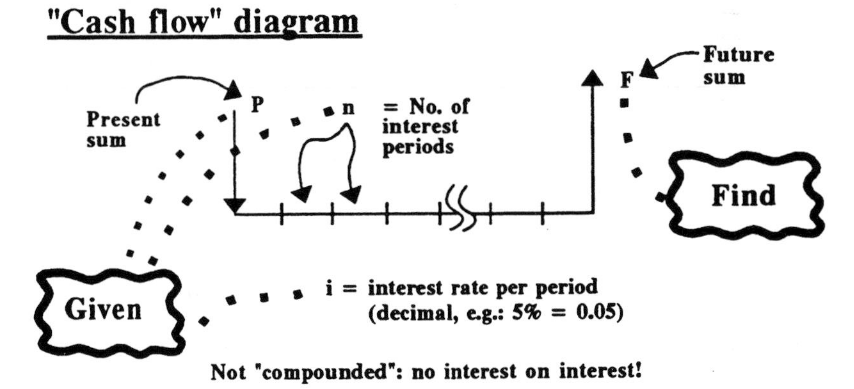

Derivation of F = f(P, n, i)

Period	Amount at beginning of period	Amount earned during period	Total amount at end of period
1	P	Pi	P+Pi = P(1+i)
2	P(1+i)	Pi	P(1+2i)
3	P(1+2i)	Pi	P(1+3i)
↓	↓	↓	↓
n	P(1+(n-1)i)	Pi	P(1+ni)

Figure 8–6 Single-payment simple-interest factor

occurring at the end of the first of n periods or at the mid-point of the first period, a different equation would have resulted. Similarly, the cash flow diagrams presented for the other seven discounting factors are explicitly tied to the resulting equations. Discounting factors should always be used with full knowledge of the cash flow diagrams used to derive them. To do otherwise is to risk illogical applications of one or more discounting factors. As suggested earlier, you are much more likely to understand and correctly solve a decision-economics problem if you first construct the cash flow diagram as a model of the problem. One reason for this observation is that by constructing the diagram you will be less likely to misuse the various discounting factors typically required to solve a problem.

Example 1

Given: $10,000 is invested at 6% for four years. Earnings occur at simple interest with a single payment at the end of the four-year period.

Find: Value of the single payment, that is, original principal plus interest earned, at the end of four years.

Solution: $F = P(1 + ni) = (10{,}000)(1.0 + (4 \times 0.06)) = (10{,}000)\ (1.24) =$ $12,400.

Note that the total interest earned is $2,400 or exactly $600 per year. In simple-interest transactions, as noted in the derivation presented in Figure 8–6, the investor does not earn interest on interest.

Single-Payment Compound-Amount Factor

Figure 8–7 presents the derivation of the equation $F = P(1 + i)^n = P(F/P\ i,n)$ where $(F/P\ i,n)$ is the single-payment compound-amount factor using the "Thuesen" notation (Thuesen et al., 1971). The key word in comparing this single-payment compound amount factor to the previously derived single-payment simple-interest factor is compound. As indicated in Figure 8–7, interest is earned on interest in the case of the single-payment compound-amount factor. As will be illustrated by some of the following examples, compounding can have a significant effect on the magnitude of the future sum.

Example 2

Given: $10,000 is invested at 6% for four years. Interest is compounded annually and a single payment is to be made at the end of four years.

Find: Value of the single payment, that is, original principal plus total interest earned (including interest on interest) at the end of four years.

Solution: $F = P(1 + i)^n = 10{,}000(1.06)^4 = (10{,}000)(1.2625) = \$12{,}625.$

Note that the total interest earned is $2,625. This is $225 more than the $600 interest earned in the simple-interest version of this situation presented as Exam-

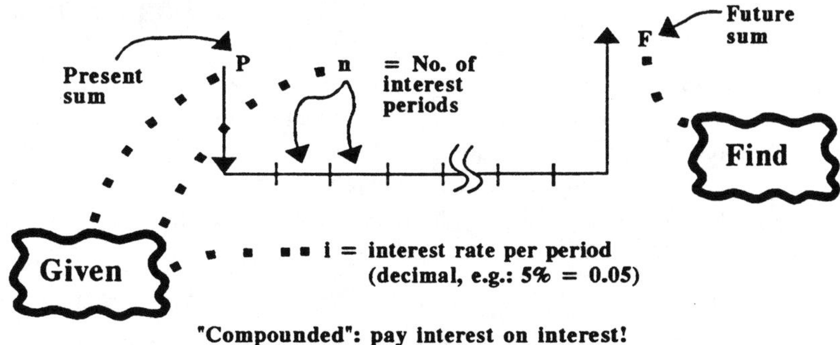

"Compounded": pay interest on interest!

Derivation of F = f(P, n, i)

Period	Amount at beginning of period	Amount earned during period	Total amount at end of period
1	P	Pi	$P+Pi=P(1+i)$
2	$P(1+i)$	$P(1+i)i$	$P(1+i)(1+i)=P(1+i)^2$
↓	↓	↓	↓
n	$P(1+i)^{n-1}$	$P(1+i)^{n-1}i$	$P(1+i)^n$

Figure 8–7 Single-payment compound-amount factor

ple 1. Obviously the future sum is greater when interest is earned on interest than when interest is earned just on the original investment.

Example 3

Given: $10,000 is invested at 6% for four years. Interest is compounded monthly and a single payment is to be made at the end of four years. Recall the convention set forth earlier in this chapter, which assumes that a given interest rate is for a one-year period unless explicitly stated otherwise. This convention has been used in Example 1 and Example 2, both of which state an interest rate, but neither of which explicitly indicates the period to which it applies. Therefore, the examples correctly assumed that the interest rate was for a one-year period. In this example, interest is to be compounded monthly.

Find: Value of the single payment, that is, original principal plus interest, at the end of four years with interest compounded monthly.

Solution: The normal practice is to calculate the monthly interest by dividing the annual interest by 12, that is, 0.06/12 = 0.005. Then calculate the future sum, using the single-payment compound-amount factor, that is, $F = 10{,}000(1.005)^{48} = (10{,}000)(1.2705) = \$12{,}705$.

The $2,705 interest earned in Example 3 for $i = 6\%$ and interest compounded monthly is $80.00 greater than the interest earned in Example 2 when $i = 6\%$ and interest is compounded only annually. In other words, for a given annual interest rate, earnings increase with the frequency of compounding. This effect is explored further in Example 4.

Example 4

Given: $10,000 is invested at 6% for four years. Interest is compounded daily, and a single payment is to be made at the end of four years.

Find: Value of the single payment at the end of four years.

Solution: Calculate daily interest as $i = 0.06/365 = 0.00016438$ and then compute $F = (10{,}000)(1.00016438)^{4\times365} = (10{,}000)(1.2711) = \$12{,}711$.

The interest earned when compounded daily is $2,711 or $6.00 more than when the same annual interest rate of 6% is compounded monthly, as previously computed in Example 3.

Example 5

Given: $10,000 is invested at 12% for four years. Interest is compounded annually, and a single payment is to be made at the end of four years.

Find: Value of the single payment at the end of four years.

Solution: $F = (10{,}000)(1.12)^4 = (10{,}000)(1.5735) = \$15{,}735$.

The earned interest of $5,735 is more than twice that earned in Example 2. This indicates that earned interest is a nonlinear function of interest rate in that doubling of interest rate more than doubles the interest earned.

Incidentally, unless clearly stated otherwise, compound interest is used in decision economics and in business transactions in general. Accordingly, you will have little use for the single-payment simple-interest factor introduced earlier in this chapter. The principal reason for introducing the single-payment simple-interest factor was to demonstrate how, all things equal, compounding produces more interest earnings than not compounding.

Example 6

Given: Money is invested at 6%.

Find: Period of time required to double the original investment.

Solution: Rearrange the single-payment compound-amount factor equation to obtain $F/P = (1 + i)^n$. Substituting $F/P = 2$ and $i = 0.06$, $2 = (1.06)^n$. Solving this, $n = 11.9$ years.

Obviously, the greater the interest rate, the shorter the doubling time. For example, at i = 12%, twice the rate used in Example 6, the period of time required to double the original investment is 6.1 years, slightly more than half that for Example 6.

Single-Payment Present-Worth Factor

Frequently you will need to calculate the present worth or value of some discrete monetary return to be received in the future or some discrete cost to be incurred in the future. For example, the salvage value of a production line at the end of its ten-year economic life has been estimated. What is it worth today? Or your community is building a pumping station that will require rebuilding at a cost of $150,000 in 15 years. What is today's liability?

See Figure 8–8 for development of the equation $P = F/(1 + i)^n = F(P/F\ i,n)$. The factor $(P/F\ i,n)$ is the reciprocal of the previously derived $(F/P\ i,n)$.

"Cash flow" diagram

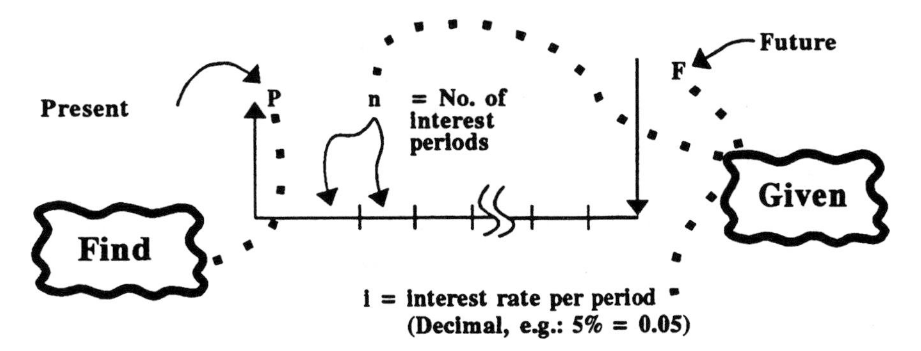

Derivation of P = f(F, n, i)

Because $F = P(1 + i)^n$

it follows that

$$P = F\left[\frac{1}{(1 + i)^n}\right]$$

Figure 8–8 Single-payment present-worth factor

Example 7

Given: As a result of a trust established for you, you will receive $75,000 in ten years.

Find: Present value of your trust on the assumption that $i = 6\%$.

Solution: $P = F/(1 + i)^n = (75{,}000)/(1.06)^{10} = (75{,}000)/(1.7908) = \$41{,}880$.

Another way of looking at this result is that $41,880 today has the same value as $75,000 ten years from now if $i = 6\%$. This observation and the Example 7 calculation could be verified by computing the value ten years from now of $41,880 invested at $i = 6\%$. This would be done, of course, using the already presented single-payment compound-amount factor ($F/P\ i,n$).

Series Compound-Amount Factor

The previously discussed factors relate to single or discrete expenses, costs, incomes, and revenues. Another class of transactions is those involving a series of equal costs or incomes. This class of transactions is referred to as *annuities;* examples of annuities are presented in Figure 8–9. Annuities are very common in professional work and in personal financial affairs. For example, as suggested by the installment-buying annuity illustrated in Figure 8–9, you want to purchase something now, but you do not have the cash. Therefore, you agree to participate in an annuity during which you will make payments at the end of each of a set number of periods. Your payments will gradually repay the original principal that you borrowed plus the interest on the ever-diminishing unpaid principal.

A sinking fund, as also illustrated in Figure 8–9, is another form of annuity. In this case, you anticipate a discrete expenditure at some point in the future and plan for it by making periodic payments, sized such that the sum of the payments plus the interest earned on payments made will equal the amount needed in the future. A retirement fund is another form of annuity, as suggested by Figure 8–9. In this case, during your working years you save in the form of an annuity and during your retirement years you withdraw in the form of an annuity.

A special set of equations is available to analyze annuities because, as suggested by the preceding, annuities are very common in engineering, business, and in personal finance. The previously derived single-payment equations could be used for annuity analysis, but the effort would be very cumbersome because of a need to repeatedly apply them.

Refer to Figure 8–10 for a portion of the derivation of the series compound-amount factor ($F/A\ i,n$). Exercise 8–1 at the end of this chapter involves completing the derivation of the equation $F = A((1 + i)^n - 1)/i = A(F/A\ i,n)$ where F is the future single value of a series of equal payments, that is, of an annuity.

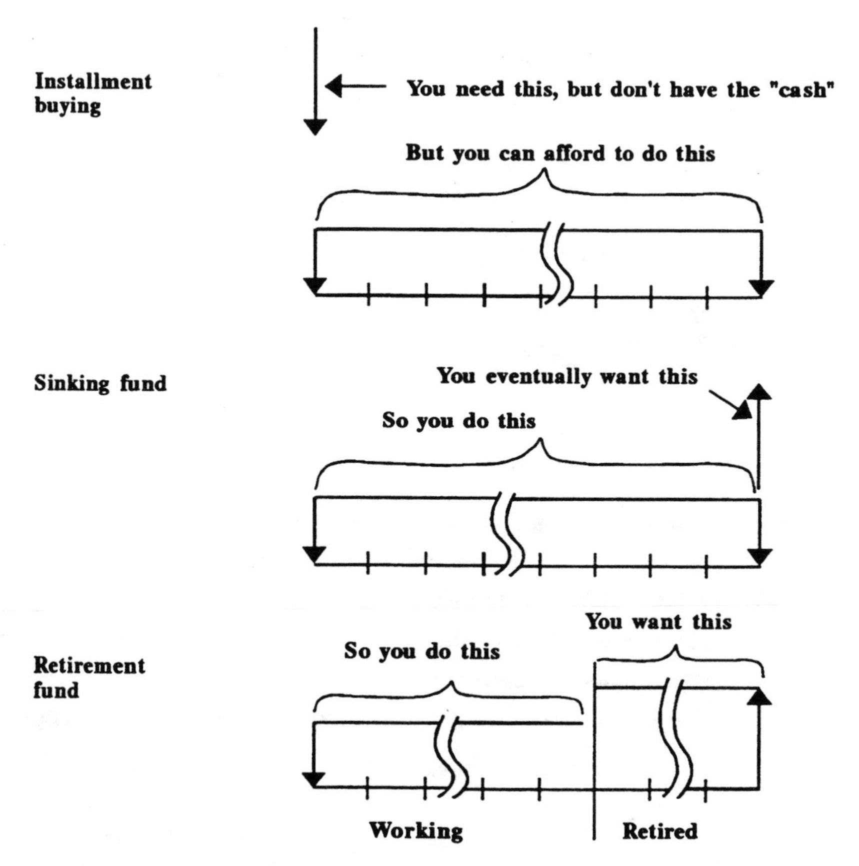

Figure 8–9 Examples of annuities–transactions involving a series of equal disbursements and/or receipts

Example 8

Given: Starting this year, you decide to put \$2,000 per year at the end of each year into a retirement plan earning 8%.

Find: Value after ten years and interest earned in ten years.

Solution: $F = (2{,}000)(1.08^{10} - 1)/(0.08) = (2{,}000)(1.1589)/(0.08) = \$28{,}973$. The interest earned is calculated by subtracting from \$28,973 the sum of the ten payments. Therefore, interest = \$28,973 − (10)(2,000) = \$8,973.

Series Sinking-Fund Factor

Sometimes you will need to compute the value of equal payments, that is, annuity payments needed to yield a single predetermined sum at a specified future time. For example, if a production line in a manufacturing plant needs to be re-

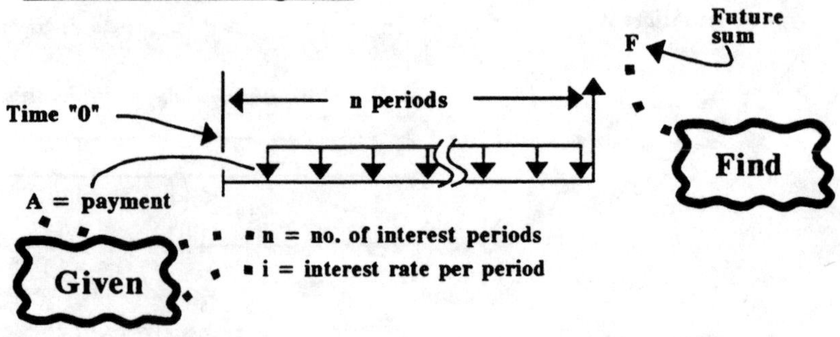

Derivation of F = f(A, n, i)

Future value of first payment $= A(1 + i)^{n-1}$

Future value of second payment $= A(1 + i)^{n-2}$

Future value of third payment $= A(1 + i)^{n-3}$

Future value of n-1 payment $= A(1 + i)^{n-(n-1)} = A(1 + i)$

Future value of n payment $= A$

$$F = A\left[(1 + i)^{n-1} + (1 + i)^{n-2} + (1 + i)^{n-3} + \ldots + (1 + i) + 1\right] \quad \text{(Equation A)}$$

Figure 8–10 Series compound-amount factor

placed four years from now and the approximate replacement cost is known, how much should be set aside each year so that the accumulated principal and interest on the principal and on interest totals the funds needed four years from now? Or consider a personal finance example. You hope to purchase a condominium three years from now and estimate that the minimal down payment and closing costs will be $8,000. How much should you save each month at 5% to accumulate $8,000 in principal and interest at the end of 36 months?

Refer to Figure 8–11 for a derivation of the equation $A = (F)\,(i)/((1 + i)^n - 1) = F(A/F\ i,n)$. Note that $(A/F\ i,n)$ is the reciprocal of the previously derived $(F/A\ i,n)$.

Example 9

Given: A small manufacturing plant projects that it will need $1,000,000 for construction of a plant expansion in ten years and can invest funds at $i = 6\%$.

Find: The amount to be set aside at the end of each of ten years to accumulate $1,000,000 at the end of ten years.

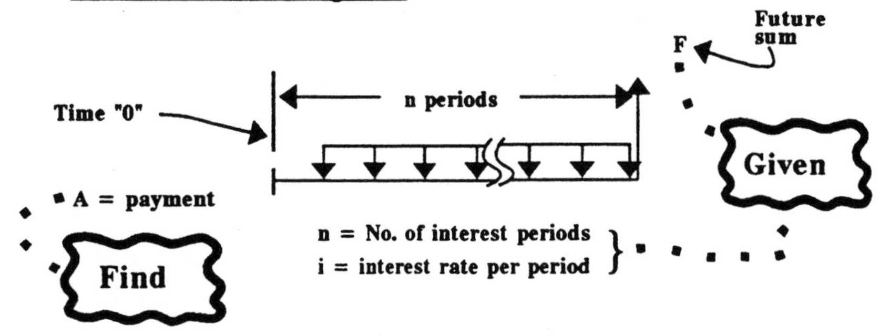

Derivation of A = f(F, n, i)

Because $F = A\left[\frac{(1+i)^n - 1}{i}\right]$

it follows that $A = F\left[\frac{i}{(1+i)^n - 1}\right]$

Figure 8–11 Series sinking-fund factor

Solution: A = (1,000,000)(0.06)/(1.06^{10} − 1) = (1,000,000)(0.06)/(0.7908) = (1,000,000)(0.07587) = $75,870.

Notice that the actual cash paid into the annuity is (10)($75,868) or $758,868. The difference between $1,000,000 and $758,868 is $241,132. This is the interest earned on principal and on interest during the ten-year period.

Series Present-Worth Factor

Assume that you purchased a new car and are "buying on time" over a period of 48 months. That is, you agreed to an annuity as a means of financing the car because you did not have the necessary funds to purchase the vehicle for cash. However, at the end of 24 months you are enjoying financial success and would like to pay off the automobile loan. Assuming that there is no prepayment penalty in your contract, how much do you owe? This situation is common in personal finance and in business, that is the determination of the present worth of an annuity or of the remainder of an annuity.

See Figure 8–12 for the beginning of the derivation of the equation $P = A((1 + i)^n - 1)/(i(1 + i)^n) = A(P/A\ i,n)$. Exercise 8.2 asks students to complete the derivation of this factor.

Example 10

Given: Annual operation and maintenance costs for a basic piece of equipment in a manufacturing plant are \$11,500, and the economic life of the equipment is 20 years.

Find: The present worth of the operation and maintenance costs for $i = 6\%$.

Solution: $P = (11{,}500)(1.06^{20} - 1)/(0.06 \times 1.06^{20}) = (11{,}500)(2.2071)/(0.1924)$ $= (11{,}500)(11.4697) = \$131{,}921$.

The total funds expended on operation and maintenance over the 20-year economic life of the facility are $20 \times \$11{,}500 = \$230{,}000$. Note that this expenditure is

"Cash flow" diagram

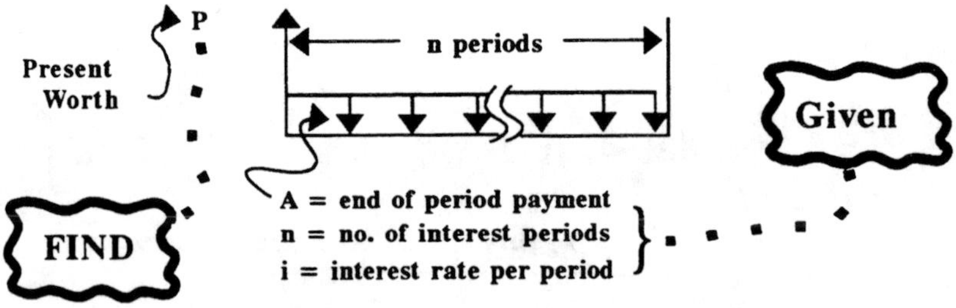

Derivation of P = f(A, n, i)

Present worth of first payment $= \dfrac{A}{(1 + i)}$

Present worth of 2nd payment $= \dfrac{A}{(1 + i)^2}$

Present worth of 3rd payment $= \dfrac{A}{(1 + i)^3}$

Present worth of n - 1 payment $= \dfrac{A}{(1 + i)^{n-1}}$

Present worth of n payment $= \dfrac{A}{(1 + i)^n}$

$$P = A\left[\frac{1}{(1+i)} + \frac{1}{(1+i)^2} + \frac{1}{(1+i)^3} + \cdots + \frac{1}{(1+i)^{n-1}} + \frac{1}{(1+i)^n}\right]$$

(Equation B)

Figure 8–12 Series present-worth factor

significantly more than the present worth of the operation and maintenance expenditures. The difference is due to the discounting of expenditures that occur in the future.

Capital-Recovery Factor

Mention was made earlier of "buying on time" or what is often referred to as "installment buying." Installment buying is illustrated in the cash flow diagram appearing in Figure 8–13. Typically, whether it is a personal matter or a business matter, the purchaser wants to acquire something now that has a value P, but does not have, or has but does not wish to spend, the necessary funds. Accordingly, the purchaser agrees to make payments A at the end of each of n periods and to pay an interest rate i for use of someone else's funds. As indicated in Figure 8–13, $A = P((i)(1 + i)^n)/((1 + i)^n - 1) = P(A/P\ i,n)$. The factor $(A/P\ i,n)$ is the reciprocal of $(P/A\ i,n)$.

As each payment is made, the payer is paying interest to the lender on the principal that remains during the just concluded period and is also paying on the

"Cash flow" diagram

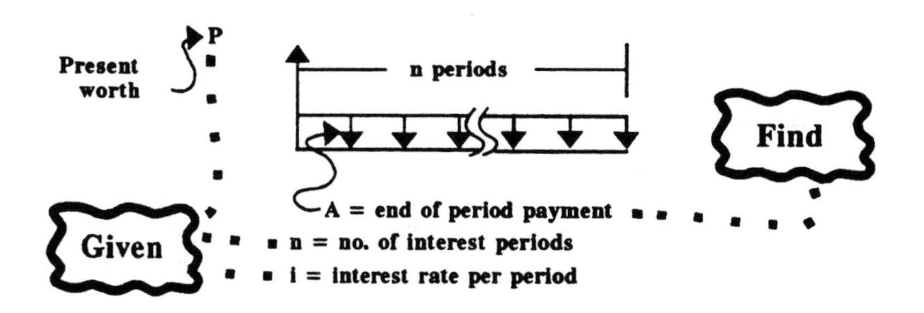

Derivation of A = f(P, n, i)

$$\text{Because } P = A\left[\frac{(1+i)^n - 1}{i(1+i)^n}\right]$$

$$\text{it follows that } A = P\left[\frac{i(1+i)^n}{(1+i)^n - 1}\right]$$

Figure 8–13 Capital-recovery factor

principal to gradually reduce it. In the early part of the annuity, most of the payment A is interest and a relatively small part is principal. As the annuity progresses, an ever-decreasing fraction of the payment A goes toward interest or "rent" on the unpaid principal and an ever-increasing fraction of the payment A goes toward repaying the principal.

Example 11

Given: You purchase a $15,000 automobile for $2,500 down and finance the balance of the purchase price for 36 months at 10%.

Find: Monthly payments and total interest paid during the life of the automobile loan.

Solution: The monthly interest rate is $i = 0.10/12 = 0.00833$. Then calculate the monthly payment as follows: $A = (12{,}500)((0.00833)(1.00833^{36}))/(1.0083^{36} - 1) = (12{,}500)(0.011229)/(0.34802) = (12{,}500)(0.032265) = \$403.31$. The total interest paid during the 36-month annuity = $(36 \times \$403.31) - 12{,}500 = \$2{,}019.16$.

Typically, installment buying involves a decision between minimizing the monthly payment by extending the payment period, which tends to increase the total amount of interest paid, versus trying to minimize the interest paid by reducing the duration of the loan, which increases the monthly payment.

Gradient-Series Present-Worth Factor

The previous four discount factors (series compound-amount factor, series sinking-fund factor, series present-worth factor, and capital-recovery factor) all involve annuities, that is, a uniform series of payments or income. Consider now one example of a gradient series, more specifically, the gradient-series present-worth factor. This gradient-series factor is presented to serve as an introduction to gradient-series factors in general. Although more gradient-series factors are available than presented in this chapter, familiarity with even one gradient-series factor is powerful because you can, through the use of transformations or superpositions, solve many problems.

Figure 8–14 presents the equation $P = G((1 + i)^{n+1} - (1 + ni + i))/(i^2)(1 + i)^n = G(\text{GSPWF})$. The derivation of the GSPWF is not presented in Figure 8–14, but is available elsewhere (e. g., James and Lee, 1971, p. 21).

Example 12

Given: Increasing operation and maintenance expenditures for a project over a ten-year period as shown in Figure 8–15. Note that these expenses are shown at the end of each of ten years. At the end of the first year,

"Cash flow" diagram

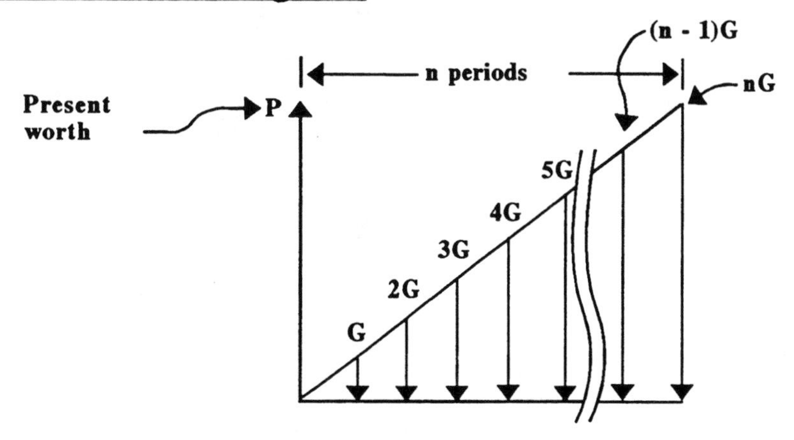

G = period-by-period increase in payment/revenue

n = no. of interest periods

i = interest rate per period

Note: Same absolute increase each year

$$P = G\left[\frac{(1+i)^{n+1} - (1+ni+i)}{i^2(1+i)^n}\right]$$

Figure 8–14 Gradient-series present-worth factor

the cost of operation and maintenance is $1,100 and then the annual increase is $100.

Find: Present worth of the operation and maintenance expenditures.

Solution: Partition the problem into an annuity and a gradient series, as shown in Figure 8–15. Then determine the present worth of the annuity and the present worth of the gradient series and sum them to get the total present worth. For the annuity portion, $P = A(P/A\ i,n) = (1{,}000)((1.06)^{10} - 1)/((0.06)(1.06)^{10}) = \$7{,}360$. For the gradient series portion, $P = G(\text{GSPWF}) = (100)((1.06^{11}) - (1.0 + (10)(0.06) + 0.06))/((0.06^2)\ (1.06)^{10}) = \$3{,}696$. Therefore, the total present worth is $7{,}360 + 3{,}696 = \$11{,}056$.

Using the principle of superposition as applied in this example, one can analyze almost any combination of discrete and continuous transactions. Exercise 8.13 provides an opportunity to use this superposition process.

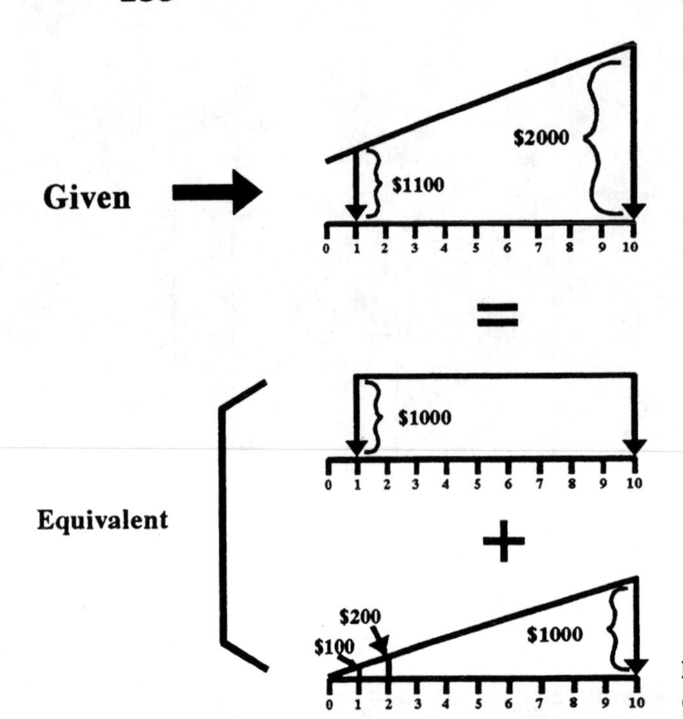

Figure 8–15 Example using gradient-series present-worth factor

As noted earlier in this chapter where annuities are introduced, special equations for annuities or uniform series (and for gradient series) greatly reduce the tedium of decision-economics computations. This observation may be illustrated by redoing the Example 12 problem by repeatedly applying the single-payment present-worth factor. This constitutes Example 13.

Example 13

Given: Same as Example 12.

Find: Same as Example 12.

Solution: The present worth of the first year O & M expense is $P_1 = F/(1 + i)^n = 1{,}100/(1.06) = \$1{,}037.74$.

The present worth of the second year O & M expense is $P_2 = 1{,}200/(1.06)^2 = \$1{,}068.00$.

The present worth of the third year O & M expense is $P_3 = 1{,}300/(1.06)^3 = \$1{,}091.51$.

The present worth of the fourth year O & M expense is $P_4 = 1{,}400/(1.06)^4 = \$1{,}108.93$.

The present worth of the fifth year O & M expense is $P_5 = 1{,}500/(1.06)^5 = \$1{,}120.89$.

The present worth of the sixth year O & M expense is $P_6 = 1{,}600/(1.06)^6 = \$1{,}127.94$.

The present worth of the seventh year O & M expense is P_7 = 1,700/(1.06)^7 = \$1,130.60.

The present worth of the eighth year O & M expense is P_8 = 1,800/(1.06)^8 = \$1,129.34.

The present worth of the ninth year O & M expense is P_9 = 1,900/(1.06)^9 = \$1,124.61.

The present worth of the tenth year O & M expense is P_{10} = 2,000/(1.06)^{10} = \$1,116.79.

Therefore, the total present worth is the sum of the ten O & M expenditures or \$11,056.35. As expected, this is identical to the Example 12 solution.

While this "brute force" method works, many more computations are typically required. Furthermore, there are many more opportunities to make mathematical errors. Clearly, the annuity and gradient series discount factors are preferable to the repeated use of discrete transaction discount factors. Incidentally, Examples 12 and 13, taken together, illustrate the technique mentioned earlier of solving each engineering economics problem two ways as a means of eliminating mathematical errors.

Summary of Discounting Factors

The eight discounting factors presented and illustrated in this chapter are summarized in Figure 8–16. Although the summary is convenient, users of the equations are advised, as noted earlier in this chapter, to first construct a cash flow diagram of the situation being analyzed and then to carefully select one or more of the equations, recalling the precise manner in which they were derived. By doing so, you will greatly reduce the likelihood of logic errors. You are also urged, as suggested earlier in this chapter, to double-check all decision-economics calculations by performing each of them in at least two different ways. This double-checking procedure will greatly reduce the likelihood of mathematical errors.

BENEFIT-COST ANALYSIS

Recall again the decision process presented in Figure 8–1 which may be used to screen, test, and evaluate alternatives. One approach to answering the "Is it economically feasible?" question is to utilize benefit-cost ratios. That is, calculate the benefit-cost ratio (B/C) for each alternative and select the alternative with the "best" B/C. Of course, in order to calculate the B/C for any alternative, the benefits and costs must first be placed on a comparable basis (e.g., present worth or

Transaction	Equation	Notation
Sum accumulated at simple interest	$F = P(1 + ni)$	—
Sum accumulated at compound interest	$F = P(1 + i)^n$	P(F/P i,n)
Present worth of an amount in the future	$P = F\left[\frac{1}{(1 + i)^n}\right]$	F(P/F i,n)
Future sum of a sinking fund	$F = A\left[\frac{(1 + i)^n - 1}{i}\right]$	A(F/A i,n)
Sinking fund payments to accumulate a sum in the future	$A = F\left[\frac{i}{(1 + i)^n - 1}\right]$	F(A/F i,n)
Present worth of an annuity	$P = A\left[\frac{(1 + i)^n - 1}{i(1 + i)^n}\right]$	A(P/A i,n)
Annuity payment equivalent to an amount now (installment buying)	$A = P\left[\frac{i(1 + i)^n}{(1 + i)^n - 1}\right]$	P(A/P i,n)
Present worth of a gradient series	$P = G\left[\frac{(1 + i)^{n+1} - (1 + ni + i)}{i^2(1 + i)^n}\right]$	G(GSPWF)

Figure 8–16 Summary of discount factors

annual) using the previously discussed discounting factors so that the ratio is meaningful. The best alternative is not necessarily the one with the largest B/C. Other considerations typically enter into the economic analysis. Three examples are given to illustrate use of the B/C in determining economic feasibility.

Example A—Alternatives with Variable Costs but Identical Benefits

The physical situation for this example is presented in Figure 8–17, and costs and benefits are summarized in Table 8–2. As indicated, the benefits are the same for all alternatives and are expressed in monetary terms. Recognize that equal mone-

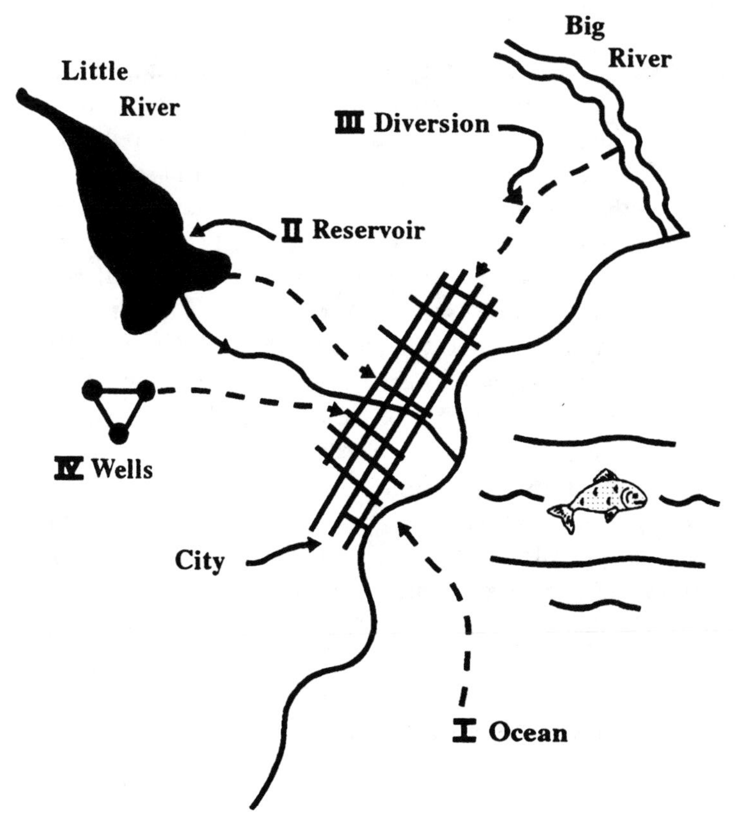

Figure 8–17 Water supply alternatives

tary benefits are unusual, but are assumed in the case of Example A to illustrate some important points. Monetary benefits for water supply might be based on factors such as reduced fire insurance premiums, additional property tax revenue by attracting new industry, and reduced water softening costs.

Alternative I would be eliminated on economic grounds because the B/C is significantly less than one. Alternative III appears to be the most advantageous

TABLE 8–2 COSTS AND BENEFITS OF WATER SUPPLY ALTERNATIVES

Alternative	Annual cost ($1,000,000)	Annual benefit ($1,000,000)	B/C
I Ocean	40.0	20.0	0.50
II Reservoir	15.0	20.0	1.33
III Diversion	10.0	20.0	2.00
IV Wells	13.3	20.0	1.50

alternative because it has the largest B/C. Note that it also has the smallest annual cost. Therefore, when benefits are fixed and all benefits are expressed in monetary terms, alternatives with B/C of less than one should be discarded. Furthermore, the alternative with the greatest B/C is clearly the most economical.

Note that the monetary value of benefits does not have to be quantified provided that decision makers feel that the benefits of the various alternatives are approximately equal and are greater than the associated costs. In this case, the decision makers simply select the least costly alternative. In fact, selection of the least costly alternative without a formal calculation of benefits is quite common in professional work decision making. Selection of the least costly alternative, without a knowledge of the quantifiable benefits, is also routinely done in personal matters. If you are planning a trip by air from location A to location B, you would typically obtain costs from several airlines and select the least costly alternative. When you do this, you are implicitly assuming that the benefits of the trip, regardless of the airline used, are about the same. Although you do not know the monetary value of those benefits, in your judgment they exceed the travel costs. You are, in effect, selecting the option with the greatest implicit B/C.

Example B—Alternatives with Variable Costs and Benefits

Figure 8–18 shows the physical situation for a variable costs and benefits example that is based on Linsley and Franzini (1979). Annualized costs and benefits for the four flood control alternatives are tabulated in Table 8–3. Benefit and cost data are presented graphically in Figure 8–19. In contrast with Example A, which had equal benefits for all alternatives, both benefits and costs vary among alternatives in Example B. In the upper part of the graph, the 45-degree line separates alternatives with benefits greater than costs (above the line) from alternatives with benefits less than costs (below the line).

Alternative I has a B/C less than 1.0 and would, on the basis of economic feasibility, be eliminated from further consideration. Alternative II has a B/C greater than 1.0 and, therefore, is economically feasible. Alternative III has an even larger B/C and would be the logical choice if the goal is a maximum rate of return, that is, maximum B-C as a percentage of C. With Alternative IV, the B/C is less than with Alternative III, but the absolute return is greatest for Alternative IV. Therefore, if the goal is maximum absolute return, Alternative IV would be the most logical choice. For Alternative V, the B/C drops further, as does the absolute value of benefits minus costs.

The preceding brief analysis suggests that maximum B/C does not necessar-

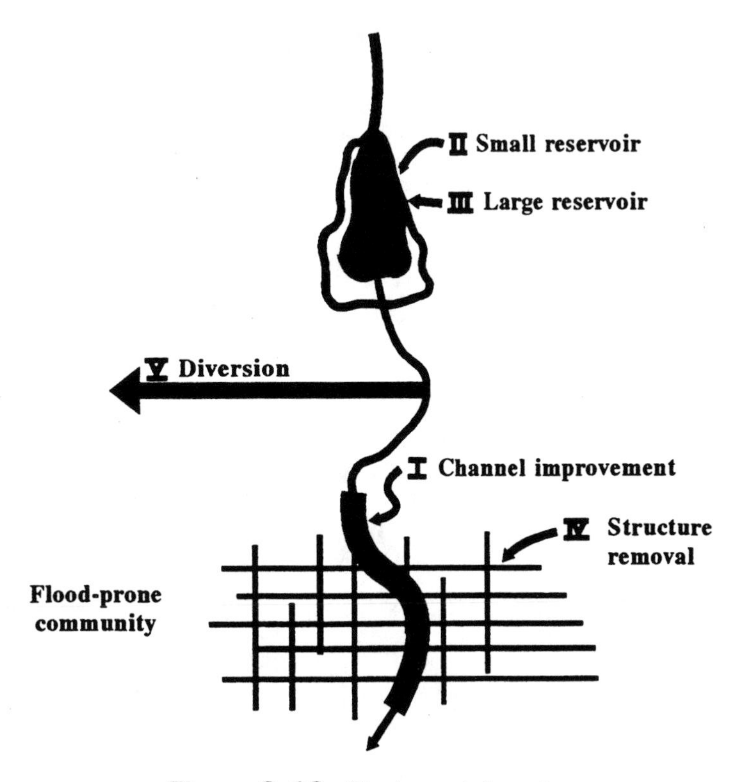

Figure 8–18 Flood control alternatives

ily determine the "best" alternative. Other factors must be considered. For example, the absolute funds available could dictate the choice. If the maximum annual cost that could be incurred is $40,000,000, then clearly Alternative II would be selected, because it satisfies the financial limits and has a B/C greater than 1.0. The owner, usually a unit of government in the case of flood control, may have other areas of concern and other investment opportunities besides flood control

TABLE 8–3 COSTS AND BENEFITS OF FLOOD CONTROL ALTERNATIVES

Alternative	Annual cost ($1,000,000)	Annual benefit ($1,000,000)	B/C
I Channel Improvement	20.0	13.0	0.65
II Small reservoir	35.0	43.0	1.23
III Large reservoir	53.0	71.0	1.34
IV Structure removal	63.0	83.0	1.31
V Diversion	80.0	98.0	1.23

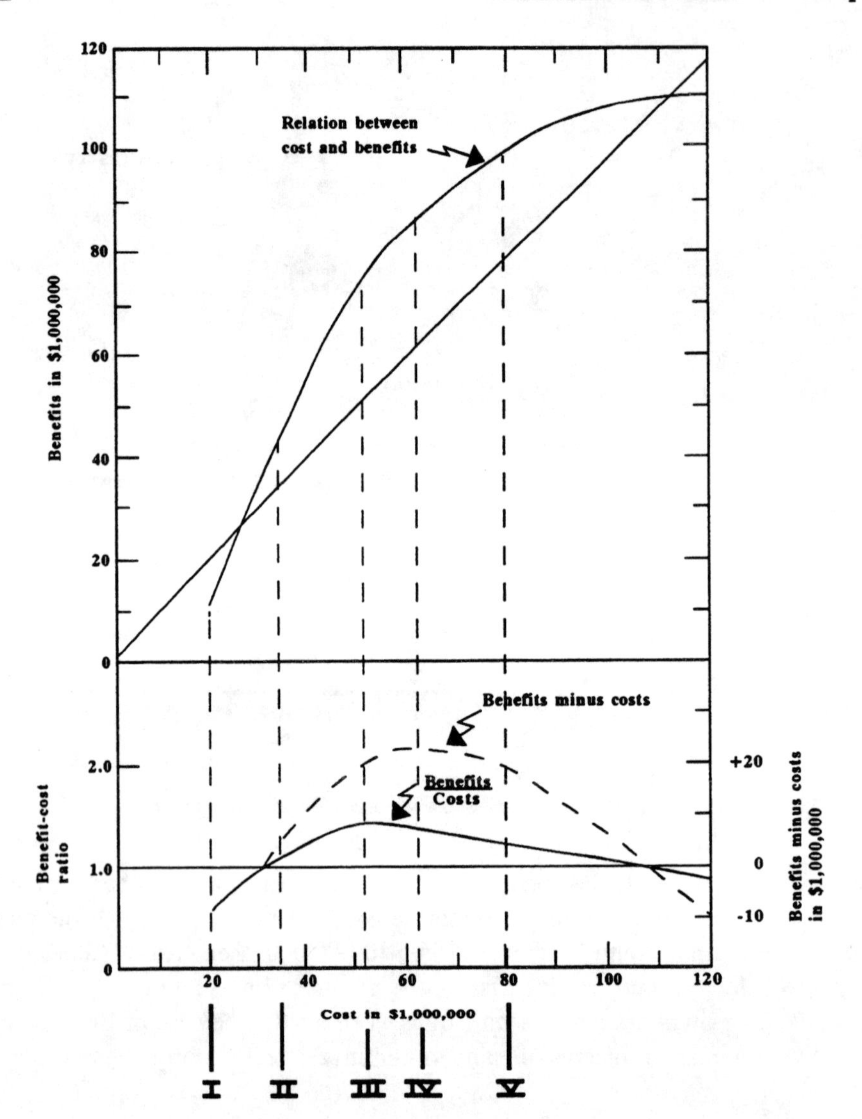

Figure 8–19 Graphical presentation of benefit and cost data (Source: Adapted with permission from R. K. Linsley, and J. B. Franzini, *Water Resources Engineering,* Third Edition, New York: McGraw-Hill, 1979.)

alternatives. For example, although Alternative III is a better choice than Alternative II based on the criterion of maximizing rate of return, the community may choose Alternative II to satisfy its flood control problems and utilize additional available funding to invest in other municipal needs that will produce an even greater return on the incremental investment.

Example C—Alternatives with Variable Costs and Benefits and with Significant Intangibles

This example (Walesh, 1989, pp. 480–487) is similar to Example B in that costs and benefits vary among alternatives. However, Example C differs from Example B in one important way, that being inclusion of important intangibles. Recall that an intangible benefit or cost is one that is important, but not quantifiable in monetary terms.

The physical situation for Example C is presented in Figure 8–20. A total of nine alternatives, seven of which were determined to be technically feasible, were developed to solve the flooding problem. Table 8–4 describes each alternative, presents monetary costs and benefits, and identifies significant nontechnical and noneconomic considerations, that is, significant intangibles. In this example, which is taken from an actual watershed planning project, the number of technically feasible alternatives gradually evolved through various combinations of early and simple alternatives. The last alternative, Alternative IX, which was recommended, includes aspects of all technically feasible alternatives.

Alternative IX was selected not solely on the basis of a benefit-cost ratio. Note that its B/C of 1.69 is clearly not the largest of all available benefit-cost ratios. It was also not selected on the basis of minimum annual cost—other less

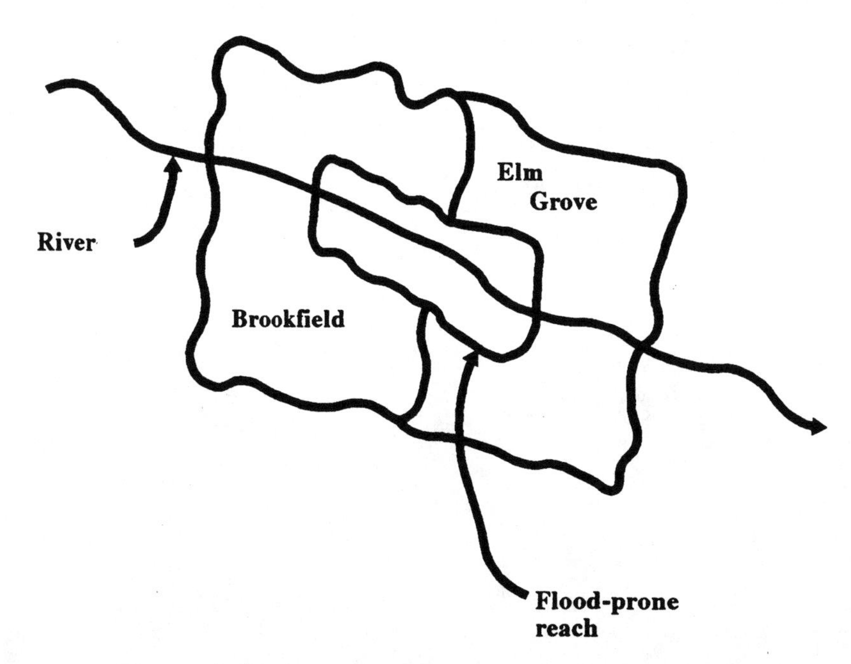

Figure 8–20 Schematic of two communities with a common flooding problem

TABLE 8–4 PRINCIPAL FEATURES AND COSTS AND BENEFITS OF FLOOD-PLAIN MANAGEMENT ALTERNATIVES FOR THE VILLAGE OF ELM GROVE, WISCONSIN

Alternative			Economic analysis (a, b)				Nontechnical and noneconomic considerations (c)		
No.	Name	Technically feasible ?	Annual cost ($1000)	Annual benefit ($1000)	Benefit-cost Ratio	Economically feasible ?	Positive	Negative	Recommended ?
1	No action	Yes	—	—	0.00	No	—	—	No
2	Detention storage	Yes	37.7	160.0	4.24	Yes	• Potential to retain public space	• Resolve only about one-half of the flood problem • May encourage new flood-prone development • Need for two communities to coordinate design, construction, and financing of storage	No
3	Structure floodproofing and removal	Yes	118.8	362.8	3.05	Yes	• Immediate partial flood relief at discretion of property owners • Most of the costs could be borne by beneficiaries	• Complete, voluntary implementation unlikely and therefore left with significant residual flood problem • Overland flooding and some attendant problems remain • Some floodproofing is likely to be applied without adequate professional advice and, as a result, structure damage may occur	No

4	Major channel modification	Yes	233.3	362.8	1.56	Yes	• Opportunity to develop an urban-oriented parkway through the business-commercial area	• Aesthetic impact in residential areas	No
5	Minor channel modification	No	—	—	—	—	—	—	No
6	Dikes and floodwalls	Yes	314.5	362.8	1.15	Yes		• Aesthetic impact of visual barrier • Pumping station operation and maintenance are critical to effective functioning of the system	No
7	Bridge and culvert alteration or replacement	No	—	—	—	—	—	—	No
8	Channelization storage composite	Yes	244.3	362.8	1.49	Yes	• Opportunity to develop an urban-oriented parkway through the business-commercial area	• Aesthetic impact of channelization component in residential areas • Need for two communities to coordinate design, construction, and financing of storage • Upstream storage may encourage new flood-prone development	No

(continued)

TABLE 8–4 PRINCIPAL FEATURES AND COSTS AND BENEFITS OF FLOOD-PLAIN MANAGEMENT ALTERNATIVES FOR THE VILLAGE OF ELM GROVE, WISCONSIN (*Continued*)

Alternative			Economic analysis (a, b)				Nontechnical and noneconomic considerations (c)		
No.	Name	Technically feasible ?	Annual cost ($1000)	Annual benefit ($1000)	Benefit cost Ratio	Economically feasible ?	Positive	Negative	Recommended ?
9	Storage—major channelization Intermediate channelization Floodproofing composite	Yes	214.2	362.8	1.69	Yes	• Potential to retain public open space • Floodproofing component would provide immediate partial flood relief at discretion of property owners	• Storage component may encourage new flood-prone development • Complete, voluntary implementation unlikely and therefore left with significant residual flood problem • Overland flooding and some attendant problems remain with floodproofing	Yes

• With floodproofing component, some costs could be borne by beneficiaries • Opportunity to develop an urban-oriented parkway through the business-commercial area	• Some floodproofing is likely to be applied without adequate professional advice and, as a result, structure damage may occur • Although less than that of major channelization, the intermediate component will have an aesthetic impact in residential areas • Erosion and attendant downstream deposition and maintenance requirements are likely for the intermediate channelization component • Need for two communities to coordinate design, construction, and financing of storage	Yes

(a) Economic analyses are based on 1975 costs, an annual interest rate of 6 percent, and a 50-year amortization period and project life.
(b) Economic analyses were not done for technically impractical alternatives.
(c) Presented only for technically and economically feasible alternatives.

(SOURCE: Adapted from SEWRPC, 1976, p. 114.)

costly alternatives were available. Intangibles, although they could not be quantified by definition, heavily influenced the selection of Alternative IX as the recommended course of action.

Concluding Thoughts

The three benefit-cost analysis examples presented in this section of the chapter are intended to illustrate the use of benefit-cost analysis and to suggest that, although benefit-cost analysis is useful, a selection of a course of action is rarely based on the notion as simple as selecting the alternative with the largest B/C. Of the three examples of benefit-cost analysis presented in this section of the chapter, Example C is by far the most realistic because it explicitly incorporates intangibles.

As an entry-level technical person serving on a team of professionals, you can assist in the client's or customer's decision-making processes by quantifying what can be quantified (e.g., certainly monetary costs and sometimes monetary benefits) and by systematically trying to identify and effectively communicate positive and negative intangibles. The mix of quantifiable and nonquantifiable factors will typically drive the decision-making process.

SENSITIVITY OF B/C TO INTEREST RATE

Most engineered products, structures, facilities, and systems incur costs early in their economic life and generate benefits later, that is, throughout their economic life. Stated differently, the "center of gravity" of costs tends to occur earlier than the "center of gravity" of benefits as suggested by Figure 8–21. Because the "center of gravity" of benefits occurs later than the "center of gravity" of costs, the interest rate selected for the analysis has a greater impact on the benefits than it does on costs. That is, the selected interest rate discounts benefits more than costs, because the benefits occur further in the future than costs.

More specifically, as the interest rate selected for the analysis is decreased, the value of future benefits increases more than the value of costs. Accordingly, an engineered product, structure, facility, or system that is uneconomic as determined by a B/C of less than 1.0 at one interest rate might become economic at a lower interest rate. Similarly, an alternative that is economic at one interest rate might become uneconomic at a higher interest rate.

The real issue at stake is the selection of the most appropriate interest rate for the situation. Engineers performing economic analyses or receiving the results of economic analyses performed by others should be aware of the great sen-

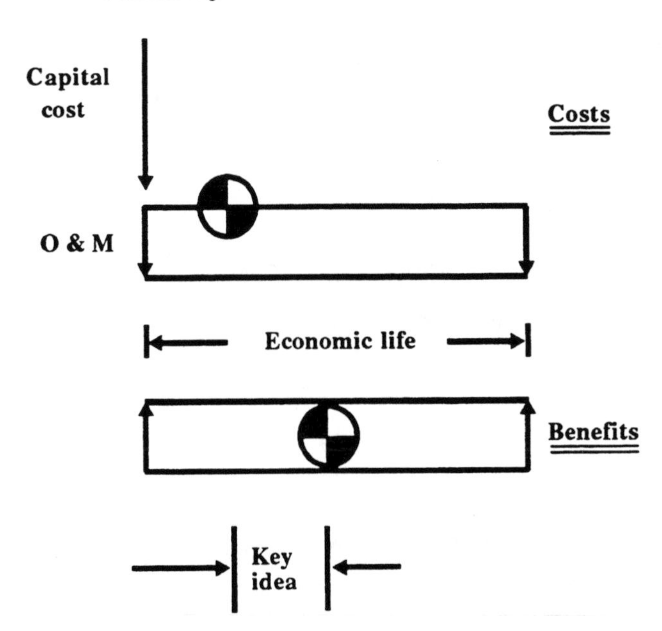

Figure 8–21 Conceptual cash flow diagram showing usual temporal position of "center of gravity" of costs and of benefits

sitivity of B/C and benefits minus costs to the selected interest rate. This sensitivity is illustrated in Example 14.

Example 14

Given: A flood control alternative, as illustrated in Figure 8–22, has been developed and the costs and benefits determined. The alternative has an economic life of 50 years.

Find: B/C for $i = 10\%$, 9%, 8%, 7%, and 6%.

Solution: For $i = 10\%$, the present worth of the benefits = P = 185,000(P/Ai, n) = $(185{,}000)((1.10^{50} - 1)/((0.10)(1.10)^{50}))$ = \$1,834,241. Therefore, B/C = 1,834,241/2,040,000 = 0.90.

For $i = 9\%$, P of benefits = \$2,027,900. Therefore, B/C = 2,027,900/2,040,000 = 0.99.

For $i = 8\%$, P of benefits = \$2,260,000. Therefore, B/C = 2,260,000/2,040,000 = 1.11.

For $i = 7\%$, P of benefits = \$2,553,000. Therefore, B/C = 2,553,000/2,040,000 = 1.25.

For $i = 6\%$, P of benefits = \$2,910,000. Therefore, B/C = 2,910,000/2,040,000 = 1.43.

Table 8–5 presents a summary of the preceding calculations in Columns (1)–(4) clearly showing how a gradual decrease in the interest rate produces a gradual increase in the B/C. In general, as the analyst gradually decreases the in-

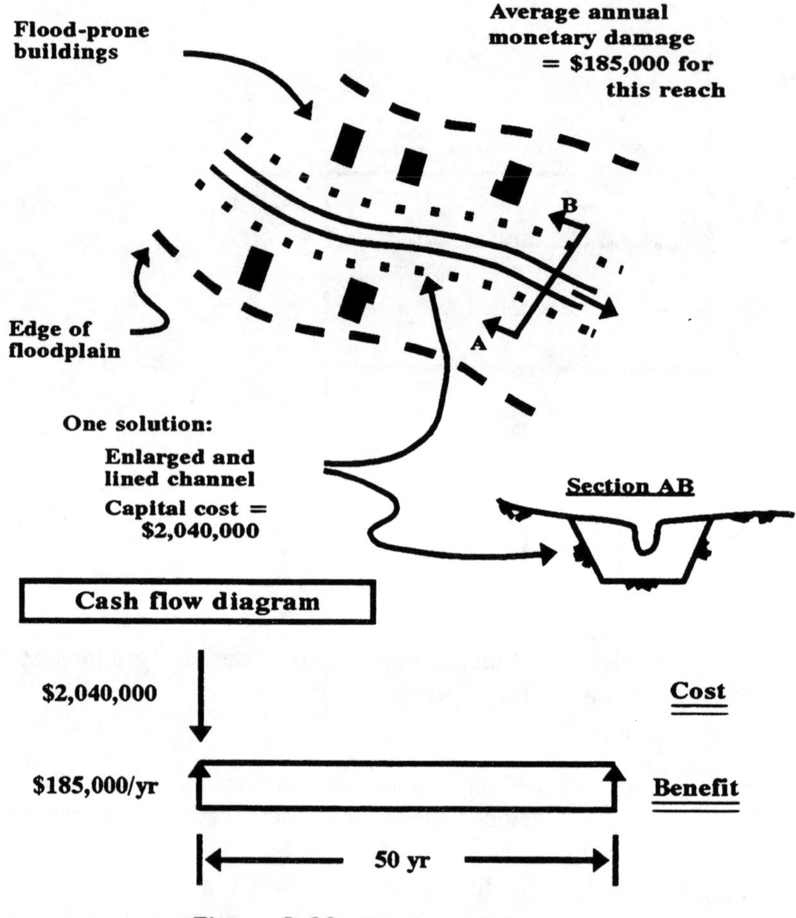

Figure 8–22 Flood control alternative

terest rate, there is usually a corresponding gradual increase in the B/C and benefits minus costs. For additional information on this topic refer to Hall and Dracup (1970, pp. 20–22) and Howe (1971, p. 70f).

SENSITIVITY OF B/C TO ECONOMIC LIFE

In addition to selecting an appropriate interest rate for an economic analysis, the engineer or other technical professional must also establish the economic life of the engineered product, structure, facility, or system. The previous discussion of the sensitivity of B/C and benefits minus costs to interest rate suggests that benefit-cost analyses are quite sensitive to interest rates within the range of inter-

TABLE 8–5 BENEFIT AND COST DATA FOR CHANNELIZATION ALTERNATIVE AS A FUNCTION OF INTEREST RATE AND ECONOMIC LIFE

(1) i	(2) Capital cost	(3) PW of benefits (n = 50)	(4) B/C (n = 50)	(5) PW of benefits (n = 100)	(6) B/C (n = 100)
%	$	$		$	
10	2,040,000	1,834,240	0.90	1,849,900	0.91
9	2,040,000	2,027,900	0.99	2,055,200	1.01
8	2,040,000	2,260,000	1.11	2,311,400	1.13
7	2,040,000	2,553,100	1.25	2,639,800	1.29
6	2,040,000	2,910,000	1.43	3,074,200	1.51

est rates typically considered in decision-economics analyses. Is economic life of similar importance in the conduct of engineering economic analyses? The answer to this question is developed in this section.

As noted in the previous section, the "center of gravity" of costs tends to occur earlier than the "center of gravity" of benefits. Because economic life determines the time span over which benefits will occur, logic suggests that as the analyst gradually increases the economic lives used in an economic analysis, there will be a corresponding gradual increase in B/C and benefits minus costs. Stated differently, "stretching" the economic life may produce a more favorable economic analysis.

One way to illustrate the impact of economic life on the results of an economic analysis is to use a specific example. Consider again the flood control alternatives presented in Figure 8–22 and the resulting economic analyses for five interest rates presented as Example 14 with the results tabulated in Table 8–5. Beginning with the results of Example 14, Example 15 is developed in which the calculations are expanded for an economic life of 100 years, or twice the 50-year economic life used in Example 14.

Example 15

Given: The results of Example 14.

Find: B/C for an economic life of 100 years.

Solution: Repeat the calculations carried out in Example 14 by substituting n = 100 for n = 50. The results are presented in Columns (5) and (6) of Table 8–5.

Consider i = 8% and n = 50 in Table 8–5 for which B/C = 1.11 as a point of reference. Note that, for i = 8%, a doubling of the economic life increases the

B/C from 1.11 to 1.13 for an absolute increase of 0.02. Although doubling the economic life increases the period of time over which benefits occur, the result of the increase in benefits is very small because benefits far out into the future—beyond 50 years—are heavily discounted. In contrast, a one percentage point decrease or increase in the interest rate used in the analysis produces a much larger increase or decrease in B/C. For example, if i is reduced from 8% to 7% with n = 50, B/C increases from 1.11 to 1.25 for an absolute increase of 0.14. This B/C increase attributed to a one percentage point decrease in i is about seven times the B/C increase attributed to a doubling of the economic life.

Although the reference point in Table 8–5 for the preceding discussion of relative sensitivity is $i = 8\%$ and $n = 50$ years, the general conclusion is the same. That is, Examples 14 and 15 illustrate that within the range of interest rates and economic lives typically considered in engineering economic analyses, the interest rate selection is much more important than the economic life selection. Stated differently, the entry-level technical professional should devote much more attention to selecting an appropriate interest rate than deciding on the appropriate economic life for engineered products, structures, facilities, or systems.

SENSITIVITY OF COSTS TO LOAD, CAPACITY, OR OTHER MEASURE OF SERVICE

The two previous sections of this chapter address factors that influence benefits and costs, mainly interest rate and economic life, and the relative impact of those factors. This section introduces the idea that, in some engineering projects, a large increase in load, capacity, or other measure of service may be gained at relatively little additional cost.

The cost of a product, structure, facility, or system, expressed on a present worth or annualized basis, is a function of its load, capacity or other measure of service. Stated differently, cost = f(load/capacity/other measure of service). Examples of various engineered products, structures, facilities, and systems and the loads, capacities, or other measures of service that determine their size are presented in Figure 8–23.

An important consideration in conducting an economic analysis is the sensitivity of cost to small changes in load, capacity, or other measure of service. Consider a personal example. Imagine that you plan to rent an apartment and, at the outset, you believe that the optimum apartment size is a two-bedroom. However, you realize that there are other options such as one-bedroom and efficiency apartments, which cost less, and three-bedroom apartments, and the penthouse which cost increasingly more. You might actually or implicitly develop a graph like that shown in Figure 8–23. If the slope of the graph in the vicinity of your initial

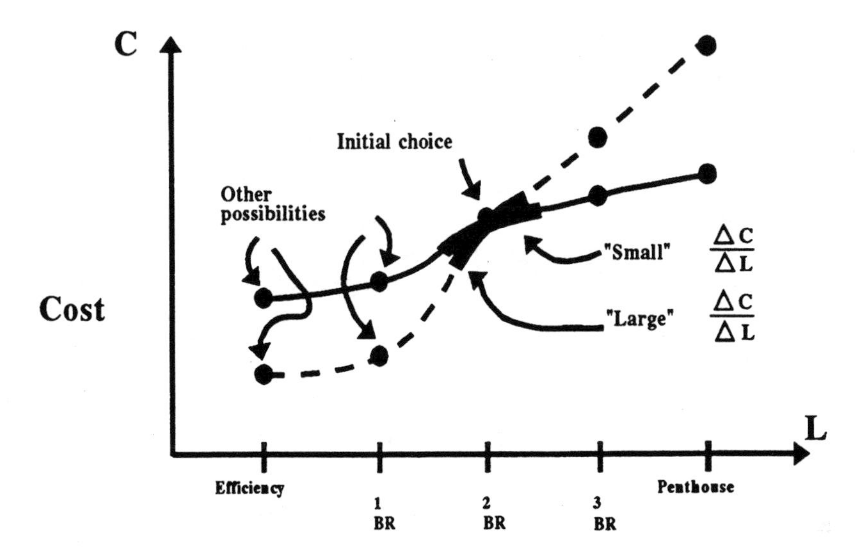

Load, capacity and other measures of service

Water main	**12 in.**	**18 in.**	**21 in.**	**24 in.**	**27 in.**
Expressway	**1 Lane**	**2 Lane**	**4 Lane**	**6 Lane**	**8 Lane**
Automobile engine	**100hp**	**125hp**	**150hp**	**175hp**	**200hp**

Figure 8–23 Cost as a function of load, capacity, or other measure of service

choice is relatively small, you might decide to rent a three-bedroom apartment rather than a two- bedroom apartment, reasoning that the marginal return is well worth the small additional marginal investment. In contrast, if the slope of the graph is relatively large in the vicinity of your initial choice, you probably would decide to stay with your initial choice. You might even decide to downgrade to a one-bedroom apartment rather than a two-bedroom apartment because of the large savings that would occur each month.

As suggested by the various alternative scales attached to the horizontal axis of Figure 8–23, the same sort of thinking on the margin can also be applied in decision economics. In most situations, engineers and other technical professionals owe those they serve, that is, their clients or customers, information or at least insight into the relationship that exists between cost and load, capacity, or other measure of service. Often, technical professionals and their clients or customers predetermine or at least have a notion of the load, capacity, or other measure of service that is desired. This may blind them to considering incremental increases or decreases and the corresponding increases and decreases in costs that might occur.

Implicit in thinking on the margin is having at least an approximate relationship between cost and load, capacity, or other measure of service, that is, having an equation or relationship of the form cost = *f*(load, capacity, or other measure of service). Ways to determine such a relationship include:

- Analytic, that is, derive equations from basic relationships.
- Computer simulation—use computer models to determine the "size" and, therefore, the cost of the product, structure, facility, or system as a function of its load, capacity, or other measure of service.
- Empirical—utilize data and information based on observation and experience as opposed to theory.
- Combinations of the preceding.

Analytic and Empirical Approach

The objective of this example is to develop a relationship between the cost per foot of gravity flow sewer (sanitary, storm, or combined sewer) and the flow it carries. That is, develop an equation of the form $u = f(Q)$, where u is the unit cost, that is, the construction cost per foot of sewer and Q is the flow-carrying capacity in cubic feet per second.

The derivation begins by obtaining a relationship between D, the sewer diameter, and Q, utilizing the Manning equation. The Manning equation is $Q = (1.49/n)(A)(R^{2/3})(S^{1/2})$ where n = the Manning roughness coefficient; R = the hydraulic radius, which is the cross-sectional area of flow (A) divided by the wetted perimeter (P); and S = the longitudinal slope of an open channel. Rearranging the equation, $Q = (1.49/n)(A)(A/P)^{2/3}(S^{1/2}) = (1.49/n)(A^{5/3})(S^{1/2})/(P^{2/3})$. This is Equation 1.

For a pipe flowing full, $A = (Pi)(D^2/4)$ and $P = (Pi)(D)$. Therefore, the term $(A^{5/3})/(P^{2/3}) = C_1D^{8/3}$, where C_1 is a constant. This is Equation 2. The same general form holds for a pipe flowing less than full with C_1 being reduced accordingly.

Combining Equation 1 with Equation 2 yields $Q = (1.49/n)(S^{1/2})(C_1D^{8/3})$. Combining the constants 1.49, n, $S^{1/2}$ and C_1 into a new constant C_2 and solving for D yields $D = (C_2)(Q^{3/8})$. This is Equation 3.

Recall that the objective of this derivation is to find a relationship between the cost per foot of sewer and the flow capacity. Although the objective has not yet been achieved, the intermediate result is of interest. Equation 3 indicates that doubling Q requires only a 30% increase in pipe diameter. Similarly, tripling or quadrupling Q requires, respectively, only 50% and 68% increases in pipe diameter. This suggests, but does not demonstrate or prove, that marginal investments in the form of increases in pipe diameter might yield significant return in terms of increased steering capacity.

The next step in the derivation is to obtain a relationship that presents the cost per foot of pipe as a function of diameter. Using national data for gravity sewer construction, Grigg (1986, pp. 90–91) indicates that unit cost or cost per foot is given by $u = C_3D$ where C_3 is a constant. This is Equation 4. Eliminating D between Equations 3 and 4, leads to $C_2Q^{3/8} = u/C_3$ or $u = (C_2/C_3)Q^{3/8}$ or $u = C_5Q^{0.375}$, where C_5 is a constant. Call this Equation 5.

The effect of various increases in Q ranging from a 50% increase in Q to a sixfold increase are calculated using Equation 5 and presented in tabular form in Table 8–6. For example, twice the flow capacity can be achieved in a sewer with only a 30% increase in construction costs. Capacity can be increased sixfold with only a doubling in construction costs. Incidentally, Hall and Dracup (1970, p. 23) report a great economy of scale for tunnels. This might be expected based on the preceding analysis, which indicates the significant economy of scale in gravity sewers.

The preceding derivation accomplishes the objective of showing how theoretical and empirical relationships can be combined to obtain a useful relationship between the cost of an engineered product, structure, facility, or system as a function of its load, capacity, or other level of service. For the particular example, that is, gravity sewers, the analysis indicates that there is a significant economy of scale in that, on the margin, great increases in flow capacity can be achieved with relatively small increases in construction costs. Clients and owners ought to have the benefit of this type of information.

TABLE 8–6 ECONOMY OF SCALE IN GRAVITY SEWER CONSTRUCTION COST

Discharge ratio	Construction cost ratio
1.5	1.16
2.0[a]	1.30[a]
4.0	1.68
5.0	1.83
6.0	1.96

[a]For example, doubling the flow capacity of the sewer would increase the cost by only 30%.

Computer Simulation Approach

In this example, computer simulation was used to find the relationship between the cost of a flood protection system as a function of the level of design provided. The level of design was determined by the recurrence interval in years of the storm that would be controlled by the system. The physical situation on which this example is based is presented in the upper part of Figure 8–24. This example is based on an actual engineering study (Walesh, 1989 p. 384). Computer simulation was used to size and then determine the costs of flood control systems that would control up to the 10-year, 50-year, and 100-year floods. The results are presented in tabular form in the lower portion of Figure 8–24.

Schematic

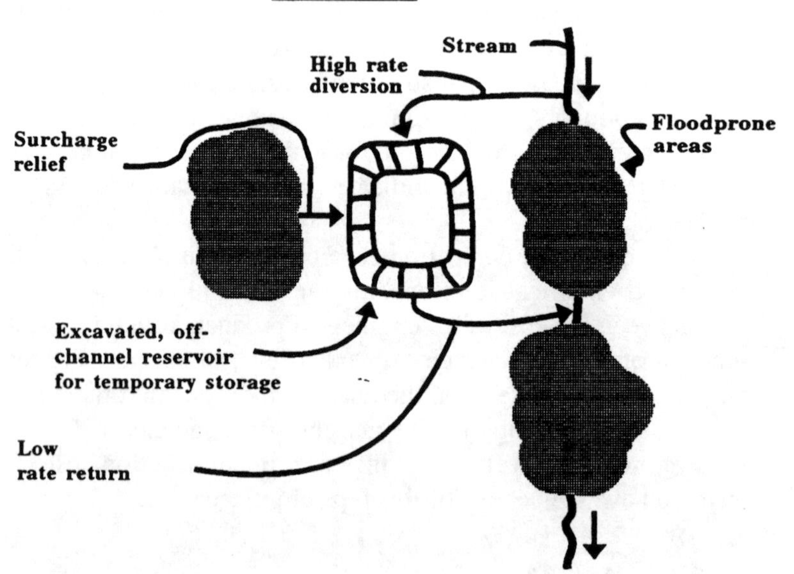

Cost as a function of recurrence interval

Recurrence interval (yr)	Probability of occurrence or exceedance in any year (%)	Capital cost (millions $)
100	1	2.90
50	2	2.65
10	10	1.82

Figure 8–24 Example of cost of flood protection as a function of design recurrence interval (Source: Adapted with permission from S. G. Walesh, *Urban Surface Water Management*, New York: John Wiley, 1989, p. 384.)

Assume that the point of reference is the second of the three designs, that is, designing to control up to the 50-year recurrence interval flood. This flood has a probability of occurring or being exceeded in any year of 2%, and the necessary protection could have been accomplished at a capital cost of $2,650,000. Note that a $250,000 or 9% increase in capital cost would produce a system that would have a probability of failure in any given year of only 1% rather than 2%. Much more protection would be provided for relatively little additional cost. On the other hand, consider the implications of reducing the design capacity from a 50-year event to a 10-year event. Capital cost would be reduced by $830,000, or 31%, but the system would now have a 10% probability of failing in any given year rather than only a 2% probability of failing. Incidentally, in the actual situation on which this example is based, the community decided to build the facility to control up to the 100-year event because they were shown the data presented in Figure 8–24 and could see the economy of scale inherent in the system.

Empirical Approach

Consider large public water utilities with the objective of finding the relationship between the annual operating cost as a function of the average annual supply and the capital cost as a function of the average annual supply. Using data based on large water utilities, Grigg (1986, p. 65) reports that annual operating costs = $C_6Q^{0.9}$, where Q is the annual supply and C_6 is a constant. Grigg also reports that capital costs = $C_7Q^{1.10}$, where C_7 is a constant. These empirical data suggest that there is no significant economy of scale in large water utilities, because annual operating costs and capital costs vary approximately linearly with Q.

Concluding Statement

You are urged to never make a recommendation on the load, capacity, or other measure of service for an engineered product, structure, facility, or system without first determining the sensitivity of cost to that load, capacity, or other measure of service within the "vicinity" of the recommendation. Give your client or customer the kind of information that he or she needs to make a good decision.

RATE OF RETURN OR RETURN ON INVESTMENT

Recall the earlier discussion of discount factors, and, in particular, the cumulative amount factor of $F = P(F/Pi,n) = P(1 + i)^n$. Recall also Example 2, where given P = $10,000, n = 4 years and i = 6%, F was calculated to be $12,625. Assume that Example 2 had been worked "backward." That is, given P = $10,000,

$F = \$12{,}625$ and $n = 4$ years, find i. Then the equation $12{,}625 = 10{,}000(1 + i)^4$ would be solved, yielding $i = 6\%$. The result $i = 6\%$ is referred to as the rate of return (ROR) or the return on investment (ROI). More specifically, rate of return (or return on investment) is defined as the single equivalent interest rate (annual) in a transaction that involves inflow and outflow of funds. This subject is treated by others including Newnan (1983, Chapter 7) and Riggs and West (1986, Chapter 6).

The methodology for calculating the ROR or ROI, as explained by Riggs and West (1986, p. 125), is to set up equations such as:

$$F(\text{investment}) = F(\text{return})$$

$$P(\text{investment}) = P(\text{return})$$

$$P(\text{receipts}) = P(\text{disbursements})$$

$$P(\text{income}) = P(\text{expenses})$$

or

$$P(\text{cost}) = P(\text{benefits})$$

where F is future value and P is present value. That is, ROR or ROI is determined by setting up an equation in which all inflows and outflows of funds are represented on a comparable basis and then solving that equation for i, the interest rate. Typically, the calculation of i is done by trial and error because an explicit expression for i usually cannot be determined. ROR or ROI is very useful because, as explained by Riggs and West (1986, p. 125), "It provides a percentage figure that indicates the relative yield and different uses of capital." Stated differently, ROR or ROI permits an analyst to "compare apples to apples" when faced with a choice among investments.

Example 16

Given: Land investment opportunity presented in Figure 8–25. An investor has an opportunity to buy a parcel of land along an existing highway immediately adjacent to a proposed interchange. The investment includes the original purchase price of the land plus annual property taxes. The projected return includes annual rental of the land and eventual sale of the land to other developers at the end of five years.

Find: ROR or ROI, assuming that the projected investment costs and returns are correct.

Solution: The basic idea is to put the investment and return on a present worth basis, equate them, and solve for i. The P of the investment = 80,000 + $850(P/A\ i,n)$ and the present worth of the return = $1{,}500(P/A\ i,n) + 150{,}000(P/F\ i,n)$.

Equating them yields $650(P/Ai,\ n) + 150{,}000(P/Fi,\ n) = 80{,}000$.

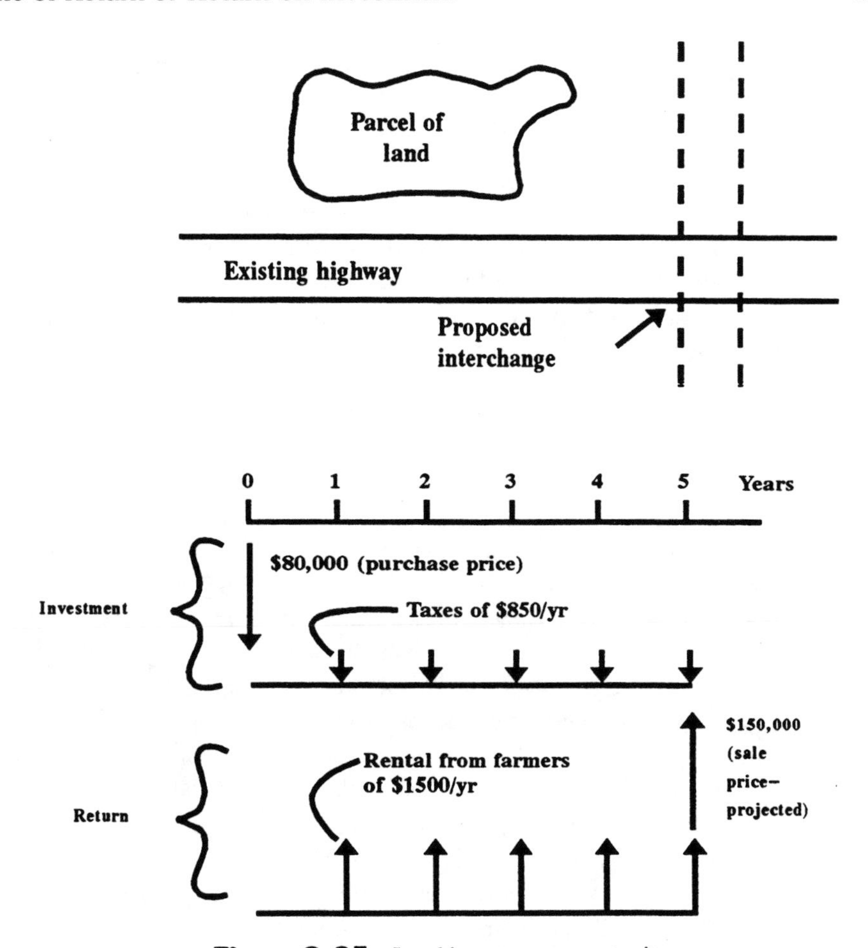

Figure 8–25 Land investment opportunity

Substituting for the discount factors leads to $650((1 + i)^5 - 1)/((i)(1 + i)^5) + 150{,}000/(1 + i)^5 = 80{,}000$.

Solve for i by trial and error. First determine if i is greater than 0. Let $i = 0\%$ and then the left side of the equation is $150,000 and the right side of the equation is $80,000. Because the return, the left side of the equation, is greater than the investment, the right side of the equation, when $i = 0$, the future is not being discounted enough. Therefore, i is greater than 0.

Arbitrarily try $i = 10\%$. The left side of the equation is $95,602 and the right side of the equation is $80,000. The future is still not being discounted enough.

Try $i = 12\%$. Then the left side of the equation is $87,457 and the right side of the equation is $80,000. The future is still not being discounted enough.

Try $i = 14\%$. Then the left side of the equation is $80,137 and the right side of the equation is $80,000. This is close enough. Accordingly, the ROI is 14%.

If you are offered an opportunity, of either a business or personal nature, consisting of one or more payments for which one or more returns are expected, calculate the ROR or ROI to quantify the value of investment opportunity. If offered or faced with two or more opportunities, calculate and compare the rates of return. Also consider the relative risks.

REFERENCES

BURKE, E., "A Letter to a Noble Lord—1795," in *Edmund Burke: Selected Writings and Speeches*, P.J. Stanlis, ed. New York: Doubleday & Company, 1963.

GRIGG, N. S., *Urban Water Infrastructure: Planning, Management, and Operations*, Chapter 4, "Waste-Water Management Systems." New York: John Wiley, 1986.

HALL, W. A., and J. A. DRACUP, *Water Resources Systems Engineering*, Chapter 2, "The Nature of Water Resources Systems." New York: McGraw-Hill Book Company, 1970.

HOWE, C. W., "Benefit-Cost Analysis for Water System Planning," *Water Resources Monograph 2*. Washington, D. C.: American Geophysical Union, 1971.

JAMES, L. D., and R. R. LEE, *Economics of Water Resources Planning*. New York: McGraw-Hill, 1971.

LINSLEY, R. K., and J. B. FRANZINI, *Water Resources Engineering*, 3rd ed., Chapter 13, "Engineering Economy in Water-Resources Planning." New York: McGraw-Hill, 1979.

NEWMAN, D. G., *Engineering Economic Analysis*, 2nd ed. San Jose, Cal.: Engineering Press, Inc., 1983.

NOVICK, D., "The Importance of Life Cycle Design for New and Rehabilitated Infrastructure," *Construction Business Review*, November/December 1991, pp. 58–61.

RIGGS, J. L. and T. M. WEST, *Essentials of Engineering Economics*, 2nd ed. New York: McGraw-Hill, 1986.

SOUTHEASTERN WISCONSIN REGIONAL PLANNING COMMISSION, *A Comprehensive Plan for the Menomonee River Watershed*, Planning Report No. 26, Volume 2, SEWRPC, Waukesha, Wis., October 1976.

THUESEN, H. G., W. J. FABRYCKY, and G. J. THUESEN, *Engineering Economy*, 4th ed. Englewood Cliffs, N.J.: Prentice Hall, 1971.

WALESH, S. G., *Urban Surface Water Management*. New York: John Wiley, 1989.

WOLOCHUK, R., "The First Cost Syndrome in Bridge Rehabilitation," Forum, *Civil Engineering*, October 1988.

EXERCISES

8.1 DERIVATION OF $(F/A\ i,n)$

Purpose

Provide students with further insight into the origin and correct application of discounting factors used in decision economics.

Tasks

Given the partial derivation of the $(F/A\ i,n)$ through Equation A presented in this chapter, complete the derivation. Suggestion: Multiply both sides of Equation A (a series) by $(1+i)$. Call the result Equation B (another series). Then subtract Equation A from Equation B on a term-by-term basis to arrive at $F = A((1 + i)^n - 1)/i = A(F/A\ i,n)$.

8.2 DERIVATION OF $(P/A\ i,n)$

Purpose

Provide students with further insight into the origin and correct application of discounting factors used in decision economics.

Tasks

Given the partial derivation of the $(P/A\ i,n)$ through Equation A presented in this chapter, complete the derivation. Suggestion: Multiply the numerator and denominator by $1 - (1/(1 + i))$ and rearrange to arrive at $P = A((1 + i)^n - 1)/(i(1+ i)^n) = R(P/A\ i,n)$.

8.3 How long from now will it take to quadruple a sum of money invested now at 9% and compounded annually?

8.4 A manufacturing plant operator anticipates a \$10,000 equipment repair expense four years from now. He/she has funds available and an opportunity to invest at 6% compounded annually. What amount should be invested now so as to have \$10,000 in four years?

8.5 A municipality's water charge income exceeds expenses by \$9,000/yr. An addition to the water treatment plant (WTP) will be needed in 10 years. If the \$9,000 yearly surplus is used to establish an annuity at 8% for 10 years, how much money will be available for WTP construction at the end of 10 years?

8.6 A power company must add a turbine in 10 years at a cost (then) of \$800,000. What amount of money should be placed into a sinking fund annually at 6% in order to accumulate \$800,000 in 10 years?

8.7 You are the city engineer and are negotiating with a land developer. The developer wants water service to a new development, and you tell him or her that service can be completed this year by city crews at a cost of \$200,000, which will be billed to the land developer and payable by the end of the year. The developer can't raise that sum now, so he/she offers to pay the city \$30,000 at the end of each of the next 10 years, which amounts to \$100,000 above what the service costs. If money is worth 8%, should you accept this offer?

8.8 A researcher is writing a proposal to a governmental agency for a five-year energy study. A portion of the proposal will be used to provide an estimated \$6,000 per year for annual data collection. The researcher plans to request a lump sum to be received one year prior to the beginning of the study and used to establish an annuity at 6%, which will pay \$6,000 at the end of each of five consecutive years. How much money should be requested to establish the annuity?

8.9 Assume that the proposal referred to in the preceding problem is approved, but that only \$15,000 is granted for establishing the annuity. What size equal payments would this provide at the end of each of five consecutive years for field data collection?

8.10 A \$17,000 car is purchased for \$3,200 down with the remainder to be financed over 36 months at 10%. Determine:

a. monthly payments

b. total interest (dollars) that would be paid in 36 months

c. cash settlement that would be due when one year remains

d. interest savings realized by paying off loan with one year remaining

8.11 An irrigator needs a motor to drive a pump.

Alternative A: \$2,500 initial cost, 10-year economic life, \$500 salvage, \$4,500/yr O & M. Buy a new one every 10 years.

Alternative B: \$3,500 initial cost, 20-year economic life, \$500 salvage, \$4,000/yr O & M. Buy a new one every 20 years.

Compute the net annual cost of each and select the most economical alternative. Use $i = 6\%$.

8.12 Redo Exercise 8.11, using a present-worth approach. Then, as a check on Exercise 8.11, convert the two present-worth costs to an average annual cost.

8.13 Operation and maintenance costs for a small power plant that is being phased out over a period of 10 years at $i = 6\%$ are shown in Figure 8–26. Determine the present worth of O & M. Do at least two ways to verify calculations.

8.14 The costs and benefits for a proposed dike flood control system are as follows:

Capital cost of dike = \$1,040,000
Capital cost of pumping station = \$475,000
Economic life of dike = 50 yr
Economic life of pumping station = 25 yr
Annual O & M = \$20,000
Benefits (average annual) = \$215,000

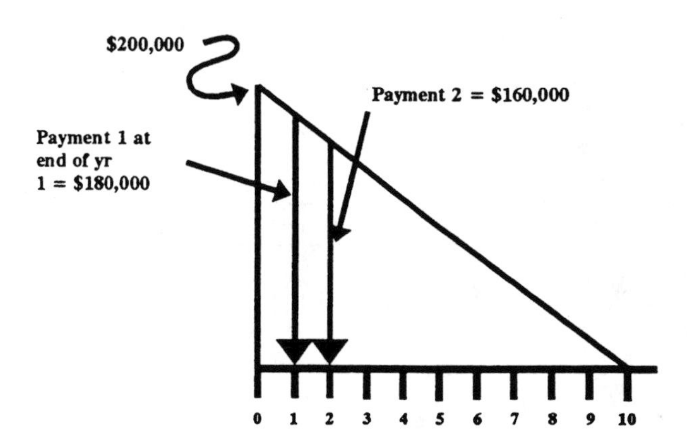

Figure 8–26 Decreasing operation and maintenance costs for power plant

If $i = 8\%$, determine the B/C ratio. Solve the problem on a present-worth basis and then on an annualized basis to verify the answer.

8.15 COMPARISON OF ALTERNATIVES ON A PRESENT WORTH BASIS. Refer to Figure 8–27. The following cost information is given:

Interest rate	10%
Alternative A—Tunnel	
Initial cost	$8,000,000
Economic life	100 years
O & M	$55,000/year
Alternative B—Canal	
Initial cost of flume	$900,000
Economic life of flume	50 years
Initial cost of canal (excluding lining)	$1,000,000
Economic life of canal	100 years
Initial cost of concrete canal lining	$550,000
Economic life of canal lining	20 years
Salvage (reuse) value of canal lining	$25,000
O & M of all components	$120,000/year

Determine:

a. Present worth of each alternative. *Note:* Do the exercise entirely on a present-worth basis, that is, not on an annualized basis. Convert each cost component (e.g., O & M of tunnel, canal lining) to its present worth and then sum to get the total present worth of each alternative.

b. Best alternative from an economic perspective.

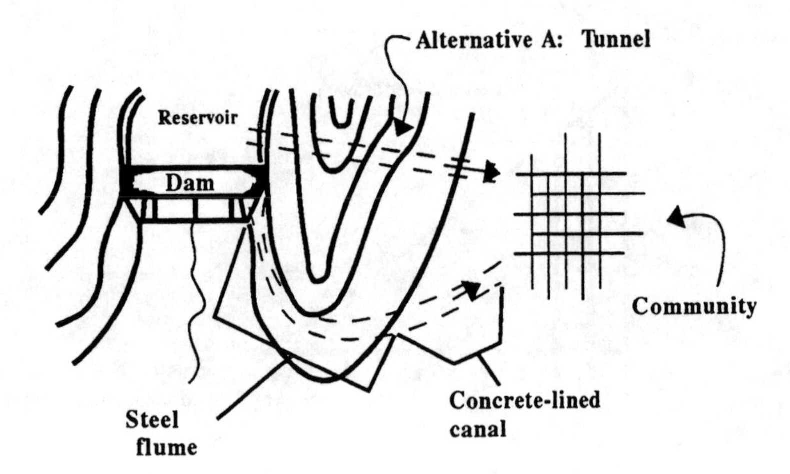

Figure 8–27 Water supply alternatives

c. Use your imagination to identify one nontechnical, noneconomic factor which might enter into the choice between the two alternatives.

8.16 COMPARISON OF ALTERNATIVES ON AN ANNUAL-COST BASIS

Repeat Exercise 8.15, except determine the annual cost of each alternative. *Note:* Do the exercise entirely on an annual-cost basis, that is, not on a present-worth basis. Convert each cost component (e.g., initial cost of tunnel, salvage value of canal lining) to its annual cost and then sum to get the total annual cost of each alternative. Convert the total annual cost of each alternative to its present worth. Present worths calculated this way should be the same as the present worths obtained as the solution to Exercise 8.15.

8.17 COMPARISON OF ALTERNATIVES ON THE BASIS OF CONVERTING ALL COSTS AND SALVAGE VALUES TO A SPECIFIED POINT IN TIME

Repeat Exercise 8.15, except determine the equivalent net cost of each alternative at a specific point in time (e.g., the end of the twentieth, fortieth, fiftieth, sixtieth, eightieth, one-hundredth year) as specified by the instructor. To verify the solution, calculate the ratio of the cost of alternative A to the cost of alternative B and compare to the solutions to Exercises 8.15 and 8.16. The ratios should be identical. To further verify the solution, calculate the present worth of each alternative and again compare them to the solutions of Exercises 8.15 and 8.16.

8.18 Select an area within engineering or a related technical field that results in products, structures, facilities, or systems. Obtain information on cost as a function of load, capacity, or other measure of service. Present cost function(s) (cost as a function of load, capacity, or other measure of service) in graphical or equation form) and comment on sensitivity of cost to the service that would be provided.

8.19 RATE OF RETURN

Given: Invest $5,000 at time $t = 0$. Receive total returns of:

$1190@ $t = 1$ yr
$1300@ $t = 2$ yr
$1200@ $t = 3$ yr
$1200@ $t = 4$ yr
$1000@ $t = 5$ yr

Determine: Rate of return.

8.20 RATE OF RETURN

One aspect of obtaining a college education is the prospect of improved future earnings compared to noncollege graduates. Sharon Shay estimates that a college education has a $32,000 equivalent cost at graduation. She believes that the benefits of her education will occur throughout 40 years of employment. She thinks she will have a $2,000 per year higher income during the first 10 years out of college, compared to a noncollege graduate. During the subsequent 10 years she projects an annual income that is $4,000 per year higher. During the last 20 years of employment she estimates an annual salary that is $6,000 above the level of the noncollege graduate. If her estimates are correct, what rate of return will she receive as a result of her investment in a college education? (Source: quoted—with different numerical values—from Newnan, 1983, p. 154.)

8.21 AMORTIZATION TABLE

Purpose

1. Provide the student with additional insight into and understanding of the principal and interest components of an annuity.
2. Expand the student's capability with spreadsheets.

Tasks

1. Refer to Table 8–7, Payment Schedule for 36-Month Financing of an Automobile, which was developed using a spreadsheet.
2. Program a spreadsheet to essentially duplicate the results of the Table 8–7 in terms of input and output. Generally follow the established format. Call the results Alternative A.
3. Then obtain results for the following alternatives:

Alternative	i (%)	Cost of auto ($)	Down payment (%)
B	10.5	16,000	10.0
C	12.5	16,000	20.0
D	12.5	16,000	10.0
E	12.5	20,000	10.0
F	12.5	20,000	20.0

4. Submit the entire output for each alternative clearly labeled A, B, C, D, E, & F.
5. Assume you financed your car under Alternative B. You subsequently and unexpectedly receive a large sum of money and decide to pay off your loan at the end of 24 months. Based on the spreadsheet, how much do you owe?
6. Verify the preceding "amount owed" using one or more discount factors for annuities.
7. Indicate the name of the spreadsheet you used.

TABLE 8–7 PAYMENT SCHEDULE FOR 36-MONTH FINANCING OF AN AUTOMOBILE

Note: (us) means user-supplied.
(c) means calculated.

i = Annual interest rate (%)	10.500	(us)
Monthly interest rate (%)	0.875	(c)
n = Finance period (months)	36	specified
Cost of auto ($)	16,000.00	(us)
Down payment (%)	20.0	(us)
Down payment ($)	3,200.00	(c)
Amount financed ($)	12,800.00	(c)
R = Monthly payment ($)	416.03'	(c)
Total paid ($)	14,977.13	(c)
Total interest paid ($)	2,177.13	(c)

(1) Month	(2) Principal owed at beginning of month ($)	(3) Payment due at end of month ($)	(4) Interest due at end of month ($)	(5) Payment on principal ($)
1	12,800.00	416.03	112.00	304.03
2	12,495.97	416.03	109.34	306.69

(continued)

TABLE 8–7 PAYMENT SCHEDULE FOR 36-MONTH FINANCING OF AN AUTOMOBILE (*CONTINUED*)

(1) Month	(2) Principal owed at beginning of month ($)	(3) Payment due at end of month ($)	(4) Interest due at end of month ($)	(5) Payment on principal ($)
3	12,189.28	416.03	106.66	309.38
4	11,879.90	416.03	103.95	312.08
5	11,567.82	416.03	101.22	314.81
6	11,253.01	416.03	98.46	317.57
7	10,935.44	416.03	95.69	320.35
8	10,615.09	416.03	92.88	323.15
9	10,291.94	416.03	90.05	325.98
10	9,965.97	416.03	87.20	328.83
11	9,637.14	416.03	84.32	331.71
12	9,305.43	416.03	81.42	334.61
13	8,970.82	416.03	78.49	337.54
14	8,633.29	416.03	75.54	340.49
15	8,292.80	416.03	72.56	343.47
16	7,949.33	416.03	69.56	346.47
17	7,602.85	416.03	66.52	349.51
18	7,253.35	416.03	63.47	352.56
19	6,900.78	416.03	60.38	355.65
20	6,545.13	416.03	57.27	358.76
21	6,186.37	416.03	54.13	361.90
22	5,824.47	416.03	50.96	365.07
23	5,459.40	416.03	47.77	368.26
24	5,091.14	416.03	44.55	371.48
25	4,719.66	416.03	41.30	374.73
26	4,344.92	416.03	38.02	378.01
27	3,966.91	416.03	34.71	381.32
28	3,585.59	416.03	31.37	384.66
29	3,200.93	416.03	28.01	388.02
30	2,812.91	416.03	24.61	391.42
31	2,421.49	416.03	21.19	394.84
32	2,026.65	416.03	17.73	398.30
33	1,628.35	416.03	14.25	401.78
34	1,226.57	416.03	10.73	405.30
35	821.27	416.03	7.19	408.85
36	412.42	416.03	3.61	412.42
	Totals:	14,977.13	2,177.13	12,800.00

9

Business Accounting Methods

> *Some know the price of everything and the value of nothing.*
>
> (Anonymous)

Accounting is the process of recording, summarizing, analyzing, verifying, and reporting in monetary terms the transactions of a business or other organization. Kamm (1989, p. 84) states that accountants are like journalists—they report what happens. In contrast with engineers and other technical professionals, who tend to make things happen, accountants normally record what has happened, at least the financial aspects. Engineering and other technical professions tend to be prospective and accounting retrospective.

Although there are major differences between the perspectives and functions of technical professionals and accountants, they need each other. The former, even entry-level technical professionals, should understand the basic terminology and concepts of accounting so that they can effectively function within the owner-consultant-contractor triangle or playing field presented in Chapter 1. The three types of organizations represented by the vertices of the triangle all utilize accounting, although the private entities place greater emphasis on it because of their need to be profitable.

The purpose of this chapter, therefore, is to provide the young professional with an introduction to basic accounting terminology and concepts. After explaining ways in which various types of technically based organizations need the services of accountants, two important financial statements—the balance sheet and the income statement—are discussed, as is the relationship between them. Building on financial statements, financial ratios are introduced. Time utilization rates, expense ratio, and the multiplier, accounting-related performance indicators for consulting engineering firms and similar service organizations, are discussed with emphasis on implications for the entry-level technical professional. The income statement is discussed further, this time as part of the business plan for a professional services firm, followed by a brief treatment of the project overruns and the implications for profitability.

WHY DO ACCOUNTING?

The general, overriding reason for doing accounting is to determine where an organization has been in financial terms and, by analysis and extrapolation, to forecast where the organization is likely to be going in financial terms. Some of an organization's accounting is done for internal reasons, and some of it is done to satisfy external needs.

Examples of internal reasons for doing accounting include general "score keeping" to determine how closely an organization is adhering to its financial plan, to provide the basis for pricing products or setting charge-out rates for professional services, and to guide internal resource allocation and optimum marginal investing. External reasons for accounting include the need to report to stockholders; providing data to lenders in support of loan applications; satisfying the requirements of federal, state, and other tax laws; and providing information to insurance companies as part of a process of securing professional liability insurance. Liability insurance premiums are based partly on annual billings and similar financial measures.

THE BALANCE SHEET: HOW MUCH IS IT WORTH?

The balance sheet is one of several financial statements that define the financial condition of a company, are prepared at regular intervals such as monthly, quarterly, and annually, and meet internal and external needs as discussed earlier. Unfortunately, the balance sheet is often referred to by other names (e.g., Clough, 1986, p. 263) such as financial statement, statement of financial condition, statement of worth, and statement of assets and liabilities. The term *balance sheet* is used in this book because the term seems appropriately descriptive and because it is widely used in engineering literature (e.g., Clough, 1986, p. 265; Kamm, 1989, p. 84).

The balance sheet shows, for a point in time, assets (A), liabilities (L), and net worth (NW) or equity (E). The basic equation for the balance sheet is

$$A - L = NW \text{ (or E)}$$

Two example balance sheets are presented and discussed in order to illustrate the features and usefulness of balance sheets. The first is a hypothetical balance sheet for a young professional's personal finances, and the second is a hypothetical balance sheet for a construction company. The personal and business examples are chosen to illustrate the broad applicability of the balance sheets, that is, their usefulness in managing one's personal finances as well as in managing business finances.

Refer to Table 9–1. This accounting of assets and liabilities could apply to a young professional and possibly his or her spouse within a year or so after graduation from college. Some observations are in order. First, this balance sheet and balance sheets in general are not as accurate as they may appear to be, particularly when line items are entered to the nearest cent. For purposes of consistency, accountants usually show all items to the nearest cent. Some line items, such as the current balance of a checking account or the amount owed on an automobile loan, can be determined and stated to the nearest cent. However, in general, the overall accuracy of a balance sheet is less than that because some values are estimates, such as the current market value of a condominium or other personal property. The accuracy inconsistency in a balance sheet is further complicated by the difficulty of getting all values to be simultaneous or coincident.

Second, while absolute values of line items, assets, and liabilities at any

TABLE 9–1 HYPOTHETICAL END-OF-CALENDAR-YEAR PERSONAL BALANCE SHEET

Assets	
Condominium	$60,000.00
Personal property (e.g., furniture)	15,000.00
XYZ stock	5,283.68
Car	10,000.00
Retirement (vested)	7,500.21
Cash/checking	893.76
Insurance (cash value)	1,012.16
Total:	$99,689.81 (A)
Liabilities	
Mortgage on condo (SPW of remaining payments)	$55,293.32
Car loan (SPW)	7,151.98
Credit cards	2,542.99
College loan (SPW)	2,016.14
Total:	$67,004.43 (L)
Net worth (or equity):	$32,685.38 (NW)

time are important, changes and trends such as a gradual increase in net worth for an individual, a couple, or a company are even more important. A balance sheet is a snapshot of net worth at a point in time. A series of balance sheets provides a moving picture of a changing and, it is to be hoped, improving situation. For example, a young professional might exhibit a negative net worth shortly after graduation from college, but as a result of sound personal financial management, quickly improve the situation so that in a few years assets exceed liabilities and the difference grows.

If you have not already done so, you should consider developing a balance sheet for your personal finances. If you owe or own anything, you have the basis for a personal balance sheet. Consider using spreadsheet software to construct your balance sheet. Update your balance sheet annually by adding a column for the end of the current year. By preparing a balance sheet now and periodically updating it you will have a measure of how well you are managing your personal finances and will also have ready access to the kinds of asset and liability data required by banks and other lending institutions in support of applications for home mortgages, automobile loans, and other common financial transactions.

The second example balance sheet is shown in Table 9–2. Note that the balance sheet represents the last day of a calendar year. This balance sheet is taken from Clough (1986, pp. 263–265) and the explanations of these various line items are based, in part, on Clough.

The asset line items in the balance sheet are as follows:

- Cash on . . . —Self-explanatory.
- Notes receivable . . . —They have apparently lent money to someone and it is due now.
- Accounts receivable . . . retainage . . . —This is for work billed and due to the contractor. Retainage (Clough, 1986, p. 152) is a portion of progress payments (usually 10%) retained by owner until all work is done by contractor. It is in addition to performance and payment bonds.
- Deposits and . . . —May include (Clough, 1986, p. 89) security deposits paid to A/E for set of plans to use in bidding.
- Inventory—E.g., pipe; might include items such as electrical materials and aggregate.
- Prepaid expenses—May be prepaid purchase price of pipe not yet delivered, dues to a contractor organization, and the like. The contractor will receive something of value.
- Notes . . . noncurrent—Future.
- Property—Fixed assets, not readily converted into cash, shown at purchase price.

TABLE 9–2 BALANCE SHEET FOR A HYPOTHETICAL CONSTRUCTION COMPANY

THE BLANK CONSTRUCTION COMPANY, INC.
PORTLAND, OHIO

BALANCE SHEET
December 31, 19__

Assets	
(a) Current Assets	
Cash on hand and on deposit	$389,927.04
Notes receivable, current	16,629.39
Accounts receivable, including retainage of $265,689.39	1,222,346.26
Deposits and miscellaneous receivables	15,867.80
Inventory	26,530.14
Prepaid expenses	8,490.68
Total Current Assets	1,679,79.31
(b) Notes Receivable, Noncurrent	12,777.97
(c) Property	
Buildings	55,244.50
Construction equipment	388,289.80
Motor vehicles	97,576.04
Office furniture and equipment	23,596.18
Total Property	564,706.52
(d) Less accumulated depreciation	422,722.51
Net Property	141,984.01
(e) Total Assets	$1,834,553.29

Liabilities	
(f) Current Liabilities	
Accounts payable	$306,820.29
Due subcontractors	713,991.66
Accrued expenses and taxes	50,559.69
Equipment contracts, current	2,838.60
Provision for income taxes	97,616.66
Total	1,171,826.90
(g) Deferred Credits:	
Income billed on jobs in progress at December 31, 19__	2,728,331.36
Costs incurred to December 31, 19__ on uncompleted jobs	2,718,738.01
Deferred credits	$9,593.35
Total Current Liabilities	1,181,420.25
Equipment Contracts, Noncurrent	7,477.72
(h) Total Liabilities	1,188,897.97
Net Worth	
(i) Common stock, 4,610 shares	461,000.00
Retained earnings	184,655.32
(j) Total Net Worth	645,655.32
(k) Total Liabilities and Net Worth	$1,834,553.29

(SOURCE: Adapted with permission from Clough, R.H., *Construction Contracting,* 5th ed., New York: John Wiley and Sons, 1986, p. 264.)

- Less accumulated depreciation.
- Total Assets—Sum of all assets.

The liability line items in the balance sheet may be explained as follows:

- Current liabilities—Debts to be completely paid within normal cycle of business.
- Income billed . . . —Contractor has billed $2,728,331.36, but work was not done. Actual work done is $2,718,738.01. Therefore, contractor "owes" $9,593.35 of work—a liability.
- Equipment contracts . . . —Contractor apparently has equipment debt.
- Total liabilities—sum of all liabilities.

The net worth portion of the balance sheet is described as follows:

- Common stock . . . —Investment in company by its stockholders—4610 shares at $100.00.
- Retained earnings—Also called *earned surplus*—available for use within corporation.
- *Note:* Book value of company = $645,655.32/4610 = $140.06 per share. The market value of stock is unknown and could be different from stock value or book value.
- Total net worth—Total assets minus total liabilities.

THE INCOME STATEMENT—INTRODUCTION

The income statement is another important type of financial statement used in business. Recall that the balance sheet had other names. Unfortunately, so does the income statement, which, as noted by Clough (1986, p. 261), is sometimes referred to as the profit and loss statement, statement of earnings, statement of loss and gain, income sheet, summary of income and expense, profit and loss summary, statement of operations, and operating statement. The term *income statement* is used in this book for simplicity and because it appears to be in common usage.

The income statement shows the type and amount of income and expense, along with the difference, over a specified period of time. Any time period is possible, but typically income statements are prepared on a monthly, quarterly, or annual basis. The basic equation for the income statement is

$$\text{Income} - \text{Expenses} = \text{Net Income}$$

As is the case with the balance sheet, two example income statements are presented—the first one applies to an individual's personal income, and the second applies to a business. Later in this chapter a second business income statement is presented as a means of further explaining the importance of the income statement in a business that provides a service.

Refer to Table 9–3, which might apply to a young professional and his or her spouse for their first full year of employment after graduation from college. This income statement might be for the hypothetical young couple whose balance sheet is presented in Table 9–1. However, there are no obvious connections between the two financial statements. Note, again, that the income statement shows income received and expenses incurred over a period of time. Contrast this with the balance sheet, which shows assets and liabilities at a point in time.

A personal income statement like that shown in Table 9–3 could be used in a postmortem mode to review income and expenses during the past year. In addition, a personal income statement could also be used in a prospective mode to plan income and its use in the near future. Retrospective and prospective uses of income statements are routinely made in the business environment.

The observation made earlier about the variation in accuracy of line items in the balance sheet also applies to the income statement. For example, while items such as salaries and interest and dividends earned can be listed to the nearest

TABLE 9–3 HYPOTHETICAL PERSONAL ANNUAL INCOME STATEMENT

	Income	
Salaries (gross)		$40,052.29
Interest/dividends		483.68
Sale of stock		1,108.23
	Total	$41,644.20 (I)
	Expenses	
Mortgage (interest and principal)		$ 9,406.16
Utilities		2,829.00
Food		3,500.00
Clothing		2,500.00
Car Payments		3,611.89
Insurance		3,421.08
Taxes		6,500.00
Entertainment/travel		3,500.00
	Total	$35,268.13 (E)
	Net income	$ 6,376.07 (NI)

penny, expense items such as entertainment/travel and clothing would be estimates unless unusually meticulous records are kept. Variation of accuracy within the income statement does not in any significant way detract from its usefulness as an analysis and planning tool.

Table 9–4 follows the same general format as the example of the personal income statement. The hypothetical business for which the income statement was developed is for the same business in the same year for which Table 9–2, the balance sheet, was developed. One indication of the relationship between the balance sheet and the income statement is "retained earnings" of $184,655.32 in Table 9–2 and the item "retained earnings, balance . . . " of $184,655.32 in

TABLE 9–4 ANNUAL INCOME STATEMENT FOR A HYPOTHETICAL CONSTRUCTION COMPANY

THE BLANK CONSTRUCTION COMPANY, INC.
PORTLAND, OHIO

INCOME STATEMENT
For the Year ended December 31, 19__

Item	Total
Project Income	$8,859,138.39
Less project costs, including office overhead expense of $239,757.04	8,705,820.15
Net Project Income	153,318.24
Other Income	
Discounts earned	23,064.93
Equipment rentals	23,758.93
Miscellaneous	12,882.64
Total Other Income	59,706.50
Net Income Before Taxes on Income	213,024.74
Federal and State Taxes on Income	97,616.66
Net income after taxes on income	115,408.08
Retained earnings	
Balance, January 1, 19__	106,127.24
Dividends paid	36,880.00
Total Retained Earnings	69,247.24
Balance, December 31, 19__	184,655.32
Earnings per share on net income	$25.03

(SOURCE: Adapted with permission from Clough, R.H., *Construction Contracting,* 5th ed., New York: John Wiley and Sons, 1986, p. 262.)

Table 9–4. However, in general, there are a few obvious connections between the balance sheet (Table 9–2), and the income statement (Table 9–4).

Note also that the accrual method of accounting is being used, not the cash method or basis. With the accrual approach, income is recognized and entered into the books as it is earned (not when the revenue is actually received, as when the cash basis is used). Expenses are recognized and entered when they are incurred (not when payments are actually made, as when the cash basis is used).

The income statement for the hypothetical construction company is taken from Clough (1986, pp. 261–263), and the explanation of various line items in the income statement is based in part on Clough. The line items in the income statement are as follows:

- Project income—Done on completed project basis. This is the sum of all income for all projects completed during the year.
- Less project costs . . . —Items like labor, fringe benefits, equipment rental, materials, subcontracts, and a portion of general overhead allocated to completed projects.
- Net project income—Difference between preceding two items. Note how small net project income is compared to project income—the former is only 1.73% of the latter. This suggests risky profitability.
- Other income—Discounts earned might be a credit or cash payment for volume purchases. Equipment rentals could mean rental of corporation's equipment to other contractors to keep it productive. Miscellaneous might be sale of building materials to contractors.
- Net income before taxes . . . —Sum of net project income and total other income.
- Federal and state taxes . . . —45% of "net income before taxes on income."
- Net income after taxes . . . —This is discretionary income to be used for purposes such as paying dividends to stockholders, raising salaries, putting back into the business, and holding onto as a cushion. What did the Blank Construction Company do with the after-tax income?
- Dividends paid—$36,880.00 of the retained earnings carried over from the beginning of the year ($106,127.24) was paid to stockholders, leaving a retained earnings balance of $69,247.24. Recall, from Table 9–2, that there are 4,610 shares, so the dividends paid per share are $36,888/4610 = $8.00.
- Balance, December 31, . . . —Sum of total retained earnings ($69,247.24) and net income . . . ($115,408.08).
- Earnings per share . . . —$115,408.08/4610 = $25.03. An index of performance.

RELATIONSHIP BETWEEN THE BALANCE SHEET AND THE INCOME STATEMENT

Although not obvious by examining them, the balance sheet and income statement for an individual or for an organization are inextricably linked. Consider a reservoir, with its continuously varying contents and fluctuating water level, as an analogy. The balance sheet for the end of a reporting period is analogous to the current contents of the reservoir. The income statement is analogous to an accounting of how much water came into the reservoir and from where and how much water went out of the reservoir and where it went during the time period. Another common analogy to the balance statement and income statement is a checking account. The end-of-month (or end-of-reporting-period) balance is like the balance sheet. The listing of deposits, checks written, fees charged, and other transactions for the month is like an income statement.

Recall the balance sheet (Table 9–2) and the income statement (Table 9–4) for the hypothetical construction company. There was one explicit connection between the two financial statements, the item "retained earnings." The income statement showed the basis for the retained earnings on the balance sheet.

Therefore, the balance sheet and income statement for an organization are completely linked. The linkage may be presented as follows:

- Balance sheet at end of period.
- Income statement for next period.
- Balance sheet at end of period.
- Income statement for next period.
- etc.

In a business or other organization, or even in one's personal financial affairs, neither the balance sheet nor the income statement are usually sufficient for a full understanding of the organizational or individual financial matters. Using the reservoir metaphor again, even if all inflows to and outflows from the reservoir were known for a year (balance sheet), the reservoir contents at the end of the year (balance sheet) could not be determined without having the reservoir contents (balance sheet) for the end of the preceding year.

There may be some exceptions to the general statement that both a balance sheet and income statement are desired. For example, a modest sole proprietorship consulting business may not require a balance sheet, because it operates on a cash basis with no significant liabilities and has little property or other assets.

FINANCIAL RATIOS

Financial professionals use various ratios to help interpret balance sheets, income statements, and other financial statements. Clough (1986, Sections 9.12 and 9.13) identifies the following four categories of financial ratios, which are shown in Figure 9–1:

- Liquidity = the ability to meet financial obligations.
- Leveraging = debt relative to assets or net worth or, stated differently, the extent to which the organization is using other people's or organization's resources.
- Activity = how well capital and other assets are being used.
- Profitability = the "bottom line."

Category	Purpose	Examples
Liquidity	Ability to meet financial obligations	1. (Current assets) / (current liabilities)
		2. (Quick assets) / (current liabilities)
Leveraging	Debt relative to assets or net worth	3. (Total liabilities) / (net worth)
Activity	How well capital and other assets are being used	4. (Project income) / (net worth)
		5. (Project income) / (net working capital)
		6. (Fixed assets) / (net worth)
Profitability	Bottom line	7. (Net project income before taxes) / (project income)
		8. (Net project income before taxes) / (net worth)

Figure 9–1 Categories of financial ratios (Source: Adapted with permission from R. H. Clough, *Construction Contracting,* 5th ed., New York: John Wiley and Sons, 1986, pp. 265–268.)

Examples of each of the eight financial ratios set forth in Figure 9–1 are presented. Data from the Blank Construction Company balance sheet (Table 9–2) and income statement (Table 9–4) are used in the eight examples and the explanations are based, in part, on Clough (1986, Sections 9.12 and 9.13).

Liquidity Ratios

- **Example 1.** (Current assets)/(current liabilities) = ($1,679,791)/($1,181,420) = 1.42

  a. Both from balance sheet.
  b. Compares liquid assets to current liabilities—the former should be greater than the latter. The larger the ratio, the better assurance that short-term debts can be paid.
  c. National median value is 1.3 for commercial building contractors.

- **Example 2.** (Quick assets)/(current liabilities) = ($1,653,261)/($1,181,420) = 1.40

  a. Both from balance sheet—"quick assets" is current assets minus inventory.
  b. Similar in purpose to previous ratio.
  c. Applicable to personal.

Leverage Ratio

- **Example 3.** (Total liabilities)/(net worth) = ($1,188,897)/($645,655) = 1.84

  a. Both from balance sheet.
  b. Also called "debt to equity ratio."
  c. Indicates relative amounts that creditors (others) and owners have invested in the business.
  d. National median value is 2.0 for commercial building contractors.
  e. From an owner's perspective, the higher the ratio, the better the owner is using someone else's money.
  f. From a creditor's perspective, the higher the ratio, the less assurance the owners (contractor, A/E) will be diligent in running business and, therefore, in repaying loans.
  g. From a potential client/customer's perspective, the higher the ratio, the less assurance the contractor or A/E will complete project.

Activity Ratios

- **Example 4.** (Project income)/(net worth) = ($8,859,138)/($645,655) = 13.7

  a. First from income statement and second from balance sheet.
  b. Indicates how well net worth is being used.

c. National median value is 9.9 for commercial building contractors.
d. If ratio is too low, funds are being underused or are stagnant, and profitability is probably going to suffer.
e. If the ratio is too high, liabilities may build up too rapidly.

- **Example 5.** (Project income)/(net working capital) = (project income)/(current assets − current liabilities) = ($8,859,138)/($1,679,791 − $1,181,420) = ($8,859,138)/($498,371) = 17.8.

  a. First from income statement and second from balance sheet.

- **Example 6.** (Fixed assets)/(net worth) = ($141,984 from balance sheet)/($498,371 from income statement) = 0.22.

Activity ratios tend to show that the construction industry and the A/E business are easy to get into in that little capital is needed; they are labor intensive, but there is no guarantee of large profitability.

Profitability Ratios

- **Example 7.** ((Net project income before taxes)/(project income)) × 100 = (($153,318 from income statement)/($8,859,138 from income statement)) × 100 = 1.73%.

  a. National median value is 1.6% for commercial building contractors.
  b. Emphasizes small profit margins in construction. For example, a 2% increase in project expense (1.02 × $8,705,820 = $8,879,936, which is greater than income) would produce a negative profitability.
  c. Consulting firms are similar, but not quite as extreme.
  d. Careful management is very important.

- **Example 8.** ((Net project income before taxes)/(net worth)) × 100 = (($153,318)/($645,655 from balance sheet)) × 100 = 23.7%.

  a. Measures efficiency of use of invested capital. Considering risks in construction, this percent should be greater than current dividends or interest rates.

- **Example 9.** ((Net project income before taxes)/(total assets)) × 100 = (($153,318)/($1,834,553 from balance sheet)) × 100 = 8.4%.

The preceding three profitability ratios may be small because of high salaries paid to officers or principals (and appearing as "project costs" on the income statement).

THE IMPACT OF TIME UTILIZATION RATE AND EXPENSE RATIO ON PROFITABILITY IN THE CONSULTING BUSINESS

Consulting firms, and most other private engineering-oriented organizations, must generate a profit. Profitability may not be their primary reason for existence when other possibilities, such as personal satisfaction and service to society, are considered. Although profitability may not be sufficient, it is necessary, as nicely explained by Peters and Waterman (1982, p. 103) who said, "Profit is like health. You need it, and the more the better. But it is not why you exist."

This profitability-oriented discussion is targeted primarily at entry-level engineers and other technical professionals who are on the staff of consulting engineering firms. More specifically, the discussion is directly applicable to businesses that generate most of their revenue by selling time as opposed to selling products, that is, offering services rather than producing goods. Profitability, or more specifically, what it is sensitive to, should obviously be of interest to engineers and others employed by consulting firms. The topic should also be of interest to government personnel, private sector personnel, and others who retain consultants so that they are in a better position to understand how the consulting firms operate.

A/E profitability (Norris, 1987) may be stated as Profitability = $f(U, R,$ other factors) where for a specified time period:

- U = utilization rate (or chargeable rate or billable rate). Consider all the hours worked by salaried and hourly personnel at all levels. Recognize that some of the hours are charged to clients. Then U = (charged time in hours) /(total time in hours). U is always less than 1.0, because not everyone can be working directly on client projects all the time.

- R = expense ratio. Let S = nonsalary costs of a business that are not billed directly to clients (e.g., Social Security, professional liability insurance, rent, utilities, entertainment). Let P = total payroll cost, that is, the dollars paid to employees for all the hours they worked as described under the description of U. Then R = expense ratio = S/P.

The preceding terms will be defined further, including why they are critical to the profitability of an A/E and who in the firm has primary control over U and R. Intuitively, raising U and decreasing R should increase profitability.

As a further introduction to time utilization rate (chargeable rate) and expense ratio, consider Table 9–5. Note that the accrual method of accounting, as described earlier in this chapter, is being used.

Review of each line item:

TABLE 9-5 HYPOTHETICAL INCOME STATEMENT FOR A CONSULTING ENGINEERING COMPANY

	Year to date	
Total revenues	$1,800,000	
Less reimbursable expenses	100,000	
Less outside consultants	500,000	
Net revenues	1,200,000	
Less direct labor	400,000	
Less nonreimbursables	10,000	
Gross income	790,000	
Less overhead	600,000	
Net income before taxes	$190,000	Profit as a % of total revenue = 10.5%
		Overhead ratio = O = 600,000/400,000 = 1.5
		Multiplier = M = 1,200,000/400,000 = 3.0

(SOURCE: Adapted with permission from Birnberg, H.G., "Communicating the Company's Operating Performance Data," *Journal of Management in Engineering-ASCE,* Vol. 1., No. 1 (January 1985) pp. 12–19.)

- Total revenues—Money paid to or due to A/E firm.

- Reimbursable expenses—Expenses (e.g., travel, printing) billed to client and paid to or due to A/E (exclude "markup," overhead, etc.). This revenue comes in, but is immediately used by the firm to reimburse expenses—no gain here for the firm.

- Outside consultants—More expenses billed to client.

- Net revenues—Revenues generated by in-house labor. But firm has to pay staff that worked on projects.

- Direct labor—Raw labor cost, that is, money paid to or due to employees who worked on projects; excludes fringes.

- Nonreimbursables—Expenses incurred as a result of projects, but not billable to client (e.g., unexpected lab test needed and not covered in contract or acceptable to client as a contract change.)

- Gross income—Income after project expenses accounted for, but before overhead. That is, all of the preceding are income and expenses directly related to projects.

- Overhead—Given as $600,000. While projects were underway, many and various expenses were being incurred within the A/E firm which were not directly billed as such to the client. Overhead = $S + P'$ = \$600,000. P' is the sum of salaries and hourly pay not billable to client, such as vacation, illness, holidays, bonuses, office staff, and business development. Recall that S is nonsalary costs not billed to client such as Social Security, unemployment insurance, workmen's compensation insurance, professional liability insurance, rent, utilities, and entertainment. Recall that P = total payroll. Then $P - P'$ = payroll paid by project income. Then overhead ratio = O = (overhead)/(direct labor cost) = $(S + P')/(P - P')$. The overhead ratio, which is usually greater than 1.0, is "burden" on direct labor—direct labor has to be "marked up" to recoup overhead.

- Net income before taxes (profit)—Note how overhead takes a big cut; gross income is reduced from \$790,000 to \$190,000 because of overhead. Net income before taxes is 10.5% of total revenues.

Note that profitability is sensitive to overhead. For example, assume that overhead increases 10% from \$600,000 to \$660,000. Then pretax profit drops \$60,000, or 31.6%. Overhead increases, dollar for dollar, go directly to and come off the bottom line.

Therefore, look further at the components of O, the overhead ratio (Norris, 1987):

- From before: $O = (S + P')/(P - P')$
- Divide numerator and denominator by P: $O = (S/P + P'/P)/(P/P - P'/P)$
- Add and subtract 1 in numerator where 1 is (P/P): $O = (S/P + P'/P + 1 - P/P)/(P/P - P'/P)$
- Rearrange: $O = (1 + S/P - (P/P - P'/P))/(P/P - P'/P)$
- Note that $S/P = R$, the expense ratio, based on earlier definition, and it is controlled largely by principals and upper management.
- $P/P - P'/P$ is an approximation of what was defined earlier as U, the utilization rate.
- Therefore, $O = (1 + R - U)/(U)$

Refer to Figure 9–2 for a graph of $O = f(U, R)$ and consider the sensitivity of O to U and R. Focus, for example, on the zone in which A/E firms tend to operate centered around $U = 0.600$, $R = 0.500$, and $O = 1.500$. These three typical values exactly satisfy the previously derived equation.

Assume that R remains constant, but U drops one percentage point from 0.60 to 0.59 (a 1.7% drop). Then $O = (1.00 + 0.50 - 0.59)/(0.59) = 1.542$. A

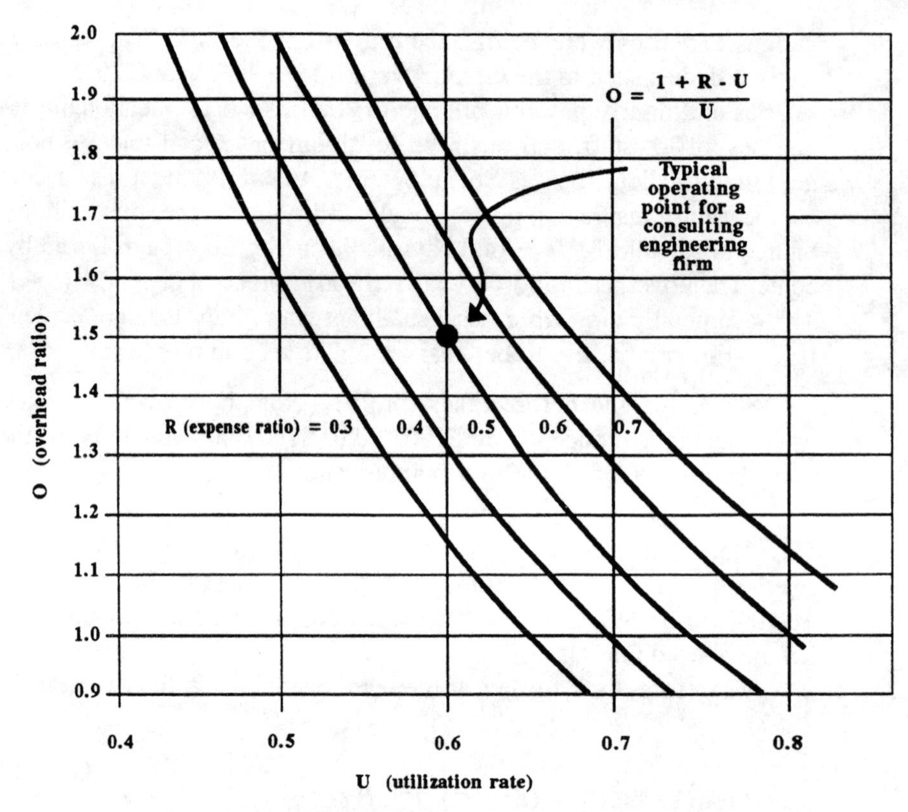

Figure 9–2 Overhead ratio as a function of the utilization rate and expense ratio (Source: Adapted with permission from W. E. Norris, "Coping with the Marketplace: A Management Balancing Act," *Journal of Management in Engineering—ASCE,* Vol. 3, No. 3 (July 1987), p. 196.)

small (one percentage point, or 1.7%) decrease in U causes a large (over four percentage points, or 2.8%) increase in O. Note how this affects the net year-to-date profit of the A/E firm whose income statement is shown in Table 9–5.

- Assume that O increases from 1.500 to 1.542 (2.8%).
- Then overhead increases from $600,000 to $616,800 or by $16,800 (2.8%).
- Therefore, profit (before tax) drops by $16,800 from $190,000 to $173,200—an 8.8% decrease.
- And percent profit drops from 10.5% to 9.6% of total revenue.
- Thus, a 1.7% decrease (1 percentage point) in U causes an 8.8% decrease in profit—about five to one.

Assume U remains constant, but R increases one percentage point from 50% to 51% (a 2% increase) for the firm whose income statement is shown in Table 9–5. Then $O = (1.00 + 0.51 - 0.60)/(0.60) = 1.517$. Therefore, a 2% (one percentage point) increase in R causes a 1.1% increase in O. See how this affects the year-to-date profit for the A/E firm whose income statement is shown in Table 9–5.

- If O increases from 1.500 to 1.517, overhead increases from $600,000 to $606,800.
- As a result, profit (before tax) drops by $6,800 from $190,000 to $183,200—a 3.50% decrease.
- Thus, a one percentage point (2%) increase in R causes an almost 4% decrease in profit—almost two to one.

What are the personal, project, and organizational management implications of the impact of time utilization rate and expense ratio? Top managers will watch R, the expense ratio, very closely. They control most of it—refer again to its components. They should be very aware that a one percentage point increase in R will cause a roughly two percent drop in profit. Top managers and all staff control U. Time is typically accounted for (e.g., logged into each employee's time sheet) to at least the nearest 0.25 hour. Time utilization is usually tracked on a weekly basis for the entire organization to the nearest 0.1%. All time legitimately worked on projects must be charged to clients—provided "the budget can take it."

Something as harmless-looking as everyone "knocking off" an hour early on Friday to clean up the office (not chargeable) would have a disastrous effect on profitability. Assume that a typical engineer normally charges 24 hours/week to projects ($U = 0.60$). Then, if $R = 0.5$, $O = (1.00 + 0.50 - 0.60)/(0.60) = 1.50$. A one-hour decrease drops this to $U = 0.575$ for a 4.2% decrease. Then, if $R = 0.5$, $O = (1.00 + 0.50 - 0.575)/(0.575)$. Therefore, O increases 7.3%. If everyone in the previously discussed hypothetical consulting firm "dropped" (failed to charge) one hour per week, overhead would increase by approximately 7% and profit would decrease by roughly 23%. Stated differently, the approximately 4% reduction in billable time as a result of knocking off early to clean up the office would diminish profit by approximately 23%.

In summary, consulting firms must be profitable. The income statement shows profit and factors leading to it. Overhead goes to bottom line, where it impacts profit. The absolute value of overhead is determined by the overhead ratio, which is a function of the utilization rate (U) and the expense ratio (R). A 1% increase in U, or decrease in R, will typically cause a several-fold percent increase in profit, with profit being more sensitive to U than R.

THE MULTIPLIER

A common term in the consulting business is the multiplier, which is a measure of a firm's efficiency. Refer again to Table 9–5, which is a hypothetical year-to-date income statement for a consulting engineering firm. The multiplier (*M*) is defined as net revenues divided by the direct cost of labor used to produce the revenues (a dimensionless parameter). Stated differently, the hours of labor that cost the firm $400,000 must generate total net revenues of $1,200,000. Therefore, the multiplier is a factor that the salary chargeable to projects must be "marked up" to cover the raw salary itself, nonreimbursables, overhead, taxes, and profit. Another way of viewing the situation is that the employer (the consulting firm) buys labor wholesale at the raw labor rate, and sells it retail, at the marked-up or multiplied rate, to its clients. Consider a young engineer employed by the hypothetical consulting firm and receiving an annual salary of $34,000. The young engineer's raw salary is $16.35 per hour, that is, $34,000 divided by 52 weeks and 40 hours per week. For each hour the young engineer works on a client's project, the client will be billed at a charge-out rate calculated as the multiplier times the young engineer's hourly rate, or 3.0 × $16.35 = $49.40 per hour.

The multiplier is one measure of cost competitiveness between consulting firms. Assuming that a particular engineering project requires a fixed number of different kinds of personnel and that raw salaries are similar between firms, the smaller the multiplier of a given firm, the less charge there will be to the client and the more cost-competitive the firm will be. Accordingly, potential clients often ask consulting firms for their multiplier, and most consulting firms try to keep their multiplier as low as possible.

Note, again, the assumptions that must be met before a multiplier comparison is useful. Each firm is assumed to require about the same number of hours of various types of professional personnel for a given project, and compensation rates for various levels of personnel are essentially the same among the consulting firms. Numerous situations may occur that invalidate these assumptions. For example, a consulting firm may have a personnel strategy under which premium levels of compensation, including benefits, are provided to existing employees and offered to prospective employees for the purpose of attracting the very best personnel. On the surface, this personnel strategy would tend to produce a high multiplier. On the other hand, this same firm might be significantly more productive because of superior personnel. That is, this firm might be able to do any given project with significantly fewer hours of labor than other firms. Even though compensation levels and, therefore, the multiplier are higher than average, the total fees paid by clients for particular projects might be less than the fees charged by most firms.

One way a firm can reduce its multiplier is to reduce its overhead. As already discussed in great detail, overhead can be reduced by increasing U, the utilization rate, or by decreasing R, the expense ratio. That is, reducing overhead costs through improved utilization of time of personnel and reduction in expenses will have the favorable effect of decreasing the multiplier. Of course, another way to reduce the multiplier, as suggested by focusing on the bottom line on the income statement shown in Table 9–5, is to reduce the profit expectation. The reduction of expected profit permits a reduction in expected net revenue which, in turn, tends to reduce the multiplier.

Consider a numerical example in which an increase in utilization rate results in a decrease in the multiplier. For the base line situation, assume the income statement presented in Table 9–5. Assume further that the hypothetical firm is able to increase its utilization rate from 0.60 to 0.61, a 1.7% increase. Based on the previously presented relationship showing the overhead ratio as a function of expense ratio and utilization rate, the stated increase in U would reduce the overhead ratio by about 2.8% or $16,800 from $600,000 to $583,200. If profit is held at $190,000, the firm can reduce total revenues and, therefore, net revenues by $16,800 to respectively, $1,783,200 and $1,183,200. Therefore, the revised multiplier is equal to $1,183,200 ÷ $400,000 = 2.96, which is a 1.33% decrease from the base line value of 3.00. In summary, a 1.7% increase in time utilization rate for the situation used in the example yields a 1.3% decrease in the multiplier with no reduction in profit.

THE INCOME STATEMENT AS PART OF THE BUSINESS PLAN FOR A CONSULTING FIRM

The preceding discussion of business income statements is retrospective. The emphasis is on the use of income statements to document what has happened in a consulting firm, construction firm, or other business. Income statements can also be used in a prospective manner. That is, the income statement can be one part of the coming year's business plan for a consulting firm, construction firm, or other business.

The income statement could also be part of the first year's business plan for a contemplated new business. Other typical elements of a business plan are market analysis, description of services to be offered, assessment of competition, and means of finance (e.g., see Fenske and Fenske, 1989.) The use of an income statement for the second purpose, that of a contemplated new business, is illustrated in Table 9–6. The example income statement applies to the first year's operation of a small—three engineers, one technician, and one secretary—consulting business. The income statement is created on a spreadsheet to facilitate running many scenarios. Careful review of the particular scenario presented in

TABLE 9–6 PLANNED ANNUAL INCOME STATEMENT FOR A NEW CONSULTING BUSINESS

YEAR: CY 199?
SCENARIO: 1
(us)

Nomenclature:
(us) user supplied
(c) calculated
1, 2, 3 . . . calculation sequence

File Name: PincSt
Date: 06/29/94

Professional and other staff salary and utilization

Category	Number in category	Annual raw salary per employee (assuming full-time employment)	Raw labor rate	Portion of work year employed	Annual raw salary for all employees	Utilization rate (U)	Billable direct labor time per employee		Billable direct labor cost for all employees
(us)	(us)	(us)	(c)	(us)	(c)	(us)	(c)	(c)	(c)
		($)	($/hr)	(%)	($)	(%)	(work weeks)	(work days)	($)
Engineer	3	30,000	14.42	100	90,000	60	31.2	156	54,000
Technician	1	20,000	9.62	100	20,000	90	46.8	234	18,000
Secretary	1	16,000	7.69	100	16,000	50	26.0	130	8,000
Total Employees	5								
		Total direct labor cost (P) =			126,000				
						Total billable direct labor cost ($P - P'$) =			80,000

Overhead			
Total nonbillable direct labor cost (P') =	46,000	(c)	
Nonbillable, nonsalary costs (S)			
Office space (5 people × 125 ft^2/person × \$10/ft^2/yr)	6,250	(us)	
Vehicle (40,000 mi × \$0.35/mi)	14,000	(us)	
Computer HW, SW, supplies, and training	4,000	(us)	
Supplies and printing	3,000	(us)	
Marketing	5,000	(us)	
Fringe benefits	31,000	(us)	
Other	750	(us)	
Subtotal (S) =	64,000	(c)	
Total overhead ($P' + S$) =	110,000	(c)	
Expense ratio = $R = S/P$ = (nonbillable, nonsalary costs)/(total direct labor cost) =	0.51	(c)	
Overhead ratio = 0 = $(P' + S)/(P - P')$ = (overhead)/(billable direct labor) =	1.38	(c)	
Income Statement			
Annual total revenue	201,000	(c)	5
Less reimbursables	0	(us)	
Less outside consultants	0	(us)	
Net revenue	201,000	(c)	4
Less billable direct labor ($P - P'$)	8,000	(c)	3
Less nonreimbursables	1,000	(us)	
Gross income	12,000	(c)	2
Less overhead ($P' + S$)	110,000	(c)	1
Net income before tax	10,000	(us)	

(continued)

TABLE 9–6 PLANNED ANNUAL INCOME STATEMENT FOR A NEW CONSULTING BUSINESS (*CONTINUED*)

Multiplier = (Net revenue)/(billable direct labor) = U =	2.51	(c)

Charge-out rates

Category (us)	Raw labor rate (c) ($/hr)	Charge-out rate (c) ($/hr)
Engineer	14.42	36.24
Technician	9.62	24.16
Secretary	7.69	19.33

Table 9–6 illustrates the application of various topics covered in this chapter, including raw labor rate; utilization rate; overhead as a sum of nonbillable direct labor cost and nonbillable, nonsalary costs; expense ratio; overhead ratio; and charge-out rates. The calculated values of parameters such as expense ratio (*R*), overhead ratio (*O*), and multiplier (*M*) when compared to values in the consulting industry provide a check on the reasonableness of each scenario. Unreasonable values of parameters would be cause for exploring additional scenarios until a workable income statement can be developed.

PROJECT OVERRUNS: IMPLICATIONS FOR PROFITABILITY AND PERSONNEL

Budgets for planning, design, and other projects typically performed by consulting firms, for construction projects carried out by contractors, and for design and production projects performed by manufacturing organizations are usually prepared as part of the process of negotiation between the firm and the client, owner, or customer. A typical, hypothetical budget estimate for a project to be carried out by a consulting engineering firm is shown in Table 9–7.

The contract or agreement between the consulting firm, contractor, or manufacturer and the client, owner, or customer typically "locks in" the total amount, which in this case is $76,000. An exception to this is successful use of the formal change-of-scope provision usually included in the agreement. Consider what happens if the budget is exceeded. An examination of a typical income statement (e.g., Table 9–5) for a consulting engineering firm indicates that the net revenue will not increase (unless a change of scope is negotiated with an additional fee). All expenses and labor costs incurred as a result of continued effort on the project

TABLE 9–7 HYPOTHETICAL PROJECT BUDGET PREPARED BY A CONSULTING ENGINEERING FIRM

Item	Quantity	Unit cost $	Total cost $
Engineer I	1000 hours	45 (1)	45,000
Engineer II	100 hours	60 (1)	6,000
Surveyor I	500 hours	30 (1)	15,000
Technician I	600 hours	25 (1)	9,000
Transportation	800 miles	0.25	200
Miscellaneous			800
		Total	$76,000

(1) Raw salary times multiplier.

may not be billed to the client and, therefore, go into overhead, with the appropriate overhead factor applied to the raw labor costs. Increases in overhead go directly to and subtract from the bottom line.

As a project gets into budget problems, particularly within consulting firms, salaried personnel may be expected to work on the project on their own time without additional compensation so that no additional labor charges are incurred by the project. You should not consider entering the consulting field if you are not willing to occasionally do this, even when "it's not your fault." While effective self-management, management of relationships with others, and project management can significantly reduce the incidence of project overruns, such overruns will occasionally occur. They must be dealt with in a profit-conscious manner.

REFERENCES

BIRNBERG, H. G., "Communicating the Company's Operating Performance Data," *Journal of Management in Engineering—ASCE,* Vol. 1, No. 1 (January 1985), pp. 12–19. (Discussed by M. D. Hensey in the *Journal*, Vol. 1, No. 3 (July 1985), p. 175.)

CLOUGH, R. H., *Construction Contracting*, 5th ed. Chapter 9, "Business Methods." New York: John Wiley and Sons, 1986.

FENSKE, S. M., and T. E. FENSKE, "Business Planning for New Engineering Consulting Firms," *Journal of Management in Engineering— ASCE,* Vol. 5, No. 1 (January 1989), pp. 89–95.

KAMM, L. J., *Successful Engineering—A Guide to Achieving Your Goals.* New York: McGraw-Hill, 1989.

NORRIS, W. E., "Coping with the Marketplace: A Management Balancing Act," *Journal of Management in Engineering—ASCE,* Vol. 3 No. 3 (July 1987), pp. 194–200.

PETERS, T. J., and R. H. WATERMAN, Jr., *In Search of Excellence.* New York: Harper & Row, 1982.

SUPPLEMENTAL REFERENCES

HOLTZ, H., *Expanding Your Consulting Practice with Seminars*, Chapter 5, "Typical Costs and Income Projections." New York: John Wiley and Sons, 1987.

MERRILL LYNCH, PIERCE, FENNER, AND SMITH, INC., "How to Read a Financial Report," 5th ed., 1984.

SHUMAN, C. H., "Managing for Profit," *Civil Engineering—ASCE,* November 1992, pp. 72–73.

STURDIVAN, M., "Trim Your Overhead to Increase Profitability," *FMG Journal*, March 1991, pp. 18–20.

SUNAR, D. G., *Getting Started as a Consulting Engineer*, Chapter 8, "Sales and Fees." San Carlos, Cal.: Professional Publications, Inc., 1986.

EXERCISES

9.1 ANALYZING FINANCIAL STATEMENTS

Purpose

Improve the student's understanding of balance sheets, income statements, and financial ratios.

Tasks

1. Obtain recent balance sheets and income statements for an actual business, preferably a technically oriented business such as a manufacturing firm, a constructor, or a consulting engineering organization. Use at least two consecutive years.
2. Calculate at least four financial ratios and comment on their meaning—that is, what do they tell you about the business and changes it is undergoing?
3. Submit your calculations, your analysis, and the financial statements to the instructor in memorandum form.
4. You may do this assignment as a group of two.

9.2 CREATING BUSINESS FINANCIAL STATEMENTS

Purpose

Improve the student's understanding of balance sheets and income statements, in general, and of the interrelationships between them in particular.

Tasks

1. Refer to the example balance sheets presented in this Chapter (Tables 9–1 and 9–2) and the example income statements (Tables 9–3, 9–4, 9–5, and 9–6).
2. Using a spreadsheet program, create a balance sheet for a hypothetical consulting, construction, or manufacturing firm as of the end of this calendar year. Assume that the firm's fiscal year is coincident with the calendar year.
3. Create an income statement for the next calendar year.
4. Use the results of Tasks 2 and 3 to create the balance sheet for the end of next year.
5. Repeat Tasks 2, 3, and 4 to simulate a "worse" year financially. Indicate what you changed and how the financial results deteriorated.
6. Repeat Tasks 2, 3, and 4 to simulate a "better" year financially. Indicate what you changed and how the financial results improved.

7. Present your spreadsheets and an interpretation of them in the form of a memorandum to the instructors.

9.3 PROJECTED INCOME STATEMENT FOR FIRST YEAR OF A NEW CONSULTING BUSINESS

1. Set up an annual income statement, similar in structure and function to Table 9–6 on a spreadsheet.
2. Assume that you and others are forming a consulting business that will begin operation on January 1 of the next year.
3. Do at least five possible trials for the first year of business. For each trial, show or state all assumptions.
4. Check each trial against "industry" patterns or typical values.
5. Assuming the project is assigned to be done by teams, make the following consulting firm assumptions for each team:
   a. Team 1: Firm will have three technical professionals plus technicians and secretaries as appropriate.
   b. Team 2: Firm will have 30 technical professionals plus technicians and secretaries as appropriate.
   c. Team 3: Firm will have 100 technical professionals plus technicians and secretaries as appropriate.
   d. Team 4: Firm will have 1,000 technical professionals plus technicians and secretaries as appropriate.
   e. Team 5: Firm will have 2,000 technical professionals plus technicians and secretaries as appropriate.
6. Prepare an executive summary memorandum, with all spreadsheets attached, that describes your analysis and sets forth your recommendations.

10

Legal Framework

As always happens in these cases, the fault was attributed to me, the engineer, as though I had not taken all precautions to ensure the success of the work. What could I have done better? (Written 152 A. D. by Nonius Datus, the Roman engineer responsible for the design and construction of a water supply tunnel through a mountain in what is now Algeria, upon visiting the construction site at which the tunnel was being excavated from both ends. He had just learned that the segments were out of alignment and had passed each other. Quoted in de Camp, 1963, p. 27)

Engineers should not practice law, just as lawyers should not practice engineering. However, just as knowing the basics of business accounting helps the young engineer practice engineering, so knowing the basics of law will help the young engineer practice engineering. The engineer should know enough about the legal aspects of engineering practice to recognize when he or she needs to take certain actions or to know when legal counsel would be prudent. The purpose of this chapter is to provide the young engineer or other technical professional with an understanding of basic legal principles and broad guidelines of law.

After citing examples of circumstances in which the entry-level professional should know some legal fundamentals, selected legal terms are explained in this chapter. Three ways in which liability is incurred are introduced, followed by examples of structural failures and lessons learned from them. The chapter concludes with a discussion of ways to minimize liability and an admonition to keep liability minimization in perspective.

THE ENTRY-LEVEL PROFESSIONAL AND LEGAL CONSIDERATIONS

Dunham et al. (1979, Chapter 1) discuss ways and describe situations in which the entry-level technical professional may need to know basic principles and broad guidelines of law. Supplementing these with additional circumstances leads to the seven items summarized in Figure 10–1. Each of the items listed in Figure 10–1 is discussed here.

1. Preparing contracts for services. The young professional may be asked to prepare or, more likely, help to prepare a contract or agreement for services between various entities such as a consulting engineering firm and a client, a construction contractor and an owner, or a manufacturer and a customer. The young technical professional may do this as a representative of the consulting firm, the manufacturer, or the owner. The desired product may be a set of plans and specifications; a constructed building, facility, or system; or a manufactured product.

2. Interpretation of contracts once a project is underway. Even a well-crafted, mutually acceptable contract or agreement will require numerous interpretations during the project. For example, the client's representative may call the project engineer at a manufacturing firm after reviewing a draft design and request that more alternatives be developed and examined. The engineer must decide if the request is reasonable, that is, within or beyond the scope of the contract.

1. **Preparing contracts for services.**
2. **Interpretation of contracts once a project is underway.**
3. **Managing to minimize personal and organizational liability.**
4. **Anticipating and/or preparing for expert-witness testimony.**
5. **Being aware of the requirements in local, state, and federal laws and rules.**
6. **Being aware of the ways in which state, federal, and other programs may provide funding for clients' projects.**
7. **Being aware of relevant pending or recent legislation and possible impacts on projects.**

Figure 10–1 Some situations in which the entry-level technical professional should know basic principles of law

Or assume that the agreement between the consulting firm and a client indicates that the latter will contract for geotechnical services with a third party, as such services may be needed. However, once the overall project is underway, someone on the owner's staff contacts the young engineer at the consulting firm and says, "Why don't you retain the geotechnical firm as subcontractor to your firm—you are more familiar with geotechnical firms anyway?" The young engineer has to determine if this is a reasonable request. The young engineer needs to or ought to determine if this request is as logical and potentially profitable as it sounds or if, because of other more important factors, prudence usually requires that the consulting firm decline the suggestion.

3. Managing to minimize personal and organizational liability. As the young engineer or other technical professional goes about his or her work, especially when doing what may appear to be relatively mundane operations and tasks, he or she should be aware of ways in which personal liability and the liability of his or her organization can be minimized. Numerous suggestions, many of which are both simple and potentially powerful, are presented later in this chapter.

4. Anticipating and/or preparing for expert witness testimony. This situation is similar to the preceding, but much more focused in that it assumes that the technical person is going to be involved as an expert witness within the litigation process. You, as an entry-level professional, are not likely to serve as an expert witness. However, you may encounter more experienced colleagues who are providing this service, and you can learn from them. Expert witness testimony is not treated in this chapter, but is discussed elsewhere (e.g., McQuillan, 1984).

5. Being aware of the requirements in local, state, and federal laws and rules. The practice of engineering and some other technical professions in the public and private sector is typically heavily influenced and constrained by the requirements of local, state, and federal laws and administrative rules. You are strongly urged to learn about those requirements in the very early stages of a project.

Consider, as an example, a consulting engineering firm responsible for planning or designing the storm water management system for a new residential development on the periphery of a growing U. S. city. The community is likely to have zoning, subdivision, drainage, and other codes and requirements that must be satisfied by the engineer's project. County and regional rules and regulations may also apply, particularly if the development will initially be outside of the corporate limits of the city. State water laws may also be applicable. For example, if the storm water system includes a detention facility, the outlet control structure may qualify as a dam under state law and require a state permit. Federal regulations and codes may also apply. Perhaps a stream passing through the area has a 100-year flood plain delineated under the flood insurance program administered by the Federal Emergency Management Agency. A storm water discharge permit may be required under the National Pollution Discharge Elimination System.

Note that all of the preceding examples apply only to the water resources portion of the project. A similar set of city, county, regional, state, and federal expectations may also exist for other aspects of the new development, such as its wastewater system, its water supply system, and its streets and highways. Unfortunately, too many technical professionals, many of whom should know better, tend to move well into a planning or design project before determining the applicable rules and regulations.

6. Being aware of the ways in which state, federal, and other programs may provide funding for clients' projects. While local through federal regulations and legislation may sometimes be reviewed as a problem because of the many requirements that must be met, some legislation, particularly at the state and federal level, also includes a "carrot" in the form of partial funding. The consulting engineering firm or other technical organization that strives to be of full service will commit the necessary resources to be aware of existing legislation that could be useful to their clients.

For example, some states have legislation that enables local communities to implement storm water management utilities which provide, through user and other fees, a means of equitably generating revenue to finance the planning, design, construction, operation, and maintenance of storm water systems. Consulting firms should know about such legislation and how to help a given community establish a utility.

7. Being aware of relevant pending or recent legislation and possible impacts on projects. In addition to knowing about already enacted legislation, you, or somebody in your organization, should be tracking pending legislation. Given the multiyear span of some planning–design–construction or manufacturing projects, legislation being debated at the beginning of the process might be enacted and available to benefit a client near the end of the process.

LEGAL TERMINOLOGY

The legal profession, like the engineering, medical, and many other professions, has a special set of terminology. Selected legal terms are defined in Figure 10–2. The four most important terms in Figure 10–2 for the young technical profes-

breach of duty violation of a right, a duty or a law, either by an act of commission or by nonfulfillment of an obligation.

caveat emptor let the buyer beware...the buyer should take pains to discover for himself any obvious defects in an article he is about to purchase.

contractor the party (either individual or organization) who undertakes for a stated price to supply goods or to perform a construction job or other project for the owner...controls the work of construction... (a)

drawings sketches and line drawing as well as...notes of explanation or instruction inserted thereon. Prints and other reproductions of drawings are generally deemed to be the equivalent of the originals from which they are made. (a)

engineer the architect or the engineer (or both) who acts for and in behalf of the owner...an engineering organization as well as an individual. (a)

fraud intentionally deceitful practice aimed at depriving another person of his rights or doing him injury in some respect.

injunction a writ issued by a court of equity ordering a person to refrain from a given course of action.

law, common those maxims and doctrines which have their origin in court decisions and are not founded upon statute. (b)

law, statute rule of conduct, enacted by the duly authorized legislative authority...represents the express, written will of the lawmaking power... (b)

liability being bound or obligated according to law or equity.

litigious prone to engage in lawsuits.

negligence breaches...duty to exercise requisite care and expertise... below the appropriate standard of care. (d)

owner individual or organization for whom something is to be built or furnished under contract...the purchaser who pays for the goods or services. (a)

privity successive (or mutual) relationship to the same property rights; a connection between parties.

statues of limitations laws that apply to architect-engineers and construction contractors and establish time beyond which these parties are no longer liable for damages arising out of completed construction projects...a typical statute provides for a three-year period for torts (negligence) and a six-year period for breach of contract. (a)

Figure 10–2 Definitions of selected legal terms

tort Essentially, "tort" is very similar to "wrong"; yet the former term is not intended to include any and all wrongful acts done by one person to the detriment of another but only those for which the victim may demand legal redress. Torts may be committed intentionally or unintentionally and with or without force. At the risk of oversimplification it may be said that tortious acts consist of the unprivileged commission (or omissions the case may be) of acts whereby another individual receives an injury to his person, property, or reputation. A tort is distinguished from a crime in that the former is a private injury on account of which suit may be brought by the affected party, while the latter is an offense against the public for which any retribution must be sought by the appropriate governmental authority. Obviously, it is entirely possible for a single act to constitute at once a tort and a crime. (c) (Note: Key ideas-wrongful act against a person or his property or reputation, intentional or unintentional, victim seeks legal remedy.)

written information information typed or printed or recorded in longhand. (a)

a. Quoted with permission from Dunham et al., 1979, Chapter 1, "Introduction."

b. Quoted with permission from Dunham et al., 1979, Chapter 2, "Law and Courts."

c. Quoted with permission from Dunham et al., 1979, Chapter 20, "Torts."

d. Quoted with permission from Dunham et al., 1979, Chapter 25, "Professional Liability of Architects and Engineers."

e. Quoted from Clough, 1986, Chapter 4, "Drawings and Specifications."

Figure 10–2 Definitions of selected legal terms (cont.)

sional and for purposes of this chapter are *breach, fraud, negligence,* and *liability.* Breach, fraud, and negligence are the three ways in which an individual professional or a technically based organization can incur liability. Each of these terms is discussed later in this chapter.

CHANGING ATTITUDES: ADDED BURDEN ON THE TECHNICAL PROFESSIONAL

Decades ago in the United States and elsewhere, people were inclined to take risks and accept the consequences whether they be favorable or unfavorable. However, in recent times citizens of many countries have become increasingly inclined to seek relief through legal means, especially when risk taking has resulted in unfavorable consequences.

A new concept of social injustice has evolved. As in the past, accidents and failures are recognized as being expensive as a result of factors such as medical costs, additional materials, and schedule delays. However, increasingly common is the idea that someone else should pay the costs, or at least part of them. More specifically, the tendency is to look for "deep pockets," that is, those individuals, organizations, or other entities having the greatest financial resources which, depending on the situation, may be a government agency, a contractor, a consulting engineer, a manufacturing firm, or an individual engineer. In effect, there has been a litigation explosion across much of society, and this includes the engineering field (Allen, 1988). As stated by DPIC (1988, p. vii) " . . . in this (litigious) environment, the private practice of a design professional can be particularly vulnerable, because of the damage that even an unfounded lawsuit can do to a reputation and financial stability." Forewarned is truly forearmed in minimizing exposure to litigation. An unfortunate result of the litigation explosion is the tendency to stifle innovation within engineering practice (Huber, 1988).

LIABILITY: INCURRING IT

As indicated in Figure 10–2, liability is defined as "being bound or obligated according to the law or equity." Liability means that individuals or organizations are responsible for doing a conscientious job and, if they do not, they will be held accountable. The concept of liability and its incorporation into laws goes back to ancient times. For example, 4,000 years ago the Babylonians developed the Code of Hammurabi, which clearly stated the importance of individual responsibility. As quoted in Petroski (1985, p. 34), the Code included these house-building provisions:

> *If a builder build a house for a man and do not make its construction firm, and the house which he has built collapse and cause the death of the owner of the house that builder shall be put to death. If it cause the death of the son of the owner of the house, they shall put to death a son of that builder. If it cause the death of a slave of the owner of the house, he shall give to the owner of the house a slave of equal value. If it destroy property, he shall restore whatever it destroyed, and because he did not make the house which he built firm and it collapsed, he shall rebuild the house which collapsed from his own property. If a builder build a house for a man and do not make its construction meet the requirements and a wall fall in, that builder shall strengthen the wall at his own expense.*

As quoted in Biswas (1970, p. 20), this dam maintenance requirement was part of the Code:

> *If any one be too lazy to keep his dam in proper condition, and does not keep it so; if then the dam breaks and all the fields are flooded, then shall he whose dam the break occurred be sold for money and the money shall replace the corn which he has caused to be ruined.*

Unfortunately, there are many and varied ways in which today's individual technical professional and technical organization can incur liability (e.g., Clough, 1986, Chapter 4; Dunham et al., 1979, Chapter 25). One way to illustrate the potential liability exposure of an engineering or similar organization is to enumerate some of the services typically provided by or within such an organization and then think through some of the possible related liabilities. This approach was used to construct Figure 10–3, which lists 18 types of services typically provided by an engineering organization and, for most of them, gives examples of potential liability in terms of fraud, breach of contract, and negligence. Recall that fraud, breach, and negligence are defined in Figure 10–2. As suggested by Figure 10–3, liability-incurring opportunities abound in the practice of engineering and related technical professions.

To elaborate, breach has little or nothing to do with intention, but nevertheless consists of violating a right, a duty, or a law. Simply failing to deliver plans and specifications on time as specified in a contract or agreement could constitute breach. Fraud, which is an explicitly intentional, deceitful action, consists of actions such as billing a client for products not delivered or falsely stating that a necessary government permit had been secured.

Negligence, the most common of the three ways in which technical professionals and their organizations incur liability, means failing to exercise care and provide expertise in accordance with the standard of the profession. For example, calculation error is likely to be considered negligence. You, as a young professional, must recognize that being honest and well-intentioned is simply not enough in avoiding negligence. You and your organization must be disciplined in the manner in which you provide services if negligence and the resulting liability are to be avoided.

The technical professional and his or her organization are not expected to be perfect (Dunham et al., 1979, p. 427). Clients contract for service, not insurance. Engineering and similar organizations cannot guarantee perfect plans and specifications or flawless products. Perfection is not expected of the technical professional. However, the professional and his or her organization are obligated to ". . . exercise ordinary professional skill and diligence and to conform to accepted . . . standards" (Dunham et al., 1979, p. 427).

1. **Participating in necessary conferences and preliminary studies.**

   Breach: contract promised X meetings; only had Y.

2. **Interpreting physical restrictions as to the use of the land.**

   Negligence: failed to allow for building setbacks.

3. **Examining the site of the construction.**

4. **Preparing and/or interpreting soil, subsoil, and hydrologic data.**

5. **Preparing drawings or verifying and interpreting existing drawings or construction.**

   Negligence: proposed water main conflicts with natural gas line.

   Breach: drawings late.

6. **Assisting in procuring of financing for the project.**

   Fraud: A/E "steers" client to certain lender in return for kickback from lender.

7. **Assisting in presentation of a project before bodies possessing approval–disapproval power.**

   Fraud: falsely claim that lower-level government units and agencies approve.

8. **Preparing drawings and specifications for architectural, structural, plumbing, heating, electrical, and other mechanical work.**

   Negligence: inadequate thermal insulation leads to fire.

9. **Assisting in the drafting of forms of proposals and contracts.**

10. **Preparing cost estimates.**

    Negligence: numerical error.

11. **Obtaining bid from contractors.**

    Fraud: alter a bid to favor a contractor.

Figure 10–3 Services typically provided by an engineering organization on a project and possible related liabilities (Source: List (the 18 points) quoted from Dunham et al., 1979, pp. 426–427. Examples of liabilities were added.) (Continued on p. 304.)

12. **Letting contracts with owner's written approval.**
13. **Inspecting the contractor's work on regular basis, including the checking of shop drawings (but without dictating the method or means by which the contractor seeks to accomplish the desired results).**

    Negligence: failing to note an unsatisfactory change (e.g., Hyatt, Kansas City failure.)

14. **Interpreting for the contractor the meaning of the drawings and specifications.**
15. **Ordering the correction or removal of all work and materials not in strict conformity with specifications.**
16. **Keeping accurate books and records.**

    Breach: failing to do and submit when required.

17. **Preparing as-built drawings that show construction changes and final locations of mechanical and electrical lines.**

    Negligence: incorrectly locating buried electric line leading to later disastrous excavation accident.

18. **Issuing certificates of payment.**

    Negligence: failing to verify that work was performed.

Figure 10–3 Continued

LIABILITY: EXAMPLES OF FAILURE AND LESSONS LEARNED

The only thing new in the world is the history you don't know.

[President Harry S. Truman quoted by Miller (1974, p. 21)]

A review of some actual failures and the resulting death, disruption, economic loss, and other consequences provides insight into the cause of the failures and the related liabilities. Although liability can be incurred as a result of breach, fraud, and negligence, an examination of failures supports the thesis that negligence is by far the dominant cause.

The focus in this section is on relatively dramatic structural failures.

Strength, or lack thereof, is typically the cause and the resulting failure is usually instantaneous and catastrophic. There can also be what might be referred to as *serviceability failures.* Examples of this much less dramatic, but more common type of failure are roof deterioration, excessive floor deflection, and vibration of structural components.

Collapse of Hotel Walkway

In 1981, two walkways suspended one above the other in the atrium of the relatively new Hyatt Hotel in Kansas City, Missouri, suddenly collapsed, killing 114 people and injuring almost 200. This description of the disaster and subsequent actions is based on data and information provided by ASCE (1989, p. 22), Goodman (1990, p. 21), and Petroski (1985, Chapter 8). The cause of the collapse is illustrated in Figure 10–4, which shows original and as-built support details. Subsequent laboratory simulation confirmed the nature of failure of the connection between the hanger rod and the box beam of the type that supported the walkway.

Two instances of negligence occurred and, unfortunately, they were additive. First, the system supporting the walkways was underdesigned in that it had only 60 percent of the strength specified by the code. However, the failure might not have occurred because of the factor of safety explicitly incorporated in the code. This negligence was attributed to the design professionals.

The second instance of negligence occurred when a design change was

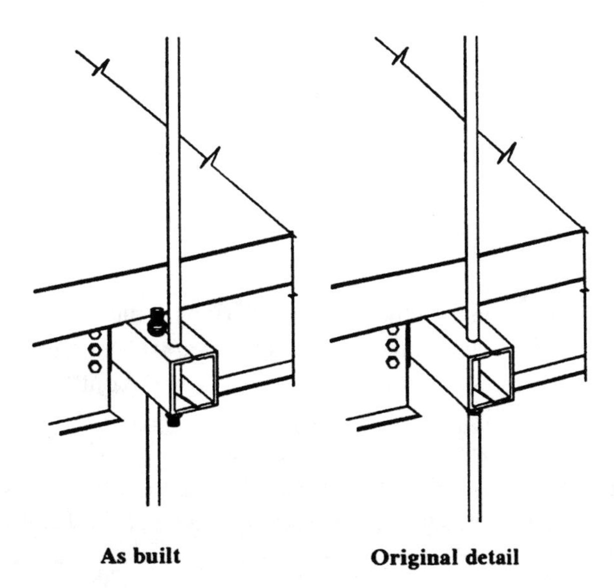

Figure 10–4. Kansas City Hyatt walkway collapse (Source: Adapted with permission from H. Petroski, *To Engineer is Human: The Role of Failure in Successful Design,* New York: St. Martins Press, 1985. Copyright 1982, 1983, 1984, 1985 by H. Petroski.)

made in the field. Instead of suspending the box beams and the walkway on single rods, which would have been very difficult from a constructability perspective, shorter rods were used. At each suspension point on the upper walkway, one rod was connected to the ceiling and terminated just below the box beam supporting the highest walkway. A second rod was connected to that box beam and extended down to and just below the box beam supporting the lower walkway. The net effect of this design change was an excessive load at the point where the upper box beams were supported by the upper rods. The change, in effect, doubled the connecting forces at this location. The designers were determined to be responsible for the unfortunate field design change in that they did not identify the increased load that resulted.

The U.S. attorney concluded that no federal or state crimes were committed, but the Missouri Attorney General successfully charged the consulting engineering firm with negligence. Two engineers lost their licenses to practice engineering, and the state certificate of incorporation of the firm was revoked.

At least three liability-related lessons may be learned from the collapse of the Kansas City Hyatt Hotel walkways. These lessons are

- A sense of responsibility and accountability must be established within an organization to minimize the probability of errors in calculations, such as what happened in the case of the walkway collapse in that the support rods were originally underdesigned.

- The responsibility of the technical professional does not necessarily end with the preparation of plans and specifications. Conscientious review and approval, as appropriate, of shop drawings may be a part of the firm's contractual responsibilities.

- An understanding of fundamentals, such as those presented in a first- or second-year college-level statics mechanics course, is essential to successful design and to the review of design changes proposed during construction.

Collapse of Supermarket Roof

In 1988, a supermarket roof, which also served as a parking structure, in Vancouver, British Columbia, collapsed and fell into the supermarket, carrying 22 automobiles with it. Somewhat miraculously, although there were about 900 shoppers in the store immediately before the collapse, only 24 people were injured. Unlike the Kansas City walkway collapse, which occurred essentially instantaneously, the supermarket roof failed slowly, giving an indication of its imminent collapse and providing adequate time to evacuate the supermarket (*ENR*, July 7, 1988, p. 14).

The failure was attributed to several factors. First, a key steel beam was underdesigned—it should have been 30 inches deep, but was only 24 inches deep. Surprisingly, although another firm had been retained to check the structural design and had identified the deficiency, they had approved it because "Mill certificates claimed the beam was 25 percent stronger than assumed." In addition, 29 critical lateral supports had been omitted. That is, they had been specified in the plans and specifications, but were not installed during construction. Finally, "Two columns were removed at the owner's request to allow room for the retail floor" (*ENR,* July 7, 1988).

At least three important lessons may be learned from this failure. They are

- Checking of assumptions and calculations, even when done by a presumably objective second or outside party, does not necessarily ensure a safe design.

- A carefully designed structure must be just as carefully constructed. Just as a faulty and undetected structural connection change was made in the Kansas City Hyatt Hotel and caused a disaster, so undetected structural omissions occurred during the construction of the Vancouver supermarket.

- Owners, customers, clients, and other constituencies clearly have a right to define their needs and engineers have an obligation to try to satisfy those needs. However, that exchange should occur during the design process, only in rare instances take place during construction, and never compromise the integrity of the product, structure, facility, or system.

Collapse of Scoreboard

As described in *ENR* (August 25, 1988), an 18-ton scoreboard normally suspended by cables over the center of a coliseum in Charlotte, North Carolina, suddenly fell to the floor, causing $1.2 million in damage. The scoreboard had been suspended by four cables so that it could be lowered from the ceiling to floor level for maintenance and changes. On the day of the collapse, the controls were incorrectly used and the load was unbalanced among the four cables, overloading at least one of them and causing failure. Incidentally, no one was injured in this failure, and a basketball game was played in the coliseum the evening of the day of the collapse.

The fundamental question raised by this case study is to what extent is the design professional responsible for anticipating and protecting against various ways in which a designed product, structure, facility, or system may be misused? Only an extremist would hold the designer of a particular model of a baseball bat accountable for injuries inflicted with the bat in a fight. On the other hand, the vast majority of reasonable citizens would be dismayed at the design of an automobile power steering system that tended to jam if it was turned too quickly or turned too far. The potential misuse of a structure, facility, or product must be considered during its planning and design and, as appropriate, preventive design changes implemented.

Collapse of Bridge Section during Construction

In 1988, as described in *ENR* (September 8, 1988), a 256-foot-long, prestressed reinforced concrete bridge segment was being placed on piers in a 1,183-foot-long bridge under construction in Aschaffenburg, Germany. The section collapsed as it was being placed; one worker was killed and five were injured; there was also significant monetary loss. Cause of the collapse: an error in the design of the temporary cable system used to construct the bridge (*ENR,* February 1, 1990).

This disastrous failure raises a question related to the issue in the preceding case study. To what extent is the designer of a product, structure, facility, or system responsible for anticipating the way the properly designed entity might be constructed, manufactured, or fabricated? If the design organization is to have such responsibility, the extent and nature of that responsibility should be clearly defined in the contract between the engineering organization and the owner.

Other Failures

Some additional failures and the lessons learned are briefly noted. Consider the 1940 collapse of the Tacoma Narrows Bridge in Washington state, which led to an improved understanding of the vortex shedding phenomenon (Petroski, 1985, pp. 164–169) and which puts the bridge designer on notice that he or she must learn from this experience and not let it happen it again. Then there is the more recent 1986 explosion of the Challenger rocket, in which seven astronauts died as the result of a faulty seal. Some believe that this disaster could have been prevented had engineers acted more responsibly (Florman, 1987, Chapter 17). In November 1989, nine children were killed in Orange County, New York when a masonry section of a cafeteria wall at a school collapsed during a storm (*ENR,* January 18, 1990). A report commissioned by the state concluded that one cause of the failure was poor communication between architects and engineers regarding lead responsibility for design of structural elements. There are lessons to be learned in each of these failures.

The engineering (e.g., Martin and Schinzinger, 1989) and popular literature contain many and varied accounts of failures of engineered works. In addition, special organizations such as the Architecture and Engineering Performance Information Center at the University of Maryland provide a repository for data and information on engineering failures (Florman, 1987, p. 102). The young professional is admonished to become a student of the history of his or her profession. By doing so, you will learn to better appreciate the legacy left by engineers and other technical professionals around the globe and also gain insight into failure and how to minimize it.

In his book *To Engineer Is Human: The Role of Failure in Successful Design*, Petroski (1985, Chapter 4) reminds design professionals that each design is not a fully tested hypothesis. Some mistakes and, yes, a few disasters are inevitable. They offer an opportunity to learn at the individual level and at the professional level and, as a result, advance the state of the art.

To reiterate a point made earlier, of the possible ways for an engineer or other technical professional to incur liability—breach, fraud, and negligence—negligence is by far the most common. Examples of preventable negligence are provided by the case studies.

LIABILITY: MINIMIZING IT

As noted at the beginning of this chapter, society is becoming more litigious, that is, more likely to take legal action. Therefore, technical professionals must be even more diligent in taking preventive and remedial action.

Insurance: Financial Protection

Although purchasing liability insurance won't directly prevent lawsuits being brought against an engineering or other technically based professional services organization, the availability of the insurance will provide some financial protection if a liability action is initiated. The insurer, in exchange for regular premium payments, agrees to make liability payments and defend suits arising out of negligence or alleged negligence in the provision of professional services by the insured.

As with most insurance policies there are exclusions, that is, actions and activities that are not covered by liability insurance. Examples are the failure of the insured to complete services on time, intentional fraudulent and other acts of the insured, and the insured providing services outside of the organization's area of expertise (Dunham et al., 1979, p. 452). As is also the case with some other forms of insurance, professional liability insurance typically has deductible provisions, that is, an initial amount of loss that is not covered by the insurance. MacLean (1982) indicates that, in the extreme, per claim deductibles may range from $2,000 to $1,000,000.

Liability insurance is expensive when annual premiums are quantified, for example, as a percentage of annual billings for a consulting engineering firm. Liability insurance premiums are of the same order of magnitude as the after-tax profit of many consulting engineering firms. In recent years, annual insurance premiums paid by consulting engineering firms have been roughly five percent of annual billings (e.g., Quick, 1991). Premiums tend to be higher for greater risk areas of service, such as structural design. Incidentally, insurance premiums are part of overhead and, therefore, go right to the bottom line on the engineering organization's income statement. There appears to be a recent tendency for liability insurance premiums measured as a percentage of annual billings to diminish slightly, perhaps reflecting improved management of risk and liability exposure by consulting engineering firms.

You should know that not all consulting firms purchase liability insurance; some "go bare." For example, a liability survey by the American Consulting Engineers Council (ACEC) in 1990 and based on responses from 1,764 consulting engineering firms revealed that 22% had no liability insurance (Quick, 1991). The survey also revealed that small firms are much more likely to go without liability insurance than large firms. For example, 43% of the firms with five or fewer employees were not insured. In contrast, only 9% of the firms with over 500 employees were without liability insurance. Firms without liability insurance are, in effect, self-insured.

Consider the disposition of claims. One insurance company reported that claims are filed against about half of the design firms insured by them in any given year. Most claims are settled or closed without payment, and only about 5% eventually go to trial and then the insurance company reported "winning" 75% of the cases (MacLean, 1982). Another study (Quick, 1991) based on 1990 data indicated that 40 claims were filed per 100 firms each year—a slightly lower rate than the earlier source. The 1990 data also revealed that some type of payment was made to claimants in 61% of the claims and that only 10% of the claims were settled in court. Figure 10–5 illustrates, in a general way based on the cited sources, the frequency and disposition of liability claims brought against consulting engineering firms.

Preventive Actions

Although liability insurance, which provides some financial protection in case of liability litigation, may be considered optional by some consulting firms and other organizations, all such organizations should aggressively and systematically undertake programs to minimize liability, particularly that incurred as a result of negligence. Figure 10–6 lists 20 ways to reduce liability, each of which is discussed here. Many of the 20 ways to reduce liability are at least partially

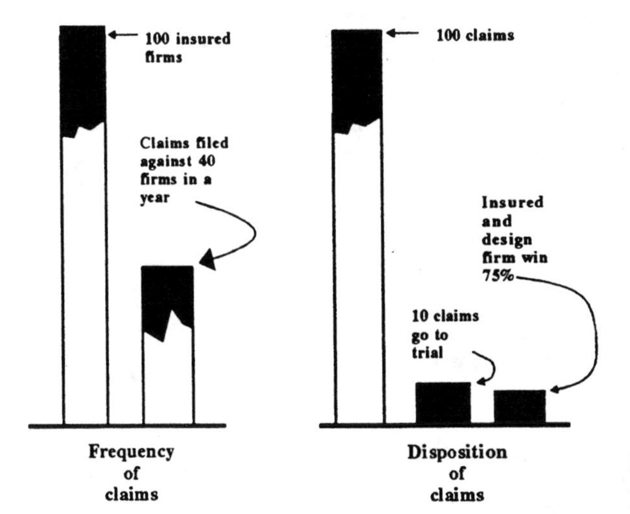

Figure 10–5. Frequency and disposition of liability claims

within the range of responsibility and authority of the entry-level technical professional. In particular, if you focus on items 4, 6, 7, 8, 9, 10, 11, 13, 14, 17, and 19, you will make a significant contribution to reducing the liability exposure of your organization.

1. Incorporate practice. Brown (1988) notes that an incorporation can shield the employees of an organization from incurring liability as a result of actions of a few members of the organization. Although plaintiffs may seek relief from others in the firm as a result of the negligence of one or a few, Brown (1988) indicates that "... incorporation places another legal barrier between the firm's principals and the risk of legal liability."

2. Limit practice to "safer" disciplines. Examples of high-litigation-potential areas of service are poorly financed client/developers, roofing projects, structural designs, hazardous waste, and geotechnical and construction inspection. A variation on this liability-reduction approach is to subcontract areas of service having higher litigation potential (Brown, 1988; Vansant, 1982).

3. Incorporate high-risk services and areas separately. For example, set up separate corporations for hazardous waste consulting or high-rise structural design. The objective is to prevent a successful claim against one part of an overall organization jeopardizing the parent or main corporation (Brown, 1988).

4. Maintain currency and competence. A proactive program of formal and informal activities is needed to maintain individual and corporate currency and competence. As discussed earlier and stated in Figure 10–2, technical services that fall below the appropriate "standard of care" may be deemed negligent

1. **Incorporate practice.**
2. **Limit practice to "safer" disciplines.**
3. **Incorporate high risk services and areas separately.**
4. **Maintain currency and competence.**
5. **Use standard contract forms.**
6. **Use tested legal language.**
7. **Develop, maintain and use written standard procedures.**
8. **Document everything.**
9. **Supplement written documentation with photographs, slides, and videotape.**
10. **Accept primary responsibility for use of computer programs and models.**
11. **Separate facts and opinions.**
12. **Hire only insured subconsultants.**
13. **Respond in a timely fashion.**
14. **Limit project comments to knowledgeable persons.**
15. **Avoid financial interest in projects.**
16. **Use peer review.**
17. **Do it right the first time!**
18. **Sign reports in corporate name only.**
19. **Communicate—communicate—communicate with those you serve.**
20. **Place liability-limiting provisions in contracts.**

Figure 10–6. Ways to reduce liability

and result in personal and organizational liability. Your and your employer's best interests require that you remain current and competent in your areas of technical specialization. Refer to the Chapter 2 sections titled "Managing Personal Professional Assets" and "Continuing Education." Clearly, being current and competent includes understanding of legal principles, with emphasis on liability, as set forth in this chapter.

5. Use standard contract forms. Consider using standard contract forms such as those produced by the Engineers Joint Contract Document Committee (EJCDC, 1992) for use by firms and constructors. Davis (1986) lists many advantages of such forms, including the all-important periodic and cooperative review of the forms by organizations representing all interests.

6. Utilize tested legal language. Consider use of tested legal language in contracts and agreements such as that developed by professional organizations like the National Society of Professional Engineers (MacLean, 1982). Certain words and phrases, even when and perhaps especially when well intended, can lead to contract conflict and claims of negligence or even breach of contract or fraud. Hayden (1987) effectively presents some commonly used words and phases to avoid and he offers suggested and safer replacements. Hayden's contribution is presented in Figure 10–7. According to Brown (1991), ". . . the big print giveth and small print taketh away." In other words, contracts and agreements, and the proposals used to obtain them, should be written and read very carefully.

7. Develop, maintain, and use written standard procedures. The availability and use of written, standard technical and even nontechnical procedures helps to prevent errors and omissions. Refer to the section titled "Written Procedures" in Chapter 7 for a detailed discussion of the advantages of developing and using written procedures for all of an organization's operations. Prudent as they may seem, very few technical organizations have written procedures (MacLean, 1982). Besides reducing the likelihood of negligence, the existence and systematic use of a procedure system within an engineering or other technical organization will be helpful in responding to claims of negligence, because the procedures will suggest that the organization is proactively involved in defining the accepted standards within the profession.

8. Document everything. Everything means essentially everything, including, but not limited to, meetings, telephone calls, field reconnaissance, and conversations. The idea is to assume that everything will someday be viewed by your peers or, worse yet, by opponents in litigation (MacLean, 1982; Vansant, 1982; Vansant, 1993). Your firm may have a uniform documentation system consisting of components such as standard filing procedures, special forms, and a project management system. If your employer doesn't have a standard documentation system, perhaps you should suggest that one be developed and offer to assist in its development. If such a suggestion is not well received, develop your own system, so that at least the work for which you are responsible is carefully documented.

9. Supplement written documentation with photographs, slides, and video tapes. Images are an extremely effective form of communication and, therefore, documentation.

10. Accept primary responsibility for use of computer programs and models. You and your organization have the primary responsibility for the correctness of computer programs or models used in your projects and for the appropriateness of the uses. Even if the software contains errors, the professional using

Do Not Use	Example Replacements
All existing information will be gathered.	Readily available information will be reviewed and collected as needed.
Coordination performed *at all times.*	Client will be apprised of approval progress.
Highly trained professionals	Professionals
Prepare summaries of *all* meetings.	Prepare summaries of monthly project status meetings with client.
Close coordination of *all stages* of the work	Perform interdisciplinary milestone review at 15%, 30%, 60%, and 90%.
All required professional support	The work will be performed by our staff.
Will *complete all* project services	Will prepare and submit for review and approval normal engineering drawings suitable for construction
Is *exceptionally* well qualified	For this opportunity we are qualified because
High quality reports	Our reports will be suitable for
Only the best	Staff selected will be appropriate for the work assigned.
The *highest level* of quality	The work performed will satisfy contract requirements.
Guarantee *successful* project completion	Committed to perform as contracted for scope, schedule, and budget.
Only experienced and qualified staff will be assigned.	Technical staff will be assigned as appropriate.
As necessary	Not less than once per
At *all* times	Will be done less than once per
Or approved equal	Similar in our opinion as to function
Insure; ensure; assure	Reasonable effort will be made
Maximum	Not less than two per month
Minimum	Not more than two per month
Periodically	Not less than once per
Supervise; inspect	Observe and report
Certify; warrant; guarantee	Statement as to our judgment based on

Figure 10–7. Expressions to use and not to use in contracts and agreements (Source: Adapted with permission from W. M. Hayden, Jr., Journal of Management in Engineering—ASCE, Vol. 4, No. 4 (October 1988), pp 284–285.

the program or model, not the outside model developer, is likely to carry all or most of the liability if a problem arises (Backman, 1993; *ENR,* October 28, 1991; Mishkin and Schwartz, 1990). Because heavy use is made of computer programs and models in technical environments, this liability principle should be publicized throughout your organization, particularly if yours is a business enterprise as opposed to a government entity. Your organization should have at least one expert associated with each computer program or model in use. This expert is an important resource for you, the entry-level professional, as you strive to be a conscientious software user.

11. Separate facts and opinions. When preparing memoranda, letters, reports, and other forms of documentation, clearly indicate when data are being presented and opinions are being expressed (Vansant, 1982).

12. Hire only insured subconsultants. This practice will tend to limit the lead firm's liability because the subconsultant's liability insurance will be available. Brown (1988) notes that another, and sometimes preferable alternative, is to have subconsultants contract directly with the owner.

13. Respond in a timely fashion. Requests from clients or customers and from subcontractors for data, information, or decisions should receive a timely response. To do otherwise is to risk a subsequent charge of negligence (MacLean, 1982).

14. Limit project comments to knowledgeable persons. Only those members of an organization familiar with its projects should make substantive comments about those projects. The principal danger being addressed with this suggestion is that one or more principals of an organization, perhaps in their zeal to serve the organization's clients, may unknowingly make promises that cannot be kept within the formal agreements that govern the conduct of the project (Lepatner and Banner, 1993).

15. Avoid financial interest in project. Dunham (1979) notes that if the design firm has a financial interest in the project, the firm may be liable for problems that arise as a result of the project, even if there is no professional negligence.

16. Use peer review. The idea is to have the technical work of one or more engineers or other technical professionals reviewed by a group of peers who, at the outset, are not familiar with the project, but are experts in the disciplines represented by the project. The effort could be as formal as retaining an outside firm to conduct the review if circumstances warranted. Another approach is to form a peer review group from within the organization. Diminished probability of errors and omissions is the obvious advantage of peer review. Potential problems or challenges inherent in peer review include increased design costs, interpersonal conflict between reviewers and reviewees, and expanding the number of potential liable individuals or organizations if a failure occurs (Preziosi, 1988; Vansant, 1982). A more elaborate but perhaps more productive approach, is to use the team building or partnering approach described in Chapter 7.

17. Do it right the first time! MacLean (1982) notes that by doing projects right the first time, not only are time and money saved in the long run for the engineering or other technical organization, but last-minute rushing and the inevitable errors resulting in liability exposure are reduced.

18. Sign reports in corporate name only. Brown (1988) suggests this, indicating that opposing attorneys will have difficulty focusing on one or a few individuals. Although this suggestion seems somewhat extreme and is atypical, it may be appropriate in certain circumstances.

19. Communicate–communicate–communicate with those you serve. All legal and other technicalities aside, there must be a clear understanding of expectations prior to formalization of an agreement between a professional services organization and a client or customer. Failure to achieve a meeting of the minds sows the seeds of conflict and litigation. Once a project is underway, ongoing communication is essential with the primary responsibility for the effectiveness of that communication lying with the project manager. Client and customer communication is important from a liability-minimization perspective, because current clients and customers—not third parties—are probably the principal source of liability for an engineering organization. Miller (1987) cites the mid-1980s experience of CH2M HILL, a large multidiscipline consulting engineering firm. He states, " . . . 65% of the significant claims and 85% of the monetary liability attributable to such claims were brought by dissatisfied clients, not by third parties or members of the public."

20. Place liability-limiting provisions in contracts. Possibilities include aggregate limits on liability, limiting potential liability to amounts recoverable with the firm's liability insurance, and restrictions on what parts of the project or who within the design firm could be subject to claims (Brown, 1988).

Danger Signals

Andrews and Ruzzo (1988) identify some client or client-related behaviors or circumstances that may indicate that serious legal problems are imminent. Examples include a reduction in the frequency of client or customer contact and a change toward the negative in the tone of such contact; involvement of new third parties, including potential plaintiffs and their attorneys; exclusion from project-related meetings and communications; or customer complaints to third parties about your organization; and termination of your services by the client or customer. Because of your contact with peers of a similar age in the client or customer organization, you may be in a position, if you are perceptive, to receive some of the danger signals. If you think you sense client or customer dissatisfaction, immediately inform your supervisor. Then the situation can be analyzed and, if a problem exists, corrective actions or at least damage control, as discussed in the next section, can be taken.

Damage Control

If litigation appears imminent, Andrews and Ruzzo (1988) suggest the following:

1. Get your "ducks in a row" by conducting internal meetings, having conversations, and reviewing project documentation, but never alter documents.
2. Assign one individual to be the spokesperson for the organization.
3. Determine decision makers and other key individuals in the client's organization.
4. Based largely on item 1 and what is being learned by the spokesperson noted in item 2, develop the position of the firm and try to formulate a solution acceptable to the firm and also likely to be acceptable to the client with whom the problem has developed. Look for an opportunity to present the proposed solution to the client, emphasizing its mutually beneficial dimensions.
5. Listen carefully to the words and for the feelings emanating from the client's organization. Recall and try to practice the empathetic listening, the highest level of listening, described in Chapter 3 of this book.

MAINTAINING PERSPECTIVE ON LIABILITY MINIMIZATION

The greatest mistake you can make in life is to be continually fearing you will make one.

(E. Hubbard)

Much of what is done by engineering and similar organizations to minimize liability exposure is also being done for one or more other reasons, some of which may even be more important—this is just good management. Your firm should guard against letting the tail wag the dog, that is, becoming overly fearful if not paranoid about liability. For example, peer review, which is presented as one way of minimizing liability, is also likely to yield a more cost-effective design as measured by life-cycle costs. Documentation, another liability minimization device, is also very useful in planning a project, coordinating a project, writing a report on a project, as a guide for future, similar projects, and as a resource for "surprise" meetings. While timely response to client requests will surely minimize liability exposure, it is also a mark of good service. Use of standard, tested contract and agreement language is another liability-minimization measure, and also a time-saving device. Finally, written technical and other procedures will certainly minimize the probability of negligence within an engineering or other technical office. Such procedures will also, in the long run, greatly reduce the time required to complete technical tasks and will serve as a very effective orientation and training tool for you, the entry-level professional, and for other new company personnel.

REFERENCES

ALLEN, C. L., "The Angry Retort Against Tort Law," *Insight*, October 31, 1988.

AMERICAN SOCIETY OF CIVIL ENGINEERS, "Court Decisions," *Civil Engineering,* January 1989, p. 22.

ANDREWS, A. S., and W. P. RUZZO, "Avoiding Litigation," *Journal of Management in Engineering*, Vol. 4, No. 1 (January 1988), pp. 8–15.

BACKMAN, L., "Computer-Aided Liability," *Civil Engineering—ASCE,* June 1993, pp. 41–43.

BISWAS, A. K., *History of Hydrology.* Amsterdam: North Holland Publishing Co., 1970.

BROWN, E. C., "Putting a Lid on Liability," *Civil Engineering*, Vol. 58, No. 7 (July 1988), pp. 66–67.

BROWN, H. J., Jr., *Life's Little Instruction Book.* Nashville, Tenn.: Rutledge Hill Press, 1991.

CLOUGH, R. H., *Construction Contracting,* 5th ed. New York: John Wiley and Sons, 1986.

DAVIS, R. O., "Advantages of Standard Contract Forms," *Journal of Management in Engineering—ASCE*, Vol. 2 No. 2 (April 1986), pp. 79–90.

DE CAMP, L. S., *The Ancient Engineers.* New York: Ballantine, 1963.

ENGINEERING NEWS RECORD, "Design Flaw Blamed for Collapse," *ENR,* January 18, 1990, pp. 11–12.

ENGINEERING NEWS RECORD, "Design Led to Downfall of Incremental Launch," *ENR,* February 1, 1990, pp. 13–14.

DPIC Companies, *Lessons in Professional Liability: A Notebook for Design Professionals.* Monterey, Cal., 1988.

DUNHAM, C. W., R. D. YOUNG, and J. T. BOCKRATH, *Contracts, Specifications, and Law for Engineers,* 3rd ed. New York: McGraw-Hill Book Company, 1979.

ENGINEERS JOINT CONTRACT DOCUMENTS COMMITTEE, "Standard Form of Agreement Between Owner and Engineer for Professional Services," National Society of Professional Engineers, American Consulting Engineers Council, and American Society of Civil Engineers, 1992.

FLORMAN, S. C., *The Civilized Engineer.* New York: St. Martin's Press, 1987.

ENGINEERING NEWS RECORD, "German Bridge Girder Fails," *ENR,* September 8, 1988, p. 18.

GOODMAN, L. J., "Revisiting the Hyatt Regency Walkway Collapse," *Civil Engineering News*, March 1990, p. 20.

HAYDEN, W. M., JR., "Quality by Design Newsletter," May 1987, A/E QMA, Jacksonville, Fla. Quoted in *Journal of Management in Engineering—ASCE,* Vol. 4, No. 4 (October 1988), pp. 284–285.

HUBER, P., "Don't Innovate, It's Dangerous," Forum, *Civil Engineering,* April 1988, p. 6.

ENGINEERING NEWS RECORD, "Key Beam Under-Designed," *ENR,* July 7, 1988, p. 14.

LEPATNER, B. B., and R. A. BANNER, "8 Tips for Avoiding Liability Claims," *American Consulting Engineer*, Second Quarter, 1993, Vol. 4, No. 2, p. 8.

MACLEAN, W. G., "Managing Liability: The Insurer's Perspective," *Managing Liability: The Individual's Challenge—The Organization's Challenge—The Manager's Challenge*, Proceedings of a Symposium sponsored by the Engineering Management Division—ASCE, J. R. King, Jr., ed., Las Vegas, Nev., April 1982, pp. 29–39.

MARTIN, M. W. and R. SCHINZINGER, *Ethics in Engineering,* 2nd ed. New York: McGraw-Hill Book Company, 1989.

MCQUILLAN, J. A., "The CE as an Expert Witness," *Consulting Engineer*, January 1984, pp. 48–50.

MILLER, D. W., "Loss Prevention: Safeguards Against Liability," *Journal of Management in Engineering—ASCE,* Vol. 3, No. 2 (April 1987), pp. 95–115.

MILLER, M., *Plain Speaking: An Oral Biography of Harry S. Truman.* New York: Berkley Publishing Corporation, 1974.

ENGINEERING NEWS RECORD, "A Million-Dollar Tilt," *ENR,* August 25, 1988, p. 17.

MISHKIN, B. and A. E. SCHWARTZ, "Architect/Engineer Liability for Use of Computer Technology in Designing Projects," Professional Ethics Report—Newsletter of the American Association for the Advancement of Science, Vol. III, No. 2 (Spring 1990), pp. 6–7.

PETROSKI, H. *To Engineer Is Human: The Role of Failure in Successful Design.* New York: St. Martin's Press, 1985.

PREZIOSI, D., "Reviewing Peer Review," *Civil Engineering*, November 1988, pp. 46–48.

QUICK, J., "Liability Crisis Is Easing but Remains a Problem," *American Consulting Engineer*, Winter 1991, Vol. 2, No. 1, pp. 48–50.

ENGINEERING NEWS RECORD, "Rugged Replacement to Rise in West Virginia," *ENR,* January 14, 1991, pp. 21–22.

ENGINEERING NEWS RECORD, "Sky Dish Collapse Probed," *ENR*, November 24, 1988, pp. 11–12.

ENGINEERING NEWS RECORD, "Training Critical in Use of Structural Software," *ENR,* October 28, 1991, pp. 29–32.

VANSANT, R. E., "Liability: Attitudes and Procedures," *Managing Liability: The Individual's Challenge—The Organization's Challenge—The Manager's Challenge*, Proceedings of a Symposium sponsored by the Engineering Management Division—ASCE, J. R. King, Jr., ed. Las Vegas, Nev., April 1982, pp. 23–28.

VANSANT, R. E., "Papering the Job Can Pay Off in the Courtroom," *The Construction Specifier*, September 1983, pp. 14–15.

WALDROP, M. M., "Collapse of a Radio Giant," *Science*, Vol. 242, November 25, 1988, p. 1120.

SUPPLEMENTAL REFERENCES

ARORA, M. L., "Writing Effective Specifications," *Civil Engineering*, March 1994, pp. 69–71.

BARON, R. M., "How to Guard Against Successor Liability," *American Consulting Engineer*, Vol. 3, No. 1, Winter 1992, p. 13.

CASSO, M. A., and J. C. KALAVRITINOS, JR., *Streamlining Civil Justice for Professionals: Six Steps Toward Reforming the Legal System.* Washington, D. C.: American Consulting Engineers Council, 1994.

FERGUSON, E. S., "How Engineers Lose Touch," *Invention and Technology*, Vol. 8, No. 3, Winter 1993, pp. 16–24.

HUTCHENS, P. E., "Risk Reduction Through Indemnification Clauses," *Journal of Management in Engineering—ASCE*, Vol. 8, No. 3 (July 1992), pp. 267–277.

MORTON, R. J., *Engineering Law, Design Liability, and Professional Ethics—An Introduction for Engineers.* San Carlos, Cal.: Professional Publications, Inc., 1983.

PETROSKI, H., *Design Paradigms: Case Histories of Error and Judgement in Engineering.* New York: Cambridge University Press, 1994.

SWEET, J., *The Legal Aspects of Architecture, Engineering and the Construction Process.* St. Paul, Minn.: West Publishing, 1994.

WARTEL, S. J., "Negotiating and Drafting the Civil Engineering Contract," *Journal of Management in Engineering—ASCE*, Vol. 5, No. 3 (July 1989), pp. 272–279.

EXERCISES

10.1 RESPONSE TO A FAILURE BY THE DESIGN FIRM

Purpose

Increase student's awareness of the need for a quick but thoughtful response to the failure of an engineered facility or system.

Tasks

1. Consider the failure of the 300-foot-diameter radio telescope at Green Bank, West Virginia in November 1988. (Waldrop, 1988, p. 1120; *ENR,* November 24, 1988; ENR, January 14, 1991, pp. 21–22).
2. Assume that your group represents the engineering firm that designed the facility. Your group includes the project manager. The chief executive of your organization is not present.
3. Assume further that a hurried meeting has been called upon learning of the collapse.
4. Conduct the meeting and write a memorandum to your chief executive, recommending a course of action.

10.2 RESPONSE TO FAILURE BY THE CONTRACTOR (CONSTRUCTOR)

Tasks

Same as Exercise 10.1, except that the group represents the contractor that constructed the facility.

10.3 RESPONSE TO FAILURE BY THE OWNER

Tasks

Same as Exercise 10.1, except that the group represents the owner of the facility.

10.4 FAILURE CASE STUDY

Purpose

Increase student's sensitivity to and awareness of methods to minimize failure of products, structures, facilities, or systems.

Tasks

1. Obtain documentation on the failure of a product, structure, facility, or system. (Do not use any of the principal examples used in this chapter).
2. Study the documentation.
3. Prepare a memorandum that:
   a. Describes the failure, including the consequences in terms such as lives lost, monetary loss, delays, ruined careers, and failure of organizations.
   b. Identify in detail the cause or causes of the failure.
   c. Indicate the lesson(s) that can be learned.

11

Ethics

> *What you are . . . thunders so that I cannot hear what you say.*
>
> (Emerson)
>
> *Reputation is what you're supposed to be; character is what you are when nobody but God is looking.*
>
> Anderson (1968, p. 3)

Television, newspapers, and magazines often report on ethical issues or, more precisely, alleged or actual unethical behavior. A wide range of society's institutions and organizations are typically involved, including government, business, academia, and religious groups. As you begin the practice of your profession, you will increasingly participate in society's institutions and organizations, and you will be faced with many and varied ethical questions and some very difficult dilemmas. The purpose of this chapter is to provide you with ideas and information, much of it from the engineering profession, to assist you in navigating through sometimes turbulent ethical waters. Frankly, you will need to develop and refine what you stand for and what you will not stand for.

Partly to set the scene for this chapter, you should know that engineers, and probably other allied professionals, are considered to be among the most ethical of all professionals. For example, a 1990 survey of executives of 200 of the largest 1,000 companies in the United States (Robert Half, 1990) asked the question "Which one of the following types of professionals do you think is most ethical?" The answers, in terms of percent of responses, were: engineers, 34%, certified public accountants, 24%; doctors, 17%; lawyers, 8%; dentists, 7%; investment bankers, 1%; and "don't know," 9%. In another U.S. survey (Modic, 1988) of "1,000 top corporate executives, business school deans, and members of

Congress," the surveyed individuals were asked to name and rank four professions in order of ethical behavior. The top four, in order of decreasing priority, were the clergy, accountants, teachers, and engineers. You are in good company.

The chapter begins with a definition of ethics and a discussion of its relationship to legal matters as discussed in Chapter 10. The need for codes of ethics is discussed, followed by an introduction to the codes developed and administered by some engineering professional organizations, other professional organizations, corporations, and government. A discussion of possible future directions in ethics as applied to the engineering and other technical professions concludes the chapter.

DEFINING ETHICS

What is ethics? Is it the same as law? Do ethics and law overlap, or are they different? Are unethical and illegal acts synonymous? Always? Sometimes? In Chapter 10, which describes the legal framework within which engineers function, breach of contract, fraud, and negligence were discussed. What is left, if anything, to consider within the context of ethics?

Tarr (date unknown) says, "Ethics is the study of conduct that is above the law." Taking a similar tack, Onsrud (1987) states, "Ethical conduct is often defined as that behavior desired by society which is above and beyond the minimum standards established by law." These two definitions suggest a benchmark concept, with behavior below the benchmark being considered unacceptable in that it is punishable by law, while behavior above the benchmark is desired, but not required by law.

On the basis of ideas presented by Vesilind (1988), ethics might be defined as the process each person uses to make value-laden decisions. This process or action-oriented perspective introduces the idea of values, which is largely a personal matter. As suggested by Figure 11–1, each individual's value profile can be composed of many individual values, each having a different relative value. Just as each snowflake is said to be different, so each person has a different value mosaic. For example, one person may rate honesty as one of the most important values and, as a result, strive for absolute honesty in all matters. In contrast, another person might also regard honesty as an important value, but honor may be valued even more highly. Each person's value profile or mosaic determines how he or she works through the process of making value-laden decisions. Because the mosaics may differ profoundly, an action considered ethical by one individual may not be considered ethical by another individual. Ethics might be defined simply

diligence	long, steady application to one's occupation or studies; persistent effort, attentive care.
efficiency	quality or property of acting or producing effectively with a minimum of waste, expense, and unnecessary effort.
equality	state or instance of being equal; especially, the state of enjoying equal rights, such as economic, and social rights.
equity	state, ideal, or quality of being just, impartial, and fair.
freedom	condition of being free of restraints; power to act, speak, or think without the imposition of restraint.
honesty	telling the truth—in other words, conforming our words to reality. (*Note*: Honesty is retrospective; it is what you say about what you've done.)
honor	esteem, respect, reverence, reputation; applicable to both the feeling and the expression of these characters.
integrity	informing reality to our words—in other words, keeping promises and fulfilling expectations. (*Note*: Integrity is prospective; it is what you do about what you said.)
knowledge	familiarity, awareness, or understanding gained through experience or study; cognitive or intellectual mental components acquired and retained through study and experience—empirical, material, and that derived by inference and interpretation.
loyalty	feelings of devoted attachment; the condition of being faithful; the unfailing fulfillment of one's duties and obligations in a close and voluntary relationship.
pleasure	enjoyable sensation or emotion; satisfaction; sometimes, though not invariably, suggests superficial and transitory emotion resulting from the conscious pursuit of happiness.
safety	freedom from danger, risk, or injury.
security	freedom from doubt; reliability and stability concerning knowledge of the future.
trust	firm reliance on integrity, ability, or character of a person or thing; implies depth and assurance of such feeling, which may not always be supported by truth.

Figure 11–1 Selected human values (Source: All values and definitions are quoted from McCuen and Wallace, 1987, pp. 34–35 except for honesty and integrity, which are quoted from Covey, 1990, pp. 195–196. Parenthetic notes were added.)

as what you do when no one is looking. Stated differently, your reputation is what others think of you; your ethics is what you really are.

In summary, ethics is

- Related to but different from and above laws. Ethical behavior is referenced to, but more than legal behavior.
- Mostly action (what you do), not knowledge (what you know).
- The personalized way you use your values profile or mosaic to make value-laden decisions.

TEACHING ETHICS

Ethics is difficult to teach in a college or university in the sense that calculus, static mechanics, and machine design can be taught. This is particularly true in light of the skeptical, but fortunately not overly cynical, perspective of many of today's college and university students. By the time today's students begin college studies, they have learned much about the unethical behavior of many private and public individuals and organizations. Young people's access to such information has increased in recent years as a result of the further improvements in global communication and the news media's increasing tendency to probe into and report on the personal lives of people holding, or wanting to hold, high positions in government, business, and academic and religious organizations. Professors and university administrators can easily make all sorts of good or lofty pronouncements about ethics, but if you perceive them to be unethical, you will give little credence to their statements. Furthermore, if you mistrust them, you may doubt what they tell you about technical matters.

As you begin your professional career, you are likely to view your supervisors and others in your organization, including the leaders of the organization, in the same skeptical manner. This is a positive perspective, provided that healthy skepticism does not slip towards unhealthy cynicism. And remember, the healthy skepticism you direct to colleagues and others is likely to be reflected back to you—at least until you all become well acquainted and prove yourselves to each other.

While this chapter cannot teach ethics, in and of itself, it can alert you, the young professional, to the need to have or develop a personalized ethical framework. Books, colleagues, professors, and friends cannot tell the young professional what his or her ethics should be, but they can tell him or her that it is necessary to consciously formulate and refine an ethical framework to use as guidance in one's professional life. You will need the guidance that only you can provide.

ACADEMIA: A CORNER ON ETHICS?

There is a presumption among many academics that the university is the last bastion of ethical behavior in society. It is presumed to be the place where the highest ethics are practiced or, if not actually practiced, at least advocated. The universities certainly have the corner on ethics rhetoric, but they do not have a monopoly on ethical conduct. As Florman (1987, p. 84) succinctly says, "For academics, there will be more conferences and publications and courses in which everything gets discussed and nothing gets decided."

The danger in all this for you, the young professional, is that you and your peers might get the erroneous idea that ethics—either personal or organizational—are irrelevant off the campus in the business and professional world. You may believe that "it is downhill from here" in that ethical expectations will be less "out there in the real world" than they are in the college and university setting. You may be conditioned to think that business ethics is an oxymoron.

To the contrary, ethical conduct is an everyday issue in the practice of the profession and in business. As a participant in the professional world, you do not have a choice as to whether or not you will be concerned with ethics. The questions are What will your rules of personal conduct be? Will you strengthen or detract from the ethical climate of your firm? What will you use as your ethical framework? What values will you hold most dear? What process will you follow? What will you stand for and not stand for? You cannot escape the domain of ethics. Your true values will be gradually revealed by many situations that arise during the normal course of the business day and your response to them.

LEGAL AND ETHICAL DOMAIN

The connection between legal and ethical areas was suggested in the preceding definitions of ethics. Figure 11–2 is a useful model of the legal and ethical domain (McCuen and Wallace, 1987; Onsrud, 1987). In this model, the position of the vertical line, which separates legal from illegal actions, is set largely by statute and common law. Accordingly, the definitive separation between legal and illegal acts is shown by a vertical solid line. The position of the horizontal

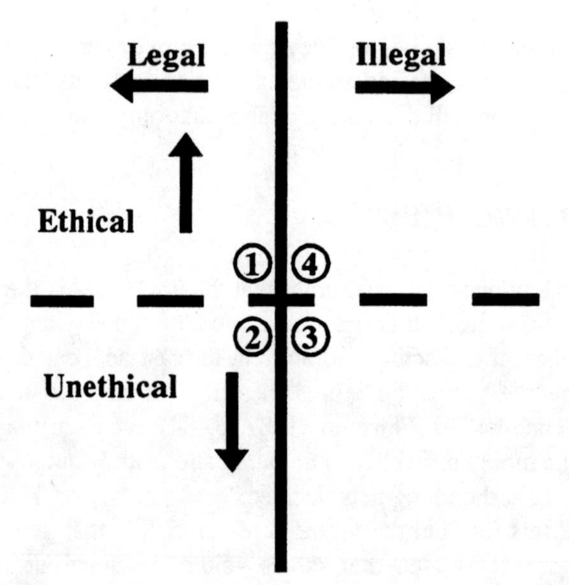

Figure 11–2 Legal and ethical domains

line, which separates ethical and unethical actions, is much less definitive, because it is based primarily on personal values as they might be informed by various codes of ethics when professional and business matters are involved. As a result, the less definitive separation between ethical and unethical lines is shown by a dashed horizontal line. Whereas most engineers would agree on the legality of an act, there would typically be much less agreement on whether or not some aspect of the act is ethical or unethical. Quadrant sizes have no meaning. The rectangular axes are simply intended to define the "space" within which all possible professional and business actions and transactions occur. Clearly, most such actions and transactions occur in the upper left, legal-ethical quadrant.

Consider some hypothetical examples of business, professional and other entries for the four quadrants, beginning with quadrant 1.

- Quadrant 1—Legal and Ethical. Most planning, design, construction, and operation projects or activities fall within this quadrant, as already noted, along with most support activities such as marketing, finance, accounting, and personnel matters.

- Quadrant 2—Legal and Unethical. An engineer successfully (he or she is "lucky") designs a structure or facility outside his or her area of competence. A consulting firm advertises its services in a highly self-laudatory manner. An individual engineer discloses confidential information developed for or with a client or former client. As will be explained later in this chapter, these hypothetical actions are usually legal, but they are unethical because they conflict with accepted codes of ethics established by engineering organizations.

- Quadrant 3—Illegal and Unethical. An example is an act of fraud, such as bid rigging or collusion with others to secure a contract for professional services. In general, illegal acts, such as fraud, breach of contract, and negligence are unethical, but there can be exceptions, as indicated in the following explanation of Quadrant 4.

- Quadrant 4—Illegal and Ethical. This somewhat problematic quadrant is best introduced by using a non-engineering or business example, such as a concerned citizen stopping at the scene of an automobile accident, putting an injured child in his or her vehicle, and exceeding the speed limit to get the child to the emergency room of a local hospital. An example taken from the world of engineering and business would be an engineer violating a signed secrecy agreement with an employer to "go public" and report on a situation that he or she believes is hazardous to the public at large.

Clearly, you are urged to be involved in activities and transactions that almost always fall in Quadrant 1. Occasionally, you may find yourself in Quadrant 4, the resolution of which may pose some serious threats to your professional career. Just as clearly, you are urged to avoid Quadrant 2 partly because to do otherwise is to risk the trust of your colleagues. Quadrant 3 is extremely dangerous ground.

CODES OF ETHICS

Essentially all engineering societies and many businesses, government entities, and other organizations have developed, adopted, and refined codes of ethics to address ambiguities. The overall purpose of such codes of ethics is to reach and document consensus, or at least the majority opinion, among members and participants particularly given the already discussed natural variation in the value profiles or mosaics of individuals. Any code of ethics can be viewed as having four purposes (McCuen and Wallace, 1987, Chapter 4). These are

- Establish goals.
- Identify values.
- Identify rights and responsibilities of the organization and its members.
- Provide guidance and inspiration.

McCuen and Wallace go on to describe some of the limitations of codes of ethics. First, they have limited power. For example, a professional society can reprimand or expel a member, but cannot necessarily keep him or her from practicing the profession. Similarly, an employer could terminate an employee who violates the employer's code of ethics but certainly can not keep him or her from seeking employment elsewhere. A university placement office could deny services to a senior student seeking employment if the student violates the code of ethics of the College Placement Council. However, the student has other means available for seeking employment. Unlike many laws, codes of ethics typically do not prescribe sanctions and punishments.

A second limitation of codes of ethics is that some form of bureaucracy, with its attendant monetary costs and procedural delays, is needed to administer them. Third, codes may appear self-serving and pompous, particularly in trying to articulate responsibilities and rights of members of a public or private organization. A fourth limitation on codes of ethics is the difficulty in obtaining consensus on all matters. Rights and responsibilities of different members of a professional society must be considered and respected. An example is an engineering faculty member being involved in consulting as viewed by academics versus consulting engineers.

The fifth limitation of codes of ethics is that they evolve over time as a result of changes within and outside the profession. Codes typically do not include a history of how the code was developed and, therefore, those who are to be governed by the code may not understand or appreciate certain provisions. An example is the change that has occurred, as discussed later in this chapter, in prohibitions against competitive bidding for professional services. The sixth and last code limitation is that individuals may be members of two or more professional organizations having conflicting elements in their codes. Lawyers and engineers are examples.

As will become evident in this section, codes are successful in that they reduce the ambiguity in the ethical arena. However, they don't eliminate all ambiguity and some, to the extent that they are unnecessarily detailed, may require considerable effort on behalf of conscientious individuals.

Engineering Professional Organizations' Codes of Ethics

Appendix B is the code of ethics of the American Society of Civil Engineers, and Appendix C is the code of ethics of the Institute for Electrical and Electronic Engineers. Note that the ASCE code consists of four fundamental principles and seven fundamental canons, with each canon being presented in considerable detail. In contrast, the IEEE code is composed of a brief introduction and ten short commitments.

Both codes make explicit reference to protecting "the safety, health, and welfare of the public." This provision appears within Fundamental Canon 1 in the ASCE code, where it is described as being "paramount," and is the first item in the IEEE code. Another feature of codes of ethics, and one that is expected, is that they typically contain some very precise and demanding provisions. For example, the ASCE code, Canon 1, Items c and d require you to act when your professional judgment is overruled and the public safety, health, and welfare are jeopardized. Similarly, the IEEE code under Item 6 mandates maintenance and improvement of "technical competence." These provisions, which are a small part of the two example codes, can put great demands on you as you progress in your professional life.

Adherence to the applicable code of ethics is a condition of membership in professional engineering and other technical organizations. Unfortunately, many professionals of all ages and levels of experience join such organizations without a detailed review of the provisions of the code. This, in turn, undoubtedly leads to some of the ethical problems that arise. You should study a professional organization's code of ethics before joining. If you cannot embrace the code, don't become a member.

As already noted, codes of ethics such as those used by engineering societies evolve in response to changing internal and external conditions. For example, many engineering societies used to include prohibitions against competitive bidding within their codes of ethics. This formerly ethical provision became a legal issue and was subsequently declared illegal by the U. S. District Court of the District of Columbia in the mid-1970s (Clough, 1986, Chapter 4.2; McCuen and Wallace, 1987, p. 95). This is an example of an ethical provision becoming an illegal act and, therefore, being removed from a code of ethics. Incidentally, in a related and somewhat compensatory manner, the Brooks Act of 1972 prohibits federal agencies from using competitive bidding to select engineering and other similar firms for design services. Some states have mini-Brooks Acts. Therefore, the intent of the former antibidding provision which appeared in and was struck from the codes of ethics of engineering organizations was, in effect, reintroduced in special circumstances by federal and state legislation.

As another example of evolving codes of ethics, consider the former provision in the code of the National Society of Professional Engineers (NSPE) that prohibited self-laudatory advertising. This provision was intended to protect the dignity of the engineering profession. The provision has been removed as a result of an agreement between NSPE and the Federal Trade Commission. The effect of this change is that NSPE can no longer try to govern the nature of its members' advertising, as long as it is truthful and nondeceptive. Self-laudatory advertising, which was until recently considered unethical but legal, is now in effect both ethical and legal (Indiana Professional Engineer, 1993).

McCuen and Wallace (1987, p. 76) remind engineers and other technical professionals who are members of professional organizations having codes of ethics that those professionals are not relieved of ethical responsibilities as defined within those codes because the professionals happen to work for organizations in which the management or culture does not support the codes. The young professional seeking his or her first employment is often already a member of one or more professional organizations through the student chapter structure. During the employment interview process, you should ask questions about the code or codes of ethics that apply within the technical organization or business you are thinking of joining. As noted later, some employers have codes of ethics that supplement or complement those of professional engineering and other organizations. If you cannot accept the potential employer's code or codes or attitude toward codes, move on to other opportunities.

Codes of Ethics of Other Professional Organizations

Organizations that are not engineering or similar organizations, but which interact with such organizations, also have codes of ethics. Young professionals who have used a college placement service may be aware of the College Placement

Council, Inc. This council, which is composed of representatives of colleges and universities and of employers of all types of college and university graduates, has established Principles for Professional Conduct for Career Services and Employment Professionals, as set forth in Appendix D.

This code is intended for use in governing the triangular relationship that includes colleges and universities, employers, and students seeking employment. The code sets forth principles for career planning and principles for placement and recruitment for each of the three parties involved in the partnership. As indicated in the introductory section, the code seeks to protect and serve the interests of all three parties. Engineering and other technical program students who use placement services at institutions belonging to the College Placement Council are expected to honor the provisions of the code.

Corporate Codes of Ethics

Business organizations that have technical professionals on their staff often adopt codes of ethics to supplement or complement the codes of ethics of various professional organizations. Appendix E presents excerpts from the Boeing Company's Business Conduct Policy and Guidelines. This code, which is presented in booklet format, has some very specific provisions dealing primarily with interactions between Boeing personnel and its government and private clients. The questions and answers appearing in Appendix E are only a very small part of the Boeing code. A review of each of the selected questions suggests the following principal ideas:

- Question 1. Ethics means good business.

- Question 2. How will you, as an employee, really know if adverse actions have been taken?

- Question 3. The focus here is on the distinction between a personal and a business relationship.

- Question 4. Boeing defers to code provisions of others if they are stricter.

- Question 5. The act, not the source of funds, is the issue.

- Question 6. Again, the source of funds is not the issue.

- Question 7. Gifts of nominal value may be accepted.

- Question 8. Many employers will gladly support community activities in a variety of ways, but to the extent that those activities involve employer's resources, the employer's permission should be obtained.

• Question 9. The importance of chargeable time within consulting firms and all other service organizations, and its impact on profitability are discussed in detail in Chapter 9 of this book. The pressure to maintain a high chargeable time utilization often creates ethical dilemmas particularly for younger engineers and other technical professionals who may be receiving mixed messages from supervisors and others within the organization.

• Question 10. Business lunches hosted by one party in a given situation are generally acceptable.

• Question 11. This and the "frequent flyer" question often arise and are usually easily resolved by simply inquiring about the corporate policy.

• Question 12. Prohibiting provision of local transportation is a very strong measure, but is presumably founded on careful consideration of circumstances.

In recent years, a growing number of engineering and other technical organizations, particularly those in the private sector, have developed corporate codes of ethics. As noted earlier, the inquisitive and well-prepared candidate for employment should ask about the existence of formal codes and, if they do not exist, inquire about the applicability of codes of ethics of one or more professional organizations.

Government Codes of Ethics

In August 1992, the Federal Office of Government Ethics (OGE) published "Standards of Ethical Conduct for the Employees of the Executive Branch" (OGE, 1992). The new standards went into effect in February 1993. The code is based on the premise that public service is a public trust. Fourteen principles are set forth in the code and are to be used by U.S. Government employees and others to define ethical conduct. The 14 principles, quoted from the *Federal Register* (OGE, 1992, p. 35042), are

1. Public service is a public trust, requiring employees to place loyalty to the Constitution, the laws and ethical principles above private gain.

2. Employees shall not hold financial interests that conflict with the conscientious performance of duty.

3. Employees shall not engage in financial transactions using nonpublic Government information or allow the improper use of such information to further any private interest.

4. An employee shall not, except as permitted by subpart B of this part, solicit or accept any gift or other item of monetary value from any person or entity seeking official action from doing business with, or conducting activities regulated by the employee's agency, or whose interests may be substantially affected by the performance or nonperformance of the employee's duties.

5. Employees shall put forth honest effort in the performance of their duties.

6. Employees shall not knowingly make unauthorized commitments or promises of any kind purporting to bind the Government.

7. Employees shall not use public office for private gain.

8. Employees shall act impartially and not give preferential treatment to any private organization or individual.

9. Employees shall protect and conserve federal property and shall not use it for other than authorized activities.

10. Employees shall not engage in outside employment or activities, including seeking or negotiating for employment, that conflict with official Government duties and responsibilities.

11. Employees shall disclose waste, fraud, abuse, and corruption to appropriate authorities.

12. Employees shall satisfy in good faith their obligations as citizens, including all just financial obligations, especially those—such as federal, state, or local taxes—that are imposed by law.

13. Employees shall adhere to all laws and regulations that provide equal opportunity for all Americans regardless of race, color, religion, sex, national origin, age, or handicap.

14. Employees shall endeavor to avoid any actions creating the appearance that they are violating the law or the ethical standards set forth in this part. Whether particular circumstances created an appearance that the law or these standards have been violated shall be determined from the perspective of a reasonable person with knowledge of the relevant facts.

The 14 principles are comprehensive and reasonable. However, their application to government activity can be very complex. The code elaborates on each of the 14 principles by setting forth, in great detail, many standards and then offering hypothetical examples to explain the applications of the principles and standards. For example, the code (OGE, 1992, p. 35046) states:

An employee may accept unsolicited gifts having an aggregate market value of $20 or less per occasion, provided the aggregate market value of individual gifts received from any one person under the authority of this paragraph shall not exceed $50 in a calendar year.

The code (OGE, 1992, p. 35046) then gives this example to illustrate application of the preceding gift provision:

An employee of the Defense Mapping Agency has been invited by an association of cartographers to speak about his agency's role in the evolution of missile technology. At the conclusion of his speech, the association presents the employee with a framed map with a market value of $18 and a book about the history of cartography with a market value of $15. The employee may accept the map or the book, but not both, since the aggregate value of these two tangible items exceeds $20.

This example drawn from the OGE code only begins to suggest the amount of individual and organizational effort that may be required to conscientiously function in accordance with very detailed and complex codes of ethics that seem to assume that most personnel require "watching." An attractive alternative, and one that may only be feasible in a small organization, is a short and strong statement of principles coupled with a very careful personnel recruitment and retention program.

LOOKING AHEAD: LESS ETHICS OR A DIFFERENT KIND?

In some ways the engineering and other technical professions may have less and less need for formal codes of ethics. Many issues and questions that used to be in the realm of codes of ethics are now in the realm of law. For example, Florman (1987, p. 89) cites federal health, safety, and environmental protection laws such as Inspection of Steam Boilers enacted in 1952, the Occupational Safety and Health Act adopted in 1970, the National Environmental Policy Act passed in 1970, the Safe Drinking Water Act enacted in 1974, the Water Pollution Control Act first adopted in 1972, and the Comprehensive Environmental Response Compensation and Liability (superfund) Act created in 1980.

In light of expansion of legislation into areas previously implicitly or explicitly within the realm of various codes of ethics, do engineering and other technical professions need to be less concerned about ethics, or should they focus on a different kind of ethics? This question is illustrated in Figure 11–3. As suggested by Figure 11–3, the legal realm appears to have enlarged and may con-

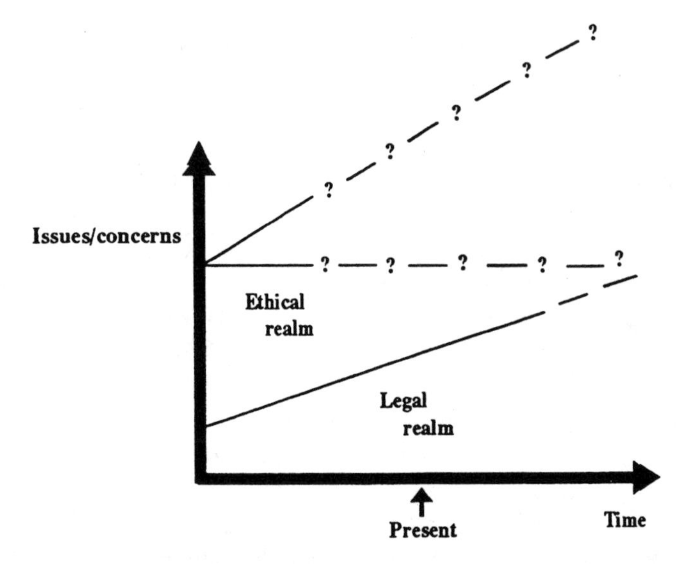

Figure 11–3 Historic changes in ethics and law in the practice of engineering

tinue to do so. Changes in the ethical realm are not as evident, and neither are future directions.

Three Possible Future Directions

Looking to the future, Vesilind (1988, p. 289) argues that ethical considerations should be expanded to include the environment in a broader sense. He goes on to say that society's environmental expectations have increased and, therefore, civil and other engineers have more to be concerned about as they go about doing their work. He cites examples such as protection of plant and animal life, preservation of unique habitat, protection of historic sites that commemorate past cultures (e.g., covered bridges, Indian mounds, forts, and mills), and aesthetics in relation to factors such as the location of facilities, the selection of materials, and preservation of vegetation. Vesilind's thesis seems to be somewhat in conflict with the idea presented earlier that laws and regulations at all levels have taken and will continue to take into account protection of the environment.

Florman (1987, Chapter 8) argues that the greatest need in the ethical realm is increased competence. He cites a review of data at the University of Maryland Architecture and Engineering Performance Information Center (AEPIC), which concludes that "... probably less than two percent of technological failures relate on ethical behavior in the conventional sense, that is, greed or intent to deceive ..." (Florman, 1987, p. 162). He also cites (Florman, 1987, pp. 102–103) a study at the Swiss Federal Institute of Technology in Zurich, which analyzed "... 800 cases of structural failure in which 504 people were killed, 592 people injured, and many millions of dollars of damage occurred. . . ." As suggested by Table 11–1, which lists the causes of the 800 cases of structural failure, competence, or rather lack thereof, is by far the dominant cause, accounting for at least 97 percent of the failures.

Florman is suggesting that the competence should receive more attention from the engineering community. Simply put, although failures are few and far between, they would be even fewer and farther between if engineers were even more careful in their work, were more thoughtful in developing solutions to problems, were more aware of new developments, were more active in sharing findings, and were more involved in public affairs.

Interestingly, Florman (1987, p. 104) clearly distinguishes between having good intentions and doing good work. While good intentions are admirable, they don't necessarily lead to competence. Good intentions do not necessarily make good engineering. Well-intentioned engineers and other technical professionals can be incompetent. Most people would rather have surgery, legal work, financial work, or engineering work done by, respectively, a conscientious and competent doctor, attorney, banker, or engineer, than a high-sounding, self-righteous doctor, attorney, banker, or engineer.

Wenk (1988, pp. 19–23), somewhat in contrast with Vesilind's environmental ethic and Florman's competence ethic, advocates a social activism ethic

TABLE 11–1 RESULTS OF STUDY BY SWISS FEDERAL INSTITUTE OF TECHNOLOGY

Researchers investigated 800 structural failures resulting in 504 deaths and 592 injuries. Where engineers were at fault, the causes were

Causes	Percent
Insufficient knowledge	36
Underestimation of influence	16
Ignorance, carelessness, negligence	14
Forgetfulness, error	13
Relying upon others without sufficient control	9
Objectively unknown situation	7
Imprecise definition of responsibilities	1
Choice of bad quality	1
Other	3
	100

(SOURCE: Adapted with permission from Florman, S.C., *The Civilized Engineer,* New York: St. Martins Press, 1987, pp. 102–103. Copyright 1987 by Samuel Florman.)

for technical professionals. He states, "Technology is more than technique; it is social process" and he goes on to argue that technology influences culture and vice versa. Technology can have adverse side effects and technology is ultimately controlled by government—it is a political process, not a market process. Therefore, Wenk argues that engineers and other technical professionals, because of their knowledge, have the responsibility to help inform and educate the public. They should go beyond answering "Can we do it?" to helping answer "Ought we do it?" and "Can we manage it?"

Key Ideas

Many issues and concerns previously in the ethical realm are now in the legal realm. Perhaps the "vacuum" in the ethical realm is or should be taken up by new ethical issues and concerns, such as an even broader environmental ethic, a competency ethic, or a social activism ethic. While pondering these future ethical issues, you as the entry-level professional, are advised to focus your early career energies on developing and refining your personal code of ethics. You will need it.

REFERENCES

ANDERSON, T., "Effort and Sacrifice," in *Here's How By Who's Who,* Jesse Grover Bell, ed. Lakewood, Ohio: Bonne Bell, Inc., 1968, p. 3.

AMERICAN SOCIETY OF CIVIL ENGINEERS, "Code of Ethics," *ASCE Official Register—1992,* New York.

BOEING COMPANY, "Business Conduct Policy and Guidelines," Seattle, Wash., June 1991.

CLOUGH, R. H., *Construction Contracting*, 5th ed. New York: John Wiley & Sons, 1986, Chapter 4.

COLLEGE PLACEMENT COUNCIL, INC., "Principles for Professional Conduct for Career Services and Employment Professionals," Bethlehem, Pa., June 1990.

COVEY, S. R., *The 7 Habits of Highly Effective People*, New York: Simon & Schuster, 1990.

FLORMAN, S. C., *The Civilized Engineer.* New York: St. Martin's Press, 1987.

INSTITUTE OF ELECTRICAL AND ELECTRONICS ENGINEERS, INC., "Code of Ethics," August, 1990.

MCCUEN, R. H., and J. M. WALLACE, eds., *Social Responsibilities in Engineering and Science: A Guide for Selecting General Education Courses.* Englewood Cliffs, N.J.: Prentice Hall, Inc., 1987.

MODIC, S. J., "Are They Ethical?" *Industry Week*, February 1, 1988, p. 20.

NATIONAL SOCIETY OF PROFESSIONAL ENGINEERS, "Engineering Ethics," *Engineering Times*, April 1988.

NATIONAL SOCIETY OF PROFESSIONAL ENGINEERS, "Engineering Ethics," *Engineering Times*, July 1988.

NATIONAL SOCIETY OF PROFESSIONAL ENGINEERS, "Engineering Ethics," *Engineering Times*, May 1989.

"NSPE Agrees to Revise Code of Ethics," *Indiana Professional Engineer*, September–October 1993, p. 1.

OFFICE OF GOVERNMENT ETHICS, "Standards of Ethical Conduct for Employees of the Executive Branch, Final Rule," *Federal Register*, 5 CFR, Part 2635, August 7, 1992, pp. 35006–35067, Washington, D. C.

ONSRUD, H. J., "Approaches in Teaching Engineering Ethics," *Civil Engineering Education—ASCE Civil Engineering Division*, Vol. IX, No. 2 (Fall 1987), pp. 11–27.

ROBERT HALF INTERNATIONAL, "Engineers and Accountants Rank Highest in Professional Ethics," press release, Menlo Park, Cal., May 10, 1990.

TARR, C. W., "Why Corporate Ethics Can Make a Difference," *The Sound Management Report*, Vol. 1, No. 6, no date.

VESILIND, P. A., "Rules, Ethics and Morals in Engineering Education," *Engineering Education*, February 1988, pp. 289–93.

WENK, E., JR., "Roots of Ethics; Wings of Foresight: New Principles for Engineering Practice," *The Bent*, Spring 1988, pp. 18–23.

SUPPLEMENTAL REFERENCES

GUNN, A. and P. A. VESILIND, *Environmental Ethics for Engineers*. Chelsea, Mich.: Lewis Publishers, 1986.

JOHNSON, D. G., ed., *Ethical Issues in Engineering*. Englewood Cliffs, N.J.: Prentice Hall, 1990.

KIRKPATRICK, D., "What Is An 'Ethical' Engineer?", *Journal of Management in Engineering—ASCE*, Vol. 5, No. 4 (October 1989), pp. 367–370.

MARTIN, M. W., and R. SCHINZINGER, *Ethics in Engineering*, 2nd ed. New York: McGraw-Hill Book Company, 1989.

MARTIN MARIETTA CORPORATION, "Code of Ethics and Standards of Conduct," Orlando, Fla., October 1993.

SUMAR, D. G., *Getting Started as a Consulting Engineer*. Chapter 11. San Carlos, Cal.: Professional Publications, Inc., 1986.

TAYLOR, J. I., "Allocating Public Funds: Morality and Constraints," *Proceedings of Highway Safety: At the Crossroads*, American Society of Civil Engineers, New York, March 1988, pp. 123–133.

EXERCISES

11.1 ETHICAL CHOICES (ADAPTED FROM ONSRUD, 1987)

Purpose

1. Illustrate the kinds of trivial to major ethical questions that must be decided by the young professional.
2. Emphasize why it is necessary for the entry-level technical professional to know his or her values and to recognize that personal value systems vary.
3. Suggest the need, at least in the business and professional environment, for guidance materials or documents such as codes of ethics.

Tasks

1. Assume that you are a young engineer employed by a city. You have been placed in charge of inspecting a sewer project that is being built for the city by a private contractor. Or assume that you are a young engineer employed by an electric utility. You have been placed in charge of a power line construction project. Because of your education and field engineering experience, you are able to suggest techniques and procedures that save the contractor both time and money. The work, however, is to be done strictly according to the plans and specifications.
2. It is quitting time on a hot summer Friday afternoon. The president of the construction firm comes to the site and offers a soft drink to each of his employees.
   a. The president offers you a soft drink. May you accept it? Why?
   b. Assume that the president hands every worker and you a case of soft drinks. Is it ethical for you to accept a case? Why?
   c. What if the president hands every worker a can of beer and offers one to you? Can you ethically accept? Why?
   d. What if the president offers every worker a bottle of Scotch? Would you accept? Why?
   e. How about a case of Scotch? Would you accept? Why?
   f. What if the president hands a joint of marijuana to each of his workers? Regardless of whether you smoke marijuana, can you ethically accept? Why?
   g. What if the president hands each worker a pen with his company name on it? Would you accept? Why?
   h. What about a hat and jacket with the company name and logo on it? Would you accept? Why?
   i. What if the construction firm buys you a new car because of the thousands of dollars you have saved their firm. Would you accept? Why?
3. Next week, you and the president meet at lunch to discuss the progress of the work. The president offers to pay the bill. Can you ethically accept this offer? Why?

11.2 A LEGAL AND UNETHICAL ACT

Purpose

Provide students with an opportunity to increase their awareness of ethics by considering the ethical and legal aspects of actual and hypothetical situations.

Tasks

1. Using the code of ethics of "your" engineering or similar society and your understanding of the legal aspects of engineering and related disciplines, identify and briefly discuss an actual (from your experience or knowledge) or hypothetical engineering or business situation that was/is both legal and unethical.
2. If you use an actual situation, please change names of people, places, and organizations as may be appropriate.

11.3 AN ILLEGAL AND UNETHICAL ACT

Purpose

Provide students with an opportunity to increase their awareness of engineering ethics by considering the ethical and legal aspects of actual and hypothetical situations.

Tasks

1. Using the code of ethics of "your" engineering or similar society and your understanding of the legal aspects of engineering and related disciplines, identify and briefly discuss an actual (from your experience or knowledge) or hypothetical engineering or business situation that was/is both illegal and unethical.
2. If you use an actual situation, please change names of people, places, and organizations as may be appropriate.

11.4 COMPANY CODE OF ETHICS

Purpose

Encourage thoughts about codes of ethics for technical organizations.

Tasks

1. Note the dilemmas posed by the kinds of scenarios presented in Exercise 11.1.
2. Assume you are the president of your own consulting firm or manufacturing organization.
3. Formulate and write your company's policy on the ethics of your employees' accepting gifts or receiving other considerations from existing clients, potential clients, contractors, or other individuals or organizations.

11.5 INTERVIEW TRIP (SOURCE: ADAPTED FROM MCCUEN AND WALLACE, 1987, P. 78.)

Tasks

1. Assume that you, as a college senior, accept an expense-paid trip to interview with an engineering or other technically based firm for potential employment upon graduation.
2. You accept the interview invitation, knowing that you will not accept a job even if it is offered.
3. Is this a violation of your professional society's code of ethics? Yes or no? Explain your answer and identify the applicable element(s) of the code.

11.6 INTERVIEW TRIP

Tasks

1. Assume that you, as a college senior, accept an expense-paid trip to interview with an engineering or other technically based firm for potential employment upon graduation.
2. You accept the interview invitation, knowing that you will not accept a job even if it is offered.
3. Is this a violation of the College Placement Council, Inc. code? Yes or no? Explain your answer and identify the applicable elements of the code.

11.7 CRITICIZING ANOTHER ENGINEER (QUOTED FROM *NSPE,* APRIL 1988)

The Situation

In a southwestern state legislature, various bills involving water supply, flood control, and production of electric power are awaiting action. The question of how to achieve the bills' goals most efficiently and economically has been debated within the legislature and in public forums for several years. A state legislative committee on public work calls a hearing to receive comments and recommendation on the various proposals.

Terry Techna, P. E., representing the state power commission, testifies that from an engineering standpoint, her team's studies point to a series of low dams as the most efficient solution. A. U. Tility, P. E., representing a private power company, testifies that according to his engineering analysis, a single high dam would produce the same results both faster and for less money. Both engineering witnesses submit voluminous engineering data in support of their positions and do not hesitate to criticize the other's analysis and findings.

What Do You Think?

Is it ethical for Techna and Tility to criticize each other's analysis and findings? Use and make specific reference to either the ASCE or IEEE codes introduced in this chapter.

11.8 EXPERT WITNESSES TESTIMONY (QUOTED FROM *NSPE*, JULY 1988)

The Situation

When damages arise from engineering failures, X. Burt Eye, P. E., is often asked to serve as technical advisor in the resulting lawsuit. He provides the contracting attorney expert analysis and consultation on the technical reasons for the alleged failure. Eye may also be called on to testify at the trial as an expert witness in support of his findings.

In the past, Eye has provided those services on a per diem basis. In one case, however, Eye is asked by the plaintiff to accept as payment a percentage of the eventual damage award—in the event the plaintiff wins the case. If the defendant comes out on top, however, Eye will not collect a dime. Eye agrees to this proposal.

What Do You Think?

Was it ethical for Eye to provide his services for a percentage of a possible court award?

11.9 CONFLICT OF INTEREST? (QUOTED FROM *NSPE*, MAY 1989)

The Situation

Abe "Buff" Borde, P. E., a principal in a consulting engineering firm, serves as chairman of the local storm drainage board. The board was established by the city council to advise on the city's general engineering needs for drainage and the related facilities. The board periodically reviews the facilities in light of economic and environmental considerations and submits to the city council recommendations for improvements. When the council determines that changes or additions to the facilities need to be designed, it directs the city's engineering staff to get statements of interest and qualifications from engineering firms. The responses are provided to the storm drainage board for review and returned with recommendations to the city council, which selects a firm to negotiate a contract.

Up to now, Borde has advised the city council and its engineering staff that his firm should not be considered because he is on the advisory board. Even so, the city engineering staff now requests Borde's firm, along with other firms, to submit its qualifications for a pending project. Borde discusses the situation openly with the members of the city council, the city attorney, and the city manager, each of whom says it would be proper for Borde's consulting firm to be considered and, if selected, to accept the project, provided that Borde does not participate in the considerations and recommendations of the advisory board.

What Do You Think?

May Borde's consulting engineering firm ethically submit its qualifications to the city advisory board if Borde disqualifies himself from the board's considerations and recommendations on selection of a firm?

11.10 GILBANE GOLD

Purpose

Increase the student's awareness of the difficulty of ethical issues in engineering practice.

Tasks

1. View "Gilbane Gold: A Case Study in Engineering Ethics," a 24-minute videotape portraying a fictitious, but realistic set of ethical quandaries. The video and study guide are available from the National Society of Professional Engineers (1420 King Street, Alexandria, VA, 22314, 703–684–2800).
2. Select one character from the following:
   a. David Jackson—young civil engineer working for Z Corp.
   b. Tom Richards—environmental engineering consultant recently fired for espousing the new test standards.
   c. Phil Port—manager, Z Corp's environmental affairs department.
   d. Frank Seedas—engineering manager, Z Corp.
   e. Diane Collins—vice president in charge of Z Corp's Gilbane plant.
   f. Lloyd Bremen—former state Commissioner of Environmental Protection. Retired and now a local farmer.
   g. Dr. Winslow Massin—professor emeritus, Hanover University, School of Engineering.
   h. Maria Renato—TV reporter for Channel 13.
   i. Dan Martin—Z Corp lawyer.
3. Identify one or more difficult questions/decisions the character had to deal with.
4. Where was it (identify quadrant) in the ethical-legal domain model (Figure 11–2) presented in this chapter?
5. What did the character do?
6. What would you have done and why?
7. Write a brief memorandum providing specific responses to tasks 2 through 6.

12

Design

> *Scientists define what is, engineers create what never has been.*
>
> (Theodore von Karman, Hungarian-American aeronautical engineer)

Designing, like navigating a vessel, playing a musical instrument, or writing your first book, has to be experienced if one is to learn how to do it and appreciate the associated height of satisfaction and thrill of achievement. Design cannot be adequately learned or fully appreciated by reading about it or hearing lectures on it—you have to do it. As a young engineer or other technical professional, you are urged to seek design opportunities very early in your career so that you are fully aware of this important aspect of the technical professions.

Although design cannot be fully appreciated without doing it, many facets or aspects of design can be defined and discussed to prepare you for design assignments and opportunities. Accordingly, the purpose of this chapter is to orient you to the design function and help you recognize and appreciate design opportunities and to grab and pursue them with gusto.

The chapter begins with a discussion of design as one of the principal functions of engineering and other technical professions. Design is then presented from a very pragmatic perspective in terms of the papers produced. The "risky business" view of design is then described, recognizing that the "designer" always risks personal failure, that is, the inability to create a successful solution. There is always the probability, small as it may be, of the failure of a product, structure, facility, or system. The chapter concludes with a view of design as a creative and satisfying process, accomplishing this, in part, by looking at some of the origins of the word *engineer*.

THE DESIGN FUNCTION

Design may be viewed as one of the four broad or principal functional areas generally recognized in technical professions, as illustrated in Figure 12–1. These broad or principal functions are

- planning
- designing
- constructing, manufacturing, fabricating, or otherwise implementing
- utilizing or operating.

Certainly there are valid arguments for identification and inclusion of other functional areas such as research and sales. A counter argument, particularly for the sake of simplicity, is that such functions are included in or part of the four fundamental functions presented in Figure 12–1.

Note how the four broad or functional areas presented in Figure 12–1 inter-

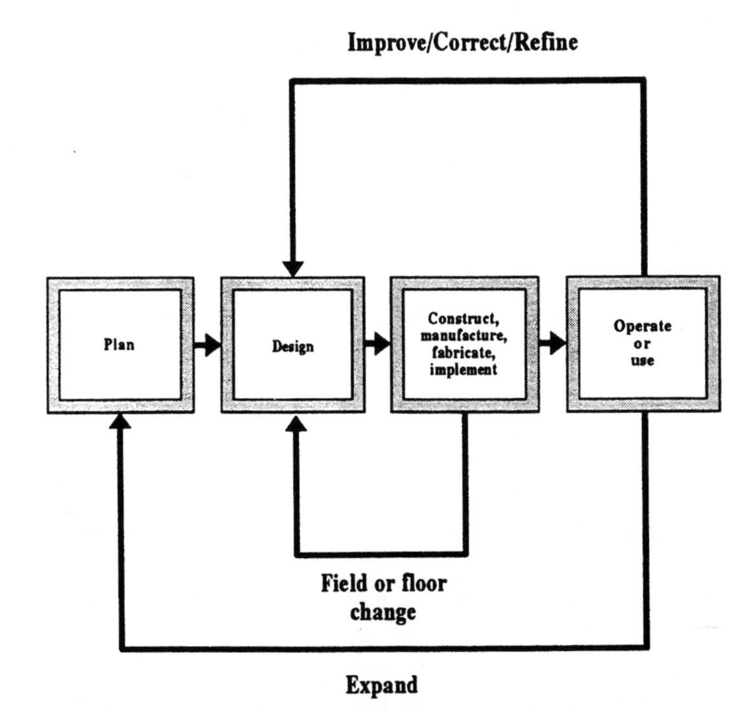

Figure 12–1. Functional areas in the creation and use of a product, structure, facility, or system

act with each other. Consider, for example, the iteration between design and construction, as illustrated by the hotel walkway collapse described in Chapter 10. In that case, the apparent failure during the design function to consider constructability led to a design change during construction which, because of inadequate review, had catastrophic results. More thought should have been given to construction during design—the two functions are inextricably connected. Or consider the interaction between design and operation illustrated by the scoreboard failure also described in Chapter 10. More thought should have been given during design to operation and maintenance. To repeat, there must be considerable interaction between the four broad or principal functional areas, including the design function, illustrated in Figure 12–1. This interaction, which is best accomplished by using a team approach, is sometimes called *concurrent* or *simultaneous* engineering (e.g., Welter, 1990) or *partnering,* as described in Chapter 7.

Iteration, more commonly referred to as trial and error, is very common within the design function. All but the most trivial designs typically involve numerous trial-and-error loops. As a young professional who was most recently in an undergraduate curriculum that focused heavily on assignments and problems requiring analysis of existing entities and usually having unique and "correct" answers, you may find design to be somewhat unsettling and unnerving. Rarely in practice is there a best or correct solution. Rather, the technical professional strives to arrive at a design that is within that "best" subset of all possible solutions as defined by factors such as those described in Chapter 8 and illustrated in Figure 8–1.

The planning function, which is one of the four broad or principal functional areas illustrated in Figure 12–1, also typically contains elements of design as design is described in this chapter. Planning differs from design in that the usual intended immediate result of design is the set of documents needed to construct, manufacture, fabricate, implement, or establish something. Planning, in contrast, is the creative process that leads to recommendations, usually in the form of a report, of what to do and why to do it, but requires subsequent steps providing more specifics.

Some distinctions between planning and design, using civil engineering as an example, are presented in Table 12–1. As indicated, within civil engineering, planning, in contrast with design, tends to have a larger areal scope, less technical detail, and less accurate cost estimates. Furthermore, planning tends to produce reports which, as already noted, focus on recommending what to do and giving the rationale. In contrast, design tends to produce plans and specifications that show and describe, in great detail, how to do what has been recommended in the planning function.

In contemplating the ultimate results of the planning and design functions, the young designer is likely to think in terms of products, structures, facilities,

TABLE 12–1 DISTINCTIONS BETWEEN PLANNING AND DESIGN IN CIVIL ENGINEERING

Item	Planning	Design	Comment
Sequence	First	Second	Unfortunately, sometimes planning is not done or it is done after design.
Areal scope	Large[a]	Small[b]	There are exceptions.
Technical detail	Broad	Deep	
Accuracy of cost estimates	Roughly ± 25%[c]	Roughly ± 1 to 5%	
Immediate product	Report[d]	Plans and specifications[e]	
Cost	Small	Large	

[a]E.g., planning of interstate highway system or planning a large residential development.

[b]E.g., design of a bridge on the interstate highway system or design of a water supply system for a large residential development.

[c]A contingency is usually added to a planning cost estimate to assure that the estimate is high rather than low.

[d]Recommends what to do and why.

[e]Shows exactly how to do it.

and systems composed of metal, concrete, plastic, and other very substantive materials. However, design can produce "soft" results. Examples include a computer model, a new way to organize a technical organization to improve utilization of human and other resources, and an improved process for manufacturing a product. As an entry-level professional you are advised to not be too myopic in your view of the scope and results of the planning–design functions. The iterative process illustrated in Figure 12–1 has wide applicability within, and even outside of, engineering and other technical fields.

DISPROPORTIONATE IMPACT OF THE DESIGN FUNCTION

One reason design is so important among the four functional areas illustrated in Figure 12–1 is that it accounts for a very small fraction of the total cost (i.e., planning, design, construction–manufacturing–fabrication, and operation) of a product, structure, facility, or system. However, design is the primary determinant of the total project cost. For example, an automobile designer stated that although design accounts for only about five percent of a new automobile's cost, it influences approximately 70 percent of the total cost of the new vehicle. By iterating between design and manufacturing, the number of parts in the 1993 Buick LeSabre grille and the time required to assemble the grille were greatly reduced compared to the 1992 model—from 52 parts and 9.6 assembly minutes to 16 parts and 3.5 assembly minutes (Moskal, 1992).

Results of a research conducted by the Construction Industry Institute (CII, 1987) provide another perspective on the impact of design on total costs and on the quality of the final product provided. Design effectiveness was measured by the following seven parameters selected from an original list of 25 possibilities:

- Final project schedule
- Constructability
- Quality of design
- Final project cost
- Plant start-up
- Performance
- Safety

After examining 40 input variables, the following ten were "identified as having the greatest impact on design effectiveness":

- Scope definition
- Owner profile and participation
- Project objectives and priorities
- Preproject planning
- Basic design data
- Designer qualification and selection
- Project manager qualification
- Construction input
- Type of contract
- Equipment sources

One conclusion of the CII research was that "Overall failure to manage design input could increase total project cost by 25 percent or more." The partnering process discussed in Chapter 7 is one mechanism for optimizing input to design.

The disproportionate or leveraging effect of design on total project costs and on the overall quality of the resulting product, structure, facility, or system is generally applicable across the technical fields. Design, or more specifically, the engineers and other technical professionals who do it, are the key to quality. As you have opportunities to participate on design teams, use the preceding list of seven design effectiveness parameters as a reminder of what is likely to be important to your organization's client, customer, or constituent. Consider the preceding list of ten design input parameters in finding ways to do effective design.

DESIGN IN TERMS OF PAPERS PRODUCED

Design can be viewed as a process that results in the production of drawings and other written information having sufficient specificity to be used by other individuals or organizations to construct, manufacture, or fabricate or otherwise implement a product, structure, facility, or system. As discussed in Chapter 1 and illustrated in Figure 1–1, the individual or organization responsible for planning and design is often not the same individual or organization responsible for constructing, manufacturing, fabricating, or implementing that which was planned or designed. Conveying the essence of the "designer's" creation to the "builder" in sufficient detail and with adequate understanding so that the latter can produce what the former intended is a monumental communication challenge. Accordingly, the full range of communication techniques discussed in Chapter 3 is typically used.

For consultants, regardless of their technical discipline, the design process typically results in the production of bidding documents, which later become contract documents. Bidding documents typically consist of a package containing the following three components, each of which is discussed in the next sections:

- Drawings
- Technical specifications
- Nontechnical provisions

Drawings

Drawings pictorially or graphically portray the type and arrangement of components that comprise the desired product, structure, facility, or system. That is, the drawings show what is to be built where. Drawings could consist of a few sheets to up to hundreds or thousands of sheets, depending on the size and complexity of the intended result. Drawings are sometimes produced by hand, but are increasingly generated by computerized drafting and design (CADD) systems.

Technical Specifications

Typically, the materials to be used in the construction, manufacturing, or fabricating process and the workmanship to be obtained are set forth in the technical specifications. These specifications describe, in great and careful detail, how the product, structure, facility, or system is to be constructed, manufactured, fabricated, or otherwise built. Technical specifications could consist of a few pages or run on to hundreds or thousands of pages. They typically are developed by combining "boiler plate" text extracted, as appropriate and applicable, from preceding or parallel projects, and original text peculiar to the design at hand.

A typical organization of technical specifications for a building project is presented in Table 12–2. Note the various engineering and other technical disciplines apparently involved in the design. The outline also suggests the breadth and depth of detail embodied in technical specifications. Each word is very important, and the specification writer must combine technical understanding with great writing skill. For example, Goldbloom (1992) advocates consistency in specification writing and discusses commonly misused, problematic words. For example, "will" is generally preferable to "shall," "inspection" to "supervision," and "consists of" to "includes."

TABLE 12–2 TYPICAL ORGANIZATION OF TECHNICAL SPECIFICATIONS FOR A BUILDING PROJECT

Specifications	
Division 1 General Requirements	
01010	Summary of work
01020	Allowances
01025	Measurement and payment
01030	Alternates/alternatives
01040	Coordination
01050	Field engineering
01060	Regulatory requirements
01070	Abbreviations and symbols
01080	Identification systems
01090	Reference standards
01100	Special project procedures
01200	Project meetings
01300	Submittals
01400	Quality control
01500	Construction facilities and temporary controls
01600	Material and equipment
01650	Starting of systems/commissioning
01700	Contract closeout
01800	Maintenance
Division 2 Site Work	
02010	Subsurface investigation
02050	Demolition
02100	Site preparation
02140	Dewatering
02150	Shoring and underpinning
02160	Excavation support systems
02170	Cofferdams
02200	Earthwork
02300	Tunneling
02350	Piles and caissons
02450	Railroad work
02480	Marine work
02500	Paving and surfacing
02600	Piped utility materials
02660	Water distribution
02680	Fuel distribution
02700	Sewerage and drainage
02760	Restoration of underground pipe-lines
02770	Ponds and reservoirs
02780	Power and communications
02800	Site improvements
02900	Landscaping
Division 3 Concrete	
03100	Concrete formwork
03200	Concrete reinforcement
03250	Concrete accessories
03300	Cast-in-place concrete
03370	Concrete curing
03400	Precast concrete
03500	Cementitious decks
03600	Grout

(continued)

TABLE 12-2 TYPICAL ORGANIZATION OF TECHNICAL SPECIFICATIONS FOR A BUILDING PROJECT *(CONTINUED)*

03700	Concrete restoration and cleaning
03800	Mass concrete
	Division 4 Masonry
04100	Mortar
04150	Masonry accessories
04200	Unit masonry
04400	Stone
04500	Masonry restoration and cleaning
04550	Refractories
04600	Corrosion-resistant masonry
	Division 5 Metals
05010	Metal materials
05030	Metal finishes
05050	Metal fastening
05100	Structural metal framing
05200	Metal joists
05300	Metal decking
05400	Cold-formed metal framing
05500	Metal fabrications
05580	Sheet metal fabrications
05700	Ornamental metal
05800	Expansion control
05900	Hydraulic structures
	Division 6 Wood and Plastics
06050	Fasteners and adhesives
06100	Rough carpentry
06130	Heavy timber construction
06150	Wood metal systems
06170	Prefabricated structural wood
06200	Finish carpentry
06300	Wood treatment
06400	Architectural woodwork
06500	Prefabricated structural plastics
06600	Plastic fabrications
	Division 7 Thermal and Moisture Protection
07100	Waterproofing
07150	Dampproofing
07190	Vapor and air retarders
07200	Insulation
07250	Fireproofing
07300	Shingles and roofing tiles
07400	Preformed roofing and cladding/siding
07500	Membrane roofing
07570	Traffic topping
07600	Flashing and sheet metal
07700	Roof specialties and accessories
07800	Skylights
07900	Joint sealers
	Division 8 Doors and Windows
08100	Metal doors and frames
08200	Wood and plastic doors
08250	Door-opening assemblies
08300	Special doors
08400	Entrances and storefronts
08500	Metal windows
08600	Wood and plastic windows
08650	Special windows
08700	Hardware
08800	Glazing
08900	Glazed curtain walls
	Division 9 Finishes
09100	Metal support systems
09200	Lath and plaster
09230	Aggregate coatings
09250	Gypsum board
09300	Tile
09400	Terrazzo
09500	Acoustical treatment
09540	Special surfaces
09550	Wood flooring
09600	Stone flooring
09630	Unit masonry flooring
09650	Resilient flooring
09680	Carpet
09700	Special flooring
09780	Floor treatment
09800	Special coatings
09900	Painting
09950	Wall coverings

(continued)

TABLE 12–2 TYPICAL ORGANIZATION OF TECHNICAL SPECIFICATIONS FOR A BUILDING PROJECT (*CONTINUED*)

Division 10	Specialties
10100	Chalkboards and tackboards
10150	Compartments and cubicles
10200	Louvers and vents
10240	Grilles and screens
10250	Service wall systems
10260	Wall and corner guards
10270	Access flooring
10280	Specialty modules
10290	Pest control
10300	Fireplaces and stoves
10340	Prefabricated exterior specialties
10350	Flagpoles
10400	Identifying devices
10450	Pedestrian control devices
10500	Lockers
10520	Fire protection specialties
10530	Protective covers
10550	Postal specialities
10600	Partitions
10650	Operable partitions
10670	Storage shelving
10700	Exterior sun control devices
10750	Telephone specialties
10800	Toilet and bath accessories
10880	Scales
10900	Wardrobe and closet specialties
Division 11	**Equipment**
11010	Maintenance Equipment
11020	Security and vault equipment
11030	Teller and service equipment
11040	Ecclesiastical equipment
11050	Library equipment
11060	Theater and stage equipment
11070	Instrumental equipment
11080	Registration equipment
11090	Checkroom equipment
11100	Mercantile equipment
11110	Commercial laundry, and dry cleaning equipment
11120	Vending equipment
11130	Audiovisual equipment
11140	Service station equipment
11150	Parking control equipment
11160	Loading dock equipment
11170	Solid waste handling equipment
11190	Detention equipment
11200	Water supply and treatment equipment
11280	Hydraulic gates and valves
11300	Fluid waste treatment and disposal equipment
11400	Food service equipment
11450	Residential equipment
11460	Unit kitchens
11470	Darkroom equipment
11480	Athletic, recreational and therapeutic equipment
11500	Industrial and process equipment
11600	Laboratory equipment
11650	Planetarium equipment
11660	Observatory equipment
11700	Medical equipment
11780	Mortuary equipment
11850	Navigation equipment
Division 12	**Furnishings**
12050	Fabrics
12100	Artwork
12300	Manufactured casework
12500	Window treatment
12600	Furniture and accessories
12670	Rugs and mats
12700	Multiple seating
12800	Interior plants and planters
Division 13	**Special Construction**
13010	Air supported structures
13020	Integrated assemblies
13030	Special-purpose rooms
13080	Sound: vibration, and seismic control
13090	Radiation protection
13100	Nuclear reactors
13120	Pre-engineered structures
13150	Pools
13160	Ice rinks

(continued)

TABLE 12-2 TYPICAL ORGANIZATION OF TECHNICAL SPECIFICATIONS FOR A BUILDING PROJECT (*CONTINUED*)

13170	Kennels and animal shelters
13180	Site constructed incinerators
13200	Liquid and gas storage tanks
13220	Filter underdrains and media
13230	Digestion tank covers and appurtenances
13240	Oxygenation systems
13260	Sludge conditioning systems
13300	Utility control systems
13400	Industrial and process control systems
13500	Recording instrumentation
13550	Transportation control instrumentation
13600	Solar energy systems
13700	Wind energy systems
13800	Building automation systems
13900	Fire suppression and supervisory systems
Division 14	**Conveying Systems**
14100	Dumbwaiters
14200	Elevators
14300	Moving stairs and walks
14400	Lifts
14500	Material handling systems
14600	Hoists and cranes
14700	Turntables
14800	Scaffolding
14900	Transportation systems
Division 15	**Mechanical**
15050	Basic mechanical materials and methods
15250	Mechanical insulation
15300	Fire protection
15400	Plumbing
15500	Heating, ventilating, and air conditioning (HVAC)
15550	Heat generation
15650	Refrigeration
15750	Heat transfer
15850	Air handling
15880	Air distribution
15950	Controls
15990	Testing, adjusting, and balancing
Division 16	**Electrical**
16050	Basic electrical materials and methods
16200	Power generation
16300	High voltage distribution (above 600 V)
16400	Service and distribtion (600 V and below)
16500	Lighting
16600	Special systems
16700	Communications
16850	Electric resistance heating
16900	Controls
16950	Testing

(SOURCE: Adapted with permission from Clough, R.H., Construction Contracting, 5th ed., New York: John Wiley and Sons, 1986, pp. 467–468.)

Nontechnical Provisions

The third and last portion of the papers produced typically produced in design may include one or more of the following: general conditions or general provisions, supplementary or special conditions, bid forms, instructions to bidders, and other items, such as supplements to bid forms, agreement forms, bonds and certificates, addenda, and modifications (Clough, 1986, p. 467).

General conditions set out procedures generally accepted in the particular area of engineering or other technical services. Examples of items typically included under the umbrella of the general conditions or provisions are descriptions of the rights and responsibilities of the owner, the technical professional, and the constructor/manufacturer/fabricator; payment and completion procedures; insurance and bonds; and means of settling disputes.

Supplemental or special conditions typically pertain to nontechnical and other idiosyncrasies of a project. Examples include special times when work must proceed, daily damages for delays, hourly wages to be paid, temporary facilities to be provided, and the need for security personnel.

The owner may require that bids be submitted in a specific format or fashion. This leads to the development, often by the technical professional, of bid forms to be completed by bidders. Similarly, special instructions to bidders may be prepared to explain steps in the bidding process, such as how to obtain a set of bidding documents, place and time to submit a proposal, withdrawal of a submitted bid, and conditions under which proposals could be rejected.

DESIGN AS RISKY BUSINESS

As noted by Florman (1987, Chapters 15–17) and Petroski (1985), design can also be viewed as risky business. Two types of failure are possible when an individual or an organization undertakes design. The first is personal failure. The second is failure of a product, structure, facility, or system.

Engineers or other technical professionals, along with other creators such as writers, composers, painters, and poets, share an apprehension or fear that they won't be able to do the task at hand. Or even if they complete a design or produce a creation, the results won't live up to their expectations and the expectations of others. This fear is probably best surmounted by recognizing and acknowledging it, by reflecting on one's depth and breadth of understanding of the problem at hand, by drawing on one's understanding of science and engineering fundamentals, by being open to new and creative approaches, and by working hard and persistently.

A less personal, but more significant kind of risk is the risk of product, structure, facility, or system failure. This risk is inherent in all technical design and should be a cause of concern because of the possibility that the failure could be catastrophic, as suggested by the descriptions of failures presented in Chapter 10. Because each nontrivial design, and most real designs are nontrivial, is new and unique, there cannot be a 100% guarantee of success. The designed product, structure, facility, or system is only as safe as its weakest link. Although the designer will naturally strive to reduce the risk, he or she cannot make it "zero," because nothing has ever been built exactly this way. Or, as argued by Petroski (1985, Chapter 4), each design is an untested hypothesis. The final testing experiment is the product, structure, facility, or system itself and how it functions. Failures can, in a cold or academic sense, be explained as disproved hypotheses. Even when failures occur, much can be learned from them, as explained in Chapter 10 of this book. Incidentally, the analysis of failures, for learning and adjudication purposes, is called *forensic engineering.* Forensic engineering is to engineering what Monday morning quarterbacking is to football—it's much easier to analyze disasters than to prevent them (Petroski, 1985, p. 188).

DESIGN AS A CREATIVE, SATISFYING PROCESS

Another way to understand and appreciate design and, to some extent, planning is to see them as part of a creative process that culminates in a tangible, often personally satisfying result. Design is much more than rote or even systematic application of science. Although technical professionals use science in design, design is much more than rote application of science. The work of the designer is very much like that of the writer, composer, painter, and poet in that bits of what is known or has been experienced are combined, typically via a trial-and-error, iterative process, in a unique and new fashion. Petroski (1985, p. 8) says it this way:

> *It is the process of design, in which diverse parts of the given-world of the scientist and the made-world of the engineer are reformed and assembled into something the likes of which Nature had not dreamed, that divorces engineering from science and marries it to art.*

Florman (1987, Chapter 2) argues that the creativity necessary and prevalent in design can be emotionally fulfilling. The anxiety associated with the fear of personal failure is more than offset by the deep and lasting satisfaction associated with the design of a product, structure, facility, or system that serves the user and society. Petroski (1985, Chapter 7) reinforces this, noting that the image of the writer staring at a blank page with a wastebasket full of false starts is completely analogous to the engineer starting a design. Likewise, the image of the writer learning of readers' enjoyment and enlightenment resulting from his or her writing or the image of the painter seeing the enjoyment of people viewing his or her work is very similar to the image of the engineer witnessing the aesthetic impact and effective functioning of his or her creation. Great personal satisfaction is often derived from that which is created, partly because of the uniqueness of the creation. However, engineers and other technical professionals often experience an even higher level of satisfaction, because the created product, structure, facility, or system is useful to society. President Herbert Hoover, who practiced engineering internationally prior to beginning public service, succinctly described the satisfaction of involvement in inherently useful creations when he said this about engineering (Fredrich, 1989, p. 546):

It is a great profession. There is the fascination of watching a figment of the imagination emerge through the aid of science to a plan on paper. Then it moves to realization in stone or metal or energy. Then it brings jobs and homes to men. Then it elevates the standards of living and adds to the comforts of life. That is the engineer's high privilege.

The Word *Engineer* and Creativity

Not only is creativity as exemplified by design one of the principal functions of engineering, but linguistically creativity and the word *engineer* are closely intertwined. Petroski (1985, p. 61) and Florman (1987, p. 42) both explore the origins of the word *engineer.* Although they follow different routes, and arrive at slightly different conclusions, both agree that *engineer* has its roots in creativity.

Petroski (1985, p. 65) states that *engineer* originally meant "one (a person) who contrives, designs, or invents." That is, engineer was synonymous with creator. This use preceded by a century the idea of an engineer as one who manages an engine. According to Petroski, the association between engineer and engine began in the mid-1800s with the emergence of the railroad as the metaphor of the industrial revolution. Petroski concludes his exploration of the origins of the word *engineer* by noting that even today there is a "confusion of the contriver and the driver of the vehicle."

Florman (1987, p. 42) traces *engineer* back to the Latin word "ingenium," which meant a clever thought or invention and was applied in about 200 A. D. to a military battering ram. That is, engineer was synonymous with that which was created. Later, in medieval times and during the Renaissance, the French, Italian, and Spanish words, respectively, *ingenieur, ingeniere,* and *ingeniero* came into use originally referring to those who designed and built military machines, such as catapults and battering rams. In English, the word progressed from the fourth through seventeenth centuries as *engynour, yngynore, ingener, inginer, enginer,* and, finally, *engineer.*

Thus, Petroski and Florman agree that *engineer* has deep roots in creativity. Petroski claims that the first emphasis was on the creative person, and Florman believes it was on what was created. However, both agree that *engineer* has its roots in contriving, inventing, designing, and creating.

CONCLUDING STATEMENT

From a very pragmatic perspective, design can be defined in terms of the papers produced, that is, the documents, needed for construction, manufacturing, fabrication, implementation, and other concrete and definitive actions. Design can also be viewed from both the individual and corporate perspective as risky business. Engineers and other technical professionals abhor failure, but learn from it. Finally, design is a highly creative, satisfying process. The very roots of the word *engineer* may be traced back to creativity.

REFERENCES

CLOUGH, R. H., *Construction Contracting*. New York: John Wiley & Sons, 1986.

CONSTRUCTION INDUSTRY INSTITUTE, "Input Variables Impacting Design Effectiveness," Publication 8–2, University of Texas at Austin, Texas, July 1987.

FLORMAN, S. C., *The Civilized Engineer*. New York: St. Martin's Press, 1987.

FREDRICH, A. J., ed., *Sons of Martha: Civil Engineering Readings in Modern Literature*. New York: American Society of Civil Engineers, 1989.

GOLDBLOOM, J., "Improving Specification," *Civil Engineering—ASCE*, Vol. 62, No. 9 (September 1992), pp. 68–70.

MOSKAL, B. S., "GM's New-Found Religion," *Industry Week*, May 18, 1992, pp. 46–53.

PETROSKI, H., *To Engineer Is Human: The Role of Failure in Successful Design*. New York: St. Martin's Press, 1985. [Note: A related film, "To Engineer Is Human," is available from Films Incorporated, 5547 N. Ravenswood, Chicago, IL, 60640–1199 (312–878–2600)].

WELTER, T. R., "How to Build and Operate a Product Design Team," *Industry Week*, April 16, 1990, pp. 35–50.

SUPPLEMENTAL REFERENCES

BEAKLEY, G. C., D. L. EVANS, and J. B. KEATS, *Engineering—An Introduction to a Creative Profession*, 5th ed., New York: Macmillan Publishing Co., 1986, Chapter 12.

FELDER, R. M., "On Creating Creative Engineers," *Engineering Education—ASEE*, January 1987, pp. 222–227.

HARRISBERGER, L., *Engineersmanship . . . The Doing of Engineering Design*, 2nd ed., Belmont, Cal.: Brooks/Cole Engineering Division, Wadsworth, Inc., 1982.

MATSON, J. V., and C. L. LOWRY, "Creativity in the Engineering Classroom—A Case Study," *Civil Engineering Education—ASEE—Civil Engineering Division*, Vol. 9, No. 1 (Spring 1987), pp. 1–16.

PETROSKI, H., *Design Paradigms: Case Histories of Error and Judgement in Engineering*. New York: Cambridge University Press, 1994.

13

Role and Selection of Consultants

> *A consultant is someone hired to come a long distance to tell you the time of the day. The consultant borrows your watch, tells you the time, keeps your watch, and sends you a bill.*
>
> (Anonymous)
>
> *The bitterness of poor quality remains long after the sweetness of low price is forgotten.*
>
> (Anonymous)

Recall the Chapter 1 discussion of the owner-consultant-contractor interaction, as illustrated in Figure 1–1, and variations on it. As suggested by that discussion, individuals and organizations requiring planning, design, management, and other engineering-related services often retain consultants to provide those services. That is, rather than have the technical work done "in-house," they choose to have it done "out-house."

What is meant by "consultant?" From a contractual, formal perspective, consultant usually means a consulting firm that enters into legal agreements with clients for the provision of services. However, on probing the matter further with owners and clients, "consultant" usually means a particular person, or perhaps a small group of professionals, on the staff of a consulting firm who have demonstrated competence and established a high-trust relationship with an individual or a small group on the staff of the client, customer, or owner. One indication of this interpretation of "consultant" is the strong allegiance shown by those who use the services of consultants to individual technical professionals when the professionals change employers, that is, move from one consulting organization to another or establish their own practice. This second interpretation of "consultant" suggests that a trustful relationship is critical in carrying out the consulting function—expertise is necessary, but clearly not sufficient. One of the highest compliments that an individual consultant can receive is to be retained on a sole-source basis by a client.

As an entry-level engineer or other technical professional on the staff of a government unit, a manufacturing organization, a contractor, or other organization, you should be familiar with the role and selection of consultants. If you are on the staff of a consulting organization, essentially all the projects you work on will be the result of the process described in this chapter. Although, as an entry-level professional, you will not play a major, formal role in the consultant-selection process, you will have opportunities to participate in it. If, for example, you are on the professional staff of an organization that utilizes consulting firms, you should observe the manner in which various firms provide services and, as opportunities arise, share your preliminary conclusions with colleagues and superiors within your organization. If you are a member of a consulting firm, you can note the variations in the expectations of your clients and share that information with colleagues and superiors. Understanding the role of consultants and being familiar with the selection process and acting on that knowledge will enable you to be a more productive young professional, regardless of where you are in the owner–consultant–contractor relationship.

After exploring reasons why consultants are retained, this chapter discusses the characteristics of successful consultants. The consultant selection process, which is sometimes extremely elaborate and costly, is described in the remainder of the chapter.

WHY RETAIN A CONSULTANT?

Consultants, in the form of consulting firms or individual professionals, are typically retained for one or more of the following reasons:

- **Temporarily acquire necessary expertise.** In this increasingly technological world, many public and private organizations, even those with engineers and other technical professionals on their staffs, do not have certain types of expertise. Although they could develop such expertise, they are often reluctant to do so unless they see a continuous need for it. Accordingly, they seek a consultant having the necessary expertise. Prudent owners, desiring to develop in-house expertise, should consider contracting with one or more consultants to provide education and training in the desired area of expertise in addition to completing the project at hand.

- **Supplement in-house staff.** Regardless of whether or not an organization has the necessary engineering and other expertise in house to carry out a project or accomplish a task, sufficient staff members may not be available at a given time to complete the project or task on schedule. The temporary use of consultants can solve the "people shortage."

• **Provide absolute objectivity.** A public or private organization, even one that has wide expertise and sufficient staff, may find itself embroiled in controversy, the resolution of which requires a high degree of objectivity. For example, a municipality may experience a heavy rainfall and resulting widespread flooding. This catastrophe may bring into question the adequacy of stormwater facilities designed by the municipality's professional staff or by consultants retained by the municipality. While the in-house professional staff or the consultant who did the original planning and design could certainly be asked to review their work, the most credible approach might be to retain an outside consultant to suggest the highest level of objectivity. In this case, the preferred "outside" consultant would probably be an individual or organization who had the necessary technical and other expertise, had never provided services to the municipality before, and perhaps had its nearest office in another community or state. Another example would be the failure of a manufactured product, such as an automobile transmission. Again, the technical professionals who carried out the original design could certainly be asked to review their work, or another professional within the manufacturing organization could be given that assignment. However, perceived and perhaps actual objectivity is likely to be the greatest when an outside consultant is retained.

• **Perform unpleasant tasks.** Carrying out unpleasant tasks or doing the "dirty work" is rarely the sole or principal purpose of a consultant, although it may be the principal purpose of the management consultant. However, engineering and other technical projects, particularly those in the public sector, often involve unpleasant and stressful tasks. For example, the long-term and frequent failure of crucial municipal facilities and services such as water supply, wastewater, transportation, and flood control can lead to deep-seated and widespread frustration among citizens of a community. Consultants are often retained to find a planning and engineering solution to such problems. Regardless of the other reasons why the consultants might be retained, such as to provide expertise, needed staff, and objectivity, the consultants are often expected to release, deal directly with, and rechannel the pent-up frustration within the community. One effective mechanism is the conduct of neighborhood information meetings shortly after the consultant is retained at which one or more representatives of the consulting organization are introduced and asked to provide an outlet for citizen frustration, optimally culminating in citizen confidence that the problems can and will be solved. One result could be a productive, ongoing program of interaction between the professionals and the public (Walesh, 1993). Similar situations involving the need to provide an outlet for pent-up frustration and instill a spirit of cooperation and optimism also occur in the private sector. Consultants may also be asked to seek cooperation among private and public entities experiencing disagreement and conflict.

In summary, consulting firms or individual consultants are typically retained for one or more of the following reasons: to provide expertise, to supplement in-house staff, to offer a high degree of objectivity, and to perform unpleasant tasks. Both parties, that is, the consultant and the client, should strive to identify and prioritize the reason or reasons for needing the consultant prior to entering into a consultant-client agreement. Mutually satisfying outcomes depend on a clear understanding of needs and expectations. Perhaps, as a result of the ideas and information presented here, you as the entry-level engineer will view your role and responsibility as being more than simply doing technical things. At minimum, as you begin work on a new project, find out why the client retained your firm to provide services.

CHARACTERISTICS OF SUCCESSFUL CONSULTANTS

One could argue that of the various employment opportunities for engineers and other technical professionals, consulting is both the most demanding and the most satisfying. The world of the consultant is typically dynamic—new problems to solve, new technologies to learn, new clients to serve, and new geographic areas to work in. These characteristics, which are likely to be viewed as positives by many young professionals, must be weighed against potential negatives, such as long hours, erratic schedules, extensive travel, and high levels of stress. Characteristics of successful consultants are discussed by others including Sunar (1986).

Success in consulting requires the characteristics and traits shown in Figure 13–1. These are

- **Inquisitiveness and currency of knowledge.** Recall that the consultant is often retained to provide expertise the client does not possess. Consultants, as individuals or as organizations, should define their areas of expertise and remain current in them. On the surface, one might think that consultants are successful primarily because of the answers they provide based on their knowledge and experience. However, the questions they ask their clients, others, and themselves based on their knowledge and experience are probably more important than the answers they give. Once key questions are asked, the consultant knows how to find the answers. The successful consultant is a perpetual student, strongly ori-

- **Inquisitiveness and currency of knowledge**

- **Responsiveness to schedules and other needs**

- **Strong people orientation**

- **Self-motivation**

- **Creativity**

- **Physical and emotional toughness**

Figure 13–1. Characteristics of successful consultants

ented towards trying to find out more about his or her area of expertise, the current task or project, and the client. Clearly, the consultant's inquisitiveness spans technical and nontechnical subjects and topics.

• **Responsiveness to schedules and other needs.** Recall that the consultant may be retained because the client does not have the personnel to complete a task or do a project. If the effort is late because of the consultant, the principal reason for retaining the consultant is negated. Responsiveness to client needs and schedules requires that the consultant have a strong service orientation.

• **Strong people orientation.** Although technical professionals plan, design, construct, fabricate, manufacture, and care for "things," they are doing this for the benefit of people. The consultant is the very important part of the interface between the needs of people and the possibilities of meeting those needs with the applications of science and technology. Accordingly, effectiveness in consulting requires a high degree of communication skills, as discussed in Chapter 3, with emphasis on listening, writing, and speaking. The successful consultant enjoys interacting with people, some of whom are not very pleasant because of their basic personality or, more often, because they are under great personal or organizational stress. The people challenges of consulting are further complicated by frequent changes in clients, potential clients, and their representatives.

- **Self-motivation.** Even though an individual consultant is "working for" a client, a client often does not know how to direct the consultant or have the time or the inclination to do so. Accordingly, most of what consultants do for clients is at the consultant's initiative within the overall framework established by the consultant–client agreement. The clients tend to assume that if they are not hearing anything from "their consultant," the consultant is proceeding with the project in a timely fashion. Moreover, the consultant will be available, on a very short notice, to answer a question, give advice, or provide a status report or other accounting of the efforts to date. Consultants must be highly proactive to the point of being intrusive in their relationships with clients.

- **Creativity.** Theodore von Karman, the Hungarian-American aeronautical engineer, said, "Scientists define what is, engineers create what never has been." Consultants must have the ability to be creative, to synthesize, and to see previously unforeseen patterns and possibilities. The typical technical project involves personnel and technical, financial, and other facets, all of which can be easily assembled in a variety of ways, most of which are suboptimal. A consultant's combination of knowledge, highly varied experience, and objectivity should enable him or her to suggest approaches and solutions not apparent to others. Recall the Chapter 12 discussion of the origin of the word *engineer,* which concludes that the word has deep roots in creativity.

- **Physical and emotional toughness.** The successful consultant needs physical and emotional strength to withstand pressure, long hours, and travel. Some of the consultant's meetings and presentations are very difficult, because they occur in situations highly charged by personality conflicts, political pressure, financial concerns, and liability issues. In addition, consultants are often not selected for projects, even though they believe they were the most qualified or had the best proposal. Frequent rejection can take its toll on conscientious and competent individuals, but is one of the realities of the consulting field.

CONSULTANT SELECTION PROCESS

The process by which an owner selects a consultant to provide planning, design, or other services is unique to each situation. However, much that is useful can be explained within the context of an overall approach or model as presented in this section. Similar and additional ideas and information on consultant selection are provided by Clough (1986, Chapter 4), Davis (1987), and Kasma (1987).

Cost versus Quality

It is unwise to pay too much, but it is worse to pay too little. When you pay too much, you lose a little money, that is all. When you pay too little, you sometimes lose everything because the thing you bought is not capable of doing the thing it was bought to do. The common law of business balance prohibits paying a little and getting a lot—it can't be done. If you deal with the lowest bidder, you might as well add something for the risk you will run and if you do that, you will have enough to pay for something better.

(John Ruskin)

Very frequently, as illustrated in Figure 13–2, the consultant selection process is driven implicitly or explicitly by the natural tension that exists between the quality of service (conformance to requirements, as defined in Chapter 7) that is likely to be received and the cost of that service. You routinely encounter the same tension in your personal life when you shop for clothing or select a restaurant for that special occasion. In a very rough way, as proposed consulting fees for a given project diminish, the quality of the resulting plan, design, or other product is likely to diminish and the capital and other costs associated with the

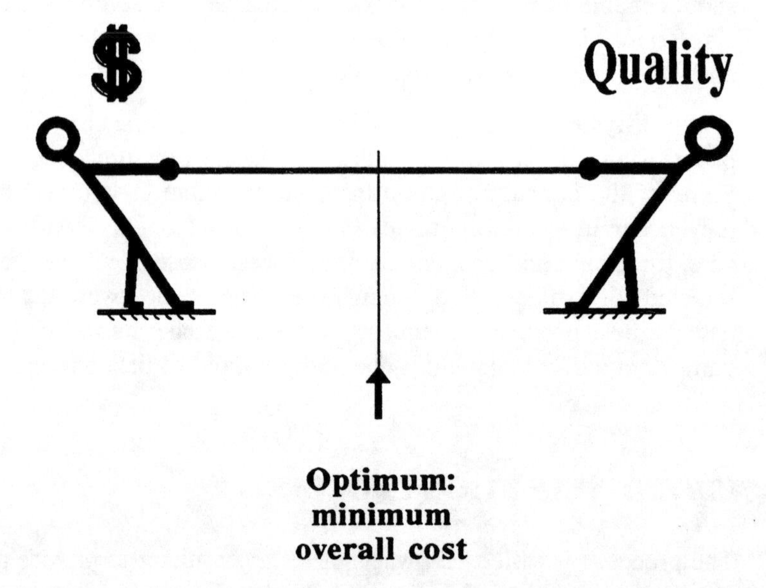

Note: **Quality is defined as conformance to requirements. Refer to Chapter 7 for a discussion of this and other definitions of quality.**

Figure 13–2. Cost-quality tension in the consultant selection process

product are likely to increase. You get what you pay for. There are, of course, exceptions.

Fees are certainly important. However, because fees can be readily quantified, relative to other selection factors such as experience and creativity, they tend to assume excessive influence. As succinctly stated by Kasma (1987), "When price is a factor in selecting consultants for negotiations, it usually becomes the deciding factor, particularly in public organizations."

Unusually low fees proposed by consultants desirous of obtaining contracts sometimes reflect a lack of experience and, therefore, awareness of all necessary aspects of a project. At other times, low fees might reflect an individual consultant or a consulting firm's desire to obtain a contract for a new type of project on which they can gain valuable experience. They are, in effect, willing to "buy" (lose money on) the assignment in exchange for the knowledge they will gain.

Most people and organizations retaining consultants know that they should avoid being penny wise and pound foolish. However, a completely rational approach does not always prevail, nor may it be possible because of insufficient information. Consultants must put themselves in the shoes of client decision makers, particularly those who are public officials subject to public scrutiny, who may be hard pressed to justify a large proposed fee over a smaller proposed fee when, at least at the surface, the resulting proposed products appear identical. Remember, at the time of consultant selection, the total cost of the project is usually not known or is not perceived as an important factor.

Refer to Table 13–1. Note that the consulting fee ultimately paid by an owner is a very small part of a total present-worth cost that will be incurred by the owner in producing a product, facility, or system. In the example, the $100,000 consulting fee is only 3.33% of the total cost of $3,000,000.

Note also that although consulting fees being proposed by various consulting firms are often known before the potential project gets underway, the remain-

TABLE 13–1 CONSULTING FEE AND TOTAL COST FOR A HYPOTHETICAL ENGINEERING PROJECT

Design fee proposed by consultant	$ 100,000
Construction/manufacturing/fabrication cost to owner (usually not known when consultant is being selected)	2,000,000
Total initial capital cost to owner	2,100,000
Operation and maintenance costs (present worth) to owner over economic life of product/structure/facility/system (usually not known when consultant is being selected)	900,000
Total (present-worth) cost to owner (usually not known when consultant is being selected)	$3,000,000

der of the total project cost and by far the largest part of the total project cost are largely unknown. Therefore, the decision makers within the client or owner organization must make a decision having incomplete fiscal information. This is one reason why some owners and clients place too much emphasis on relative magnitudes of the proposed consulting fees and not enough emphasis on the likely total cost that they will incur as a result of the consultant they select. The matter is further complicated by the wide range in consulting fees—sometimes an order of magnitude—being proposed on ostensibly the same project.

Note in Table 13–1 that a $10,000 or a 10% savings in consulting fee reduces the total cost to the owner by only 0.333% for the hypothetical example. Of course, this assumes that the consulting firm charging a $90,000 fee can produce the same quality product as the consulting firm charging a $100,000 fee. If they can't, because of insufficient knowledge, experience, or attention, the resulting increases in capital or operation and maintenance costs could easily more than offset their lower up-front fees.

Even much larger savings in up-front consulting fees will tend to result in only small savings in total project costs. For example, if one consulting firm proposes to do the hypothetical project for $50,000, a 50% savings relative to the consulting firm proposing to do the project for $100,000, the reduction in total project cost is only 1.67%. Again, this assumes, probably not realistically, that the "low-cost" firm can produce a product of similar quality to the "high-cost" firm.

Ideally, clients and owners should select consultants with the goal of minimizing their total costs, that is, the sum of the consulting fee, initial capital cost, and present worth of operation and maintenance cost. This ideal selection concept is illustrated in Figure 13–3. The fees proposed by potential consultants A, B, and C vary widely, with the largest fee being approximately twice the smallest fee. Similarly, there are significant variations, although not as dramatic in a relative sense, in the present worth of the construction/manufacturing/fabrication costs and operation and maintenance cost that would be incurred by the owner. Of course, as already noted, these total costs and even relative values of these costs are not likely to be known by the owner at the consultant selection stage. If they were known or could be known, the owner would obviously determine the total cost associated with each of the three potential consultants and their likely products and select the consultant who would offer the lowest total cost. This would be consultant B in Figure 13–3.

Recall the discussion in Chapter 12 on the disproportionate impact of the design function and, therefore, the competence of the design professionals in determining total project costs. Fees saved as a result of retaining a consulting firm with inadequate knowledge and experience may result in much larger increases in total project costs.

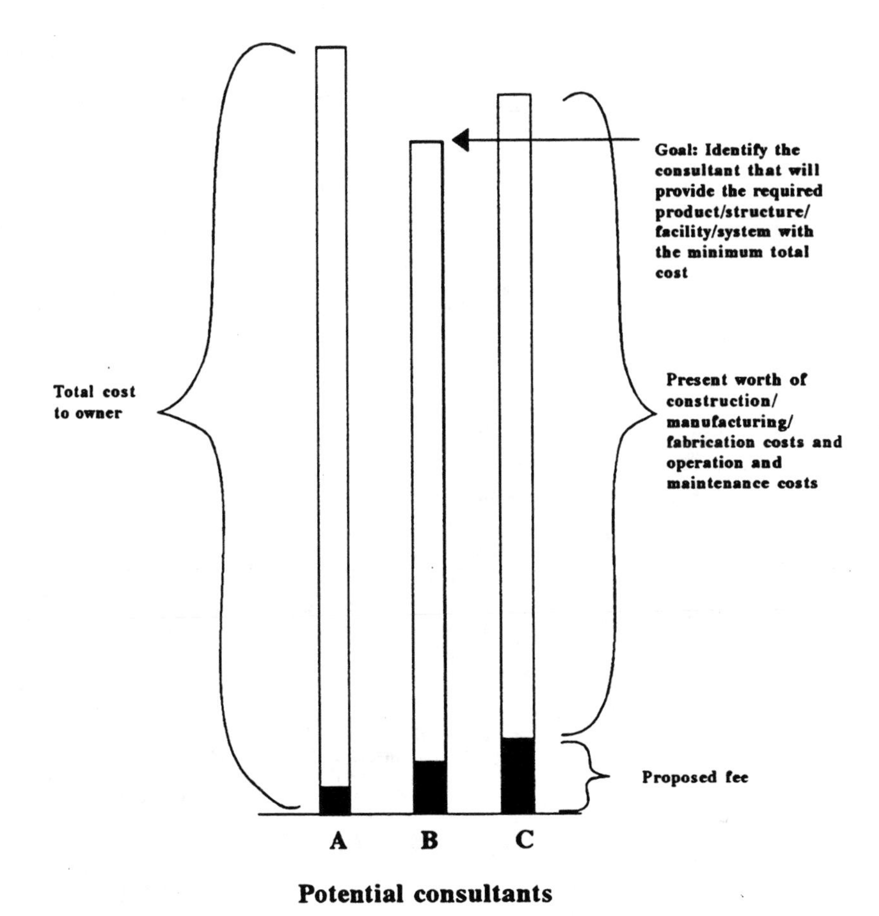

Figure 13–3. Ideal cost-based consultant selection decision

Steps in the Selection Process

The detailed process used by an owner to select a consultant to provide planning, design, and other services is unique to each situation. However, steps common to the consultant selection process can be identified and linked together as shown in Figure 13–4, for the benefit of the entry-level technical professional. Beginning with "Start," the most formal and involved selection process is the series of steps 1 through 12 proceeding clockwise around the figure. Various optional shortcuts are possible and are frequently used by owners, especially those in the private sector, as they select consultants. The full and formal 12-step process consists of the following steps:

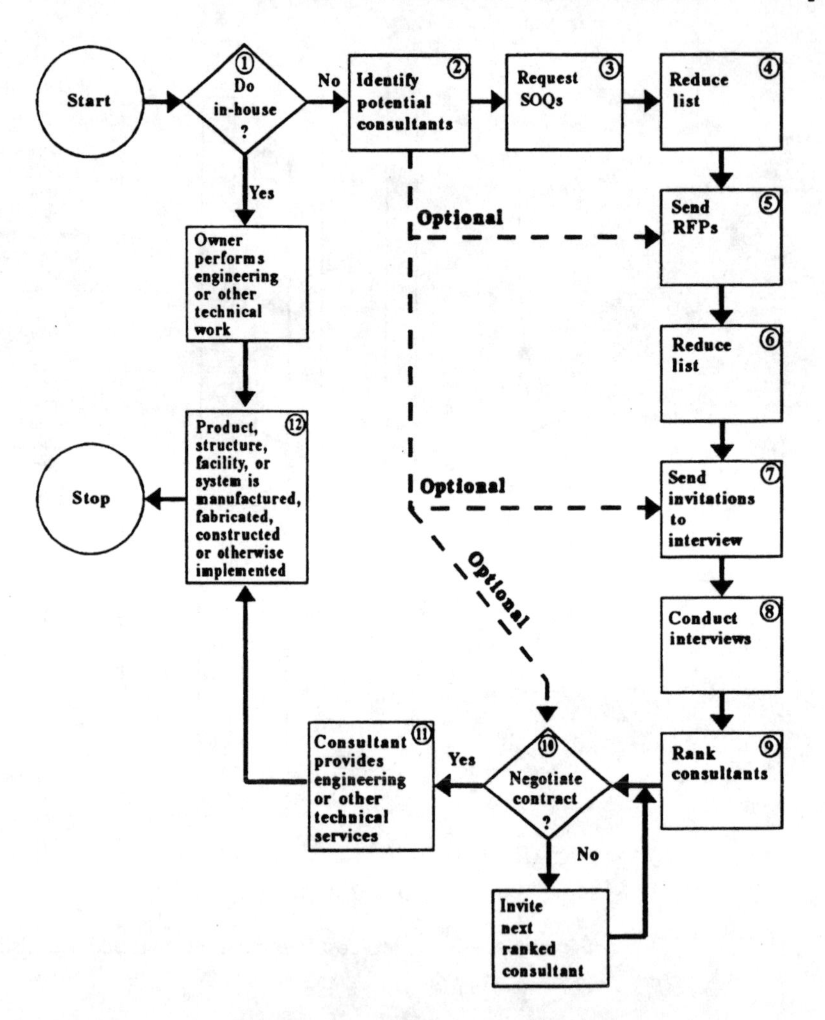

Figure 13–4. Consultant selection process

• Step 1. The owner or client determines whether or not a consultant will be retained for a task or project. Recall the four basic reasons to retain a consultant, as discussed earlier in this chapter. Assume that a consultant is to be utilized.

• Step 2. The client identifies potential consultants. The client may informally or formally create a selection committee. A list of potential consultants might be assembled using personal and other first-hand knowledge, referrals from colleagues at other organizations, and formal listings such as those appearing in directories of engineering organizations, business cards appearing in professional publications, and even the yellow pages in telephone books.

• Step 3. After screening the list, the client requests statements of qualifications (SOQs) from consultants who presumably have the ability to provide the necessary services. In some cases, usually government organizations, a request for SOQs is published in a newspaper and any consulting firm may respond. SOQs are usually standard, off-the-shelf items or documents readily assembled from standard text and graphics maintained on computer systems. An SOQ focuses on qualifications of the consulting firm with emphasis on its experience with projects similar to that being considered by the owner. An SOQ typically does not address the manner in which the consultant, if selected, would approach the client's specific project. However, respondents may decide to include some project-specific ideas, and information. SOQs typically include basic information on the consulting firm (e.g., size, office location or locations, services offered, project experience with emphasis on projects similar to that about to be undertaken by the client, resumes of selected professional and support staff with emphasis on relevant project experience, and references).

The checking of a firm's references, that is, representative list of current and past clients, would seem to be a very effective way to screen consultants. After all, who is in a better position to comment on the quality of a firm's services than those who have received those services? Of course, and as noted, the references must be truly representative. One way to ensure this is for the client to ask the candidate consultant to provide the names of all organizations receiving certain services (e.g., machine design, highway planning, management assistance) over the past few years along with permission to contact any or all organizations on this list and inquire about any aspect of the services received. A massive, unusual and enlightening example of the use of systematic reference checking early in the consultant selection process was provided by the early 1990s actions of the San Francisco Airport Commission (Escobedo, 1993). Consultants were sought for a ten-year, $2.4 billion airport expansion project. The commission's February 1992 solicitation letter specified that interested firms must submit the names of all clients they served from 1989 to 1991. Candidate firms also "had to sign an affidavit agreeing that they would not take legal action against clients that provided information or attempt to learn what clients testified about them to the Commission." Reference checking was used to reduce the original list of 617 firms submitting preliminary information, including the required list of references, to 80 finalists. Qualifications of proposed project personnel were used by the commission to select firms from the short-listed 80 for contracts. Oral presentations by consultants, which are common as discussed in step 8 of the 12-step process described in this chapter, were not permitted. Heavy reliance on thorough reference checking might be a replacement for, or at least an offset to, reliance on projected fees.

• Step 4. Client reduces list. One or more professionals representing the client review the SOQs. They match the perceived needs of their project with their interpretation of the experience and ability of each consulting firm. Firms which, in the judgment of the client representatives, do not have adequate qualifications are eliminated from further consideration.

• Step 5. Client sends requests for proposals (RFPs). The client now invites consulting firms remaining on the eligibility list to describe how they would go about doing the contemplated specific project and often asks interested firms to include an estimate of the cost of their services. The RFP package sent from the client to each consultant typically includes items such as a letter of explanation and invitation; a description of the project; an explanation of the scope of services required (e.g., feasibility study, preliminary engineering, preparation of plans and specifications, construction management, startup); a project schedule; a list of related reports, studies, and investigations available from the client; a description of available data and information available from or known by the client; the name of a contact person; an indication of whether or not the proposers should provide an estimate of fee for services; a description of minority business enterprise (MBE), women's business enterprise (WBE), and disadvantaged business enterprise (DBE) requirements; and the due date for the proposal. In some cases, the owner is prohibited from requesting an estimate of project fees. For example, recall the discussion of the Brooks Act and similar state provisions in Chapter 11.

The client may conduct, as part of the RFP process, a single explanatory meeting for all consultants intending to submit a proposal. Kasma (1987) suggests that this meeting serves two purposes. First, all consultants receive the same information, including answers to questions. Second, the client has an opportunity to meet representatives of the consulting firms.

Each consultant receiving the RFP typically revisits its initial decision to pursue the project as indicated by sending its SOQ. In the interim, any given consultant may have learned much more about the project and on the basis of that new knowledge may decide not to pursue the project further. The initial decision to pursue the project is revisited at this time because the effort required to prepare a proposal in response to the RFP is typically several magnitudes greater than the effort required earlier to assemble and submit the SOQ. As already noted, the SOQ consists of largely preprepared materials. In contrast, a proposal that is submitted in response to a RFP is a largely original document requiring considerable time and effort, including that of high-level and, therefore, costly professional personnel. In fact, the likelihood of a consultant's successfully and profitably completing a project, assuming that they are ultimately selected, depends on the care used to prepare the proposal, because the typical proposal must be prepared with a clear understanding of what the consultant will do, how the consultant will do it, how long it will take, and what it will cost. In a sense, the project is worked out "on paper" as part of the proposal preparation process.

• Step 6. Client reduces list of consultants. Using the project-focused information provided by consultants receiving the RFPs, representatives of the client eliminate some consultants from further consideration. Factors considered by a client in forming a reduced list may include one or more of the following: responsiveness to the RFP; indications of creativity, including too much or too little; specific personnel to be assigned to the project; experience or lack thereof on similar projects; results of reference checks; list of deliverables; and, of course, the proposed fee.

• Step 7. Client invites remaining consultants to interview for the project. Each consulting firm receiving an invitation to interview is likely to accept. However, as was the case when a firm was invited to prepare a proposal in response to the RFP, consulting firms typically revisit or should revisit their initial decisions to pursue the project. In the time that has passed since submitting the proposal, additional information may have been obtained about the potential project or the client that might cause a reversal of the original decision. This revisiting of earlier decisions to pursue the project is prudent because of the additional time that will now be required to prepare for the interview. Although, in some cases, an interview is a relatively informal affair requiring minimal preparation, in other situations an interview is a very formal event requiring a major investment by the consulting firm. The labor and expenses invested in assembling SOQs, preparing proposals, and getting ready for interviews all add to the consulting firm's overhead, as discussed in Chapter 9.

• Step 8. Conduct interviews. Typically, each interview is conducted in private; that is, representatives of the client meet with representatives of the consultant with no one else in attendance. The consulting firm's team usually consists of a principal of the firm, the person who would manage the project, and one or more members of the designated project team, possibly including a specialist with expertise specifically related to the project. Incidentally, consultants sometimes send a team to the interview that is not representative of the team—particularly the project manager and key members of the project team—that would actually work on the project. This approach, while it might enhance the interview, is a very poor business practice. The best approach to follow is "What you see is what you get." Additional text, tables, figures, and other printed material might be provided to client representatives by the consulting team prior to or at the interview. In addition, the consulting team may use audiovisual materials such as posters, 35-mm slides, transparencies, videotapes, and computer demonstrations. After some sort of formal presentation by the consulting team, the client team typically asks questions and a general discussion ensues.

The consulting team tries to develop rapport with the client's selection team, and the client's selection team tries to determine if a good working relationship could be established with the consulting team. Although difficult to measure and sometimes denied, interpersonal "chemistry" probably becomes a significant factor at this point in the overall selection process, because the owner and each potential consultant are now interacting with each other in a manner that roughly approximates the working relationship that would exist on the project. Interviews often conclude with a closing statement by someone on the consulting team.

• Step 9. Client ranks consultants. Largely on the basis of the interview, but perhaps on additional consideration of the proposal received prior to the interview, the client representatives typically rank the competing consultants. This is a difficult task for the client because of the voluminous amount of quantitative and qualitative information that is now available.

• Step 10. Client and first-choice consultant try to negotiate a contract. The first-ranked consultant is invited to prepare a contract for professional services and present it in draft form to the client as the basis for negotiation. Typically the consultant will revisit the proposal submitted earlier, convert it to contract language, make modifications based on ideas and information obtained during and subsequent to the interview, and submit the new document to the client. Somewhat self-laudatory language and other terminology that typically appears in a proposal should not appear in a draft contract or agreement. This matter is discussed in detail in Chapter 10. After a draft agreement has been sent to the client, typically one or more representatives of the client meet with one or more representatives of the consulting firm to review the document in detail and arrive at a mutually agreeable contract. Although rare, occasionally the client and consulting firm are not able to arrive at a mutually acceptable agreement, in which case the client enters into negotiations with the second-ranked consulting firm. As noted earlier, a carefully prepared proposal in response to RFP is crucial to the successful and profitable completion of a project, assuming the consultant is selected. The value of a carefully prepared proposal usually becomes evident during the contract negotiation phase, because the consulting firm is in a position to convince the client that the scope of services will meet client's needs and other project requirements and that the fee is reasonable.

• Step 11. Consultant provides engineering or other technical services.

• Step 12. Product, structure, facility, or system is manufactured, fabricated, constructed, or otherwise implemented. Sometimes the consulting organization is involved in this step. For example, a mechanical engineering consulting firm that designed a new manufacturing process might be retained to supervise the installation and start-up of the process. A civil engineering consulting firm that designed a high-rise structure may be retained to monitor, but not supervise, its construction so that the owner knows the degree to which the construction is conforming to the plans and specifications. Some consulting firms provide even broader services with respect to engineered structures, facilities, and systems. For example, design–build firms do both design and construction; other firms offer operation and maintenance services; and a few organizations offer all or most of the preceding plus finance services.

As indicated earlier and as suggested by Figure 13–4, numerous optional, shorter, and simpler selection processes are possible. For example, the client might move directly from step 2, identifying potential consultants, to step 5, sending RFPs, thus eliminating requesting and reviewing SOQs. An even shorter version of the overall process is to move directly from step 2, identifying potential consultants, to step 7, extending invitations to interview. This shortcut might apply in situations where a client is very familiar with the qualifications of a set of consulting firms and wants to focus immediately on how any one of those firms would go about doing a particular project. Sometimes, most of the process is omitted, and this is most likely to happen in the private sector, when the client goes from step 2, identifying potential consultants, to step 10, negotiating a contract. The client makes a predetermination regarding which consulting firm is most likely to provide the desired services at an acceptable fee and invites that firm to meet with the client to learn about the project and negotiate a contract. As noted at the beginning of this chapter, one of the highest compliments that an individual consultant or consulting firm can receive is to receive an invitation to be retained on a sole-source basis by a client. Other options, besides those shown in Figure 13–4, are possible.

Figure 13–4, might also be viewed as a client-consultant communication process. The five components of communication (listening, speaking, writing, use of graphics, and use of mathematics) discussed in Chapter 3 are typically used throughout the consultant selection process. The most qualified firm may not be selected and the client may be denied the best possible services because of inadequate communication skills on behalf of either the consulting firm or the client. A final thought is in order on the consultant selection process in the context of marketing of engineering services. As long and complex as the process may be, it is only one component—and a small one at that—of marketing. Chapter 14 is a comprehensive treatment of marketing and provides further context for the consultant selection process.

REFERENCES

CLOUGH, R. H., *Construction Contracting*, 5th ed. New York: John Wiley & Sons, 1986.

DAVIS, B. H., "How and Why to Hire a Consultant," *The Whole Non-Profit Catalog*, Winter 1987/1988, pp. 7–9.

ESCOBEDO, D., "San Francisco Procurement Takes A/E's by Surprise," *Engineering Times*, August 1993, p. 16.

KASMA, D. R., "Consultant Selection," *Journal of Management in Engineering—ASCE*, Vol. 3, No. 4 (October 1987), pp. 288–296.

SUNAR, D. G., *Getting Started as a Consulting Engineer*, San Carlos, Cal.: Professional Publications, Inc., 1986.

WALESH, S. G., "Interaction with the Public and Government Officials in Urban Water Planning," presented at Hydropolis—The Role of Water in Urban Planning, Wageningen, The Netherlands, March 1993.

SUPPLEMENTAL REFERENCES

BARTHOLOMEW, C. L., "Consulting by Civil Engineering Faculty," *Journal of Professional Issues in Engineering—ASCE*, Vol. 114, No. 2 (April 1988), pp. 157–161.

HOLTZ, H., *Expanding Your Consulting Practice with Seminars*. New York: John Wiley & Sons, 1987.

KANIGEL, R., "The Endangered Professional," *Johns Hopkins Magazine*, Vol. 40, No. 3 (June 1988), pp. 17–43.

14

Marketing Technical Services

> *I don't care how much you know until I know how much you care.*
>
> (Anonymous)

The expression "marketing technical services" often engenders negative reactions or connotations. Images of brash, high-pressure car salespeople may come to the mind of the young professional. You, as a young, entry-level engineer or other technical professional considering joining a consulting firm or other private-sector technical organization, may be repulsed by the thought of "wasting" your professional education by doing "sales" work or even being on the receiving end of any aspect of marketing.

Worse yet, marketing may suggest unscrupulous individuals. You may not have a well-informed understanding of various marketing philosophies and, therefore, you may have an aversion to marketing based on general ethical grounds. Marketing may seem unsavory based on your experiences outside of your technical field because, as perceived by you, it requires misrepresentation if not outright dishonesty. If so, you should be at least receptive to the positive marketing model presented in this chapter. To the extent that you learn to view marketing as earning trust and meeting client needs, which is the essence of the model offered in this chapter, you may conclude that not only is it an ethical process, but also a very satisfying and mutually beneficial one.

Incidentally, codes of ethics, some of which are presented in Chapter 11, contain many provisions directly applicable to marketing as marketing is presented in this chapter. For example, Fundamental Canon 2 of the Code of Ethics of the American Society of Civil Engineers states, "Engineers shall perform services only in areas of their competence." For additional discussion of the crucial role of ethics in marketing of engineering services, refer to Snyder (1993, Chapter II).

The very positive perspective on marketing of technical services offered here is supported by research and various writings and, much more importantly, by the admirable success of some consulting and other technical organizations. That positive perspective is presented in this chapter with the hope that it will be of interest and value to you, the entry-level technical professional, whether you are employed by a consulting engineering firm, a manufacturing organization, a government agency, or some other entity that provides or uses the services of consulting firms. There is little doubt that you, as a young professional, will have the opportunity to participate in marketing. The question is will that participation be positive and productive or negative and destructive?

This chapter begins with a review of the financial motivation for marketing. Marketing is then defined, followed by a synopsis of marketing research and case studies, most taken from the consulting field, which suggest some principles of marketing, including the principle that everyone in an organization should be part of the marketing effort. Presentation of a suggested working model for planning and implementing a marketing program concludes the chapter.

This chapter focuses on the marketing of engineering and other technical services to public and private clients as is typically done by consulting firms. Such firms, particularly the larger ones, typically employ civil, electrical, mechanical and other engineers as well as nonengineering disciplines. The range of disciplines on the consulting firm team is another reason this chapter should be of interest to all technical professionals, regardless of their area of specialization.

The marketing of engineered products is not explicitly treated in this chapter. However, many of the underlying principles as well as the tool and techniques are directly applicable to the marketing of engineered products. In fact, many of the principles, tools, and techniques set forth in this chapter are applicable to any marketing effort.

FINANCIAL MOTIVATION FOR MARKETING TECHNICAL SERVICES

Income for consulting firms emanates from essentially one source: clients who need and are willing to pay for services. At what rate does a consulting firm's marketing program have to generate sales? The answer, of course, depends on many factors but is heavily influenced by the size of the consulting firm in terms of total number of professional and other employees.

Consider the case of a small consulting firm having ten full-time employees, most of whom are technical professionals. Assume, for simplification purposes, that the overall time utilization (U) is 0.7, the firm's multiplier (M) is 3.0, and the overall raw labor rate is \$15 per hour. Let P_w be the total payroll cost per week. Then P_w = (40 hours per week) × (10 employees) × ($15 per hour per employee) = $6000. Given this weekly payroll cost, the necessary weekly net revenue is $P_w \times U \times M$ = ($6000) × (0.7) × (3.0) = $12,600. In other words, this small, hypothetical consulting firm needs to successfully market an average of $12,600 of new contract work per week, or $2,520 per day based on a five-day work week.

Assume that the total value of work under contract that has yet to be done is $50,000. Given that the necessary weekly net revenue is $12,600, the quotient of the value of contracted work to be done and the necessary weekly net revenue is referred to as the backlog and in this case is 3.97 weeks. That is, if there are no additional contracts as a result of the firm's marketing efforts, there will be no revenue to pay salaries and cover expenses one month from now.

Now, for emphasis, consider the case of a 4,000-person, multidiscipline firm similar to any one of several such firms in the United States. Assuming the values for *U, M,* and raw labor used in the preceding small firm example, P_w, the necessary weekly net revenue, is $5,040,000. The firm needs to generate an average of $1,008,000 per day of new work based on a five-day work week. Assuming seven percent of new revenue is spent on all aspects of marketing, which is a representative consulting industry value, the average daily marketing cost or investment is $70,560.

The preceding hypothetical but realistic examples illustrate the idea that marketing, to the extent that it leads to contracts for consulting services, is essential to the financial health of a consulting or other business. Marketing is a major expense. Therefore, the marketing effort must be carefully managed, as must other aspects of the consulting business.

DEFINITIONS OF MARKETING AND SOME OBSERVATIONS

According to Cronk, "Marketing is creating the climate that will bring in future business" (Cronk quoted by Smallowitz and Molyneux, 1987). Kolter and Fox (1985, p. xiii) state, "Marketing is the effective management by an institution of its exchange relations with its various markets and publics." According to W. Coxe, "Marketing is to selling as fishing is to catching" (Coxe, quoted by Smallowitz and Molyneux, 1987). This definition suggests that making the sale, that is, selling, is only one small aspect of marketing, just as actually catching fish is only one small part of the overall fishing process. This definition further suggests that an individual or organization must undertake a range of activities under the general umbrella of marketing in order to achieve sales, just as one must carry out certain activities using specialized equipment under the general umbrella of fishing in order to actually catch fish. Or stated differently, if an individual or organization attempts to simply sell without seeing selling in the context of other activities, the individual or organization is likely to be unsuccessful, just as if one seeks to catch fish without doing the related activities and using proper equipment, one is not likely to be successful. Drucker succinctly states, "The aim of marketing is to make selling superfluous" (Drucker, quoted by Kolter and Fox, 1985, p. 7). This definition, like the previous one, reinforces the idea that selling is only one part of marketing and suggests that if marketing is done well, sales will occur naturally.

Some characteristics of marketing that follow from or are consistent with the preceding definitions, based in part on Kolter and Fox (1985, pp. 7–8), are:

- Marketing is much more than selling.
- Marketing is a managed process, not a collection of random actions, that involves analysis, planning, implementation, and control.
- Marketing involves mutually beneficial exchanges of needs and services or products.
- Marketing requires broad participation by employees—it is not the responsibility of a few, nor is it to be done unilaterally by individuals or units within an organization.

MARKETING RESEARCH AND CASE STUDIES

The literature contains the results of some marketing or related studies that help interpret the relevance of and the desired approach to marketing of products and services. For example, System Corporation studied 3,000 industrial salespeople in about 1988. This study, which included ratings by the customers of the sales representatives, concluded that "the ability to establish trust" was more important than likability (Sheridan, 1988.) Rabeler (1991) cites references that, like the System Corporation study, emphasize the prime role of trust in marketing. The focus on trust in the cited studies is noteworthy and is discussed again later in this chapter.

Hensey (1987) conducted a study in 1986 of 25 of the most profitable U. S. consulting engineering firms. Two significant findings were the following:

- The average percentage of repeat work for clients was 90% compared to 60% for all firms nationwide.
- 85% of these firms announced their strategic marketing plan and financial results to all staff.

This study suggests the impact of repeat work on profitability, raises the question of how the high repeat work percentages were obtained, and suggests that high levels of repeat work and profitability are achieved by having a marketing plan and by employee knowledge of that plan.

Since 1990, the Top Firms Survey has been conducted among U. S. engineering and architectural firms nominated by professional societies as being very successful (Salwen Business Communications, 1993). The 1993 survey generated completed questionnaires from 39 firms. Five key success patterns emerging from the survey were

- Market by listening more than talking.
- Watch costs like a hawk.
- Never take money-losing jobs.
- Niche market—don't try to be all things to all clients.
- Establish a corporate vision.

Wahby (1993) effectively communicates the importance of giving attention to and taking care of existing clients when he says, "The best firms understand that if a project is an apple, a client is an apple tree." This concept is illustrated in Figure 14–1. According to Wahby, appreciative firms " . . . spend the bulk of their marketing efforts judiciously cultivating, pruning, and harvesting their 'orchard' of past and current clients."

As an entry-level engineer or other technical professional, you may be inclined to think that you are outside of your firm's marketing efforts—it is not part of your job description. On the contrary, to the extent that you have contact with clients on projects you have been assigned to, you are in the marketing arena. More than that, you are on the playing field and in a position to score. You can

Figure 14–1. If a project is an apple, a client is an apple tree

carry out your marketing responsibilities by helping your firm meet your client's needs on the current project and by being alert to their other needs and plans.

Assuming, as suggested by the preceding sections of this chapter, that marketing is a vital, broad function of a consulting firm and of other similar organizations and one that requires a major management effort, consideration of some marketing models or case studies is in order. That is, in designing a marketing program for a particular firm, much can be learned from the positive and negative experiences of other firms. Most of the approaches and case studies described in the literature apparently deal with successful programs. This is understandable, given the natural reluctance to document or report failed or less than satisfactory efforts. However, there are probably many undocumented failures—either failures of marketing programs or failure to have a marketing program.

Dupies (1979) describes the then-new approach taken by the Milwaukee, Wisconsin area office of Howard, Needles, Tammen, and Bergendoff (HNTB) in the late 1970s. This large firm, with many somewhat autonomous offices throughout the United States, tended to serve a few large clients. The Milwaukee office of HNTB was apparently experiencing business problems such as declining backlog, changing markets, increased overhead, and declining profits. Accordingly, the Milwaukee office of HNTB successfully developed and implemented a marketing plan. The HNTB experience provides a useful case study.

According to Dupies (1979) some key steps in the preparation and implementation of a marketing plan were

- Determining strengths and weaknesses of personnel.

- Reviewing marketing-related performance over the past five years. Included were fees by type of service, clients and their location, cost of marketing, marketing tools (e.g., brochures, reprints, photographs available) and interviews with clients and community leaders. One somewhat surprising finding is that whereas HNTB in Milwaukee thought of itself as a multidiscipline firm offering quality services for a solution of local problems, many clients viewed HNTB as an out-of-state highway and bridge firm interested only in big projects.

- Researching future opportunities, including areas of population growth and increased economic activity, relevant state and federal legislation, and newly growing service areas, such as bridge design and environmental services.

- Establishing long-range (5-year) objectives by area of service and by private and public sector. The overall annual growth rate was set at 20% with large variations among the various areas of service and client types within the Milwaukee office.

- Solving immediate problems that impeded long-range goals, such as conflicts in marketing and project assignments, inadequacy of marketing materials, insufficient monitoring, and poor participation by the professional staff in community service organizations.

• Forming a marketing committee consisting of the office's marketing director plus directors of the four major service areas (structures, transportation, environment, and architecture). In addition to their technical responsibilities, directors assumed marketing responsibilities.

• Conducting in-house seminars on marketing with emphasis on the intelligence-gathering process.

The 5-year marketing plan was apparently successful. At the three-year point, new fees, staff size, and success ratios were up and above plan, while overhead was down and at plan.

Smallowitz and Molyneux (1987) advocate a marketing model that consists of finding answers to three questions. The three questions are

- What do we do well?
- Within these areas of service, what is needed in the marketing place?
- How can we be hired to meet those needs?

This simple model resonates with the earlier Kolter and Fox (1985, p. xiii) definition of marketing as the " . . . effective management by an institution of its exchange relations with its various markets and publics."

Another approach is suggested by Versau (1986), who asserts that long-term success in the consulting engineering marketplace is achieved by focusing on three key elements. These elements are

• **Positioning.** This element or first step in planning and implementing a marketing program is defined as "the art of standing apart from the crowd that offers increasingly numerous choices to the buyer of professional services." Choices include being highly specialized, being a low-cost provider of a wide range of basic services, or offering something unique. This element or step is similar to the answer to Smallowitz's and Molyneux's first question.

• **Bias-toward action.** Meet, define, document but, in the process, assign responsibilities, set time-tables, and monitor results. In other words, work systematically and rationally, but avoid analysis paralysis.

• **People.** Demonstrate interest in the client and his or her needs and demonstrate interest in employees by creating a high-expectations–high-support environment within the consulting firm.

Concluding Observations

Marketing research and case studies, like those presented here, suggest that a consulting engineering firm or other organization will be successful in marketing if it:

- Earns the trust of potential clients.
- Knows and focuses on its strengths.
- Peers into the future and proactively prepares for it—rather than reacts.
- Develops a marketing plan, shares the plan with all members of the organization, and expects all of them to assist with implementation.
- Provides training.
- Appreciates and takes care of existing clients.

SUGGESTED WORKING MODEL FOR PLANNING AND IMPLEMENTING A MARKETING PROGRAM

Covey (1990, pp. 255–257) notes that the Greek philosophy for what might now be called win/win interpersonal and interorganizational relations was based on ethos, pathos, and logos. According to Covey:

> - Ethos is your personal credibility, the faith people have in your integrity and competency. It's the trust that you inspire . . . .
> - Pathos is the empathic side—it's the feeling. It means you are in alignment with the emotional thrust of another's communication.
> - Logos is the logic, the reasoning part of the presentation.

Covey emphasizes that these three elements of win/win interpersonal and interorganizational relations must occur in the indicated order. That is, trust must be established first; then needs must be understood, and then a logical follow-up occurs. The ethos–pathos–logos view of marketing as a mutually beneficial process is illustrated in Figure 14–2. The rational tendency in interpersonal relations, however, is to start with logos, which usually leads to less than satisfactory results.

The point is that the sequential ethos–pathos–logos process is an excellent basis for marketing. That is, first establish trust, then understand needs, and then followup logically. Engineering and other technical organizations must first earn the trust of potential clients. Once that trust is established, potential clients are

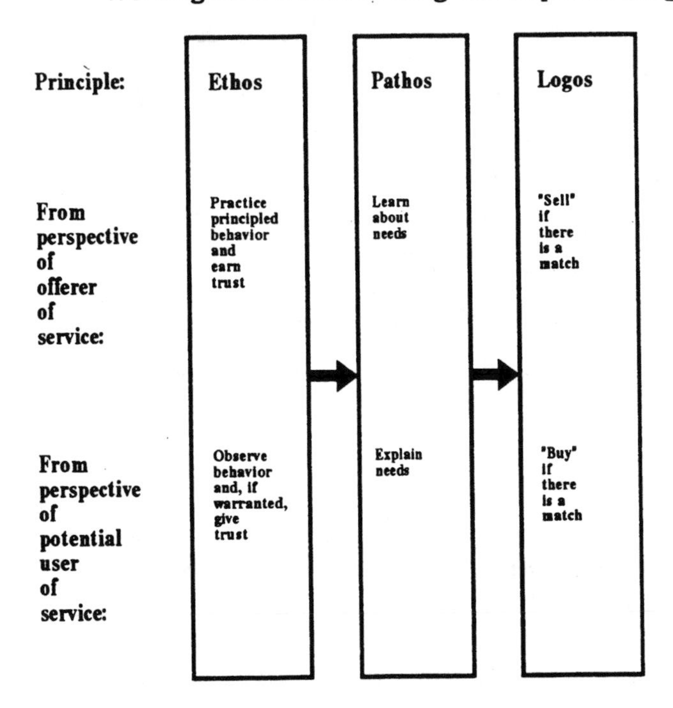

Figure 14–2. Ethos–pathos–logos view of marketing

likely to reveal their needs to the organization. If there is a match, that is, if the organization understands and can meet those needs, then a logical follow-up in the form of a contract or agreement is likely to occur. If a match does not develop between the offerer of services and the potential user of services, then the former should provide assistance by referring the latter to another engineering firm or other organization.

To reiterate, an excellent foundation for individual or corporate marketing effort is

- Ethos—earn trust
- Pathos—learn needs
- Logos—close deal

Presented now is a model for planning and implementing a marketing program that includes five largely sequential steps and is based on the ethos–pathos–logos philosophy. The model is illustrated in Figure 14–3. There are many possible feedback loops from the second, third, fourth, or fifth steps back to any earlier step.

In step 1, the firm's expertise, strengths, and niche are determined. Using

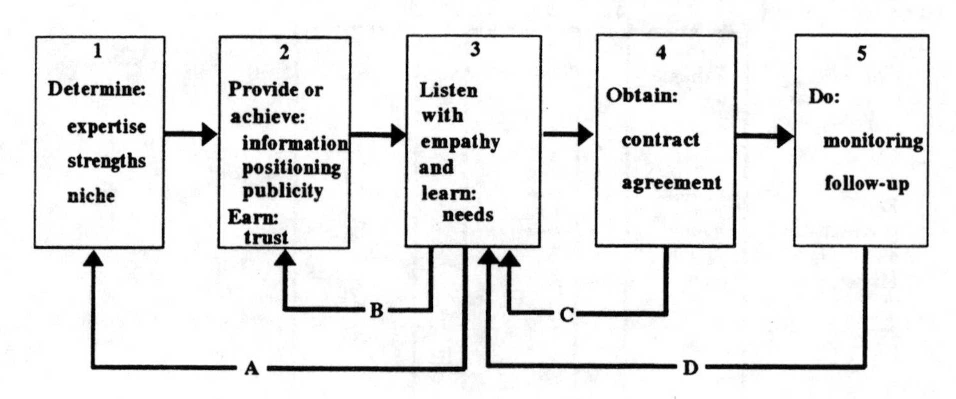

Figure 14–3. Steps in preparing and implementing a marketing plan

these results, step 2 involves directly or indirectly sharing information about the firm's expertise, strengths, and niche with the existing and potential clients and doing it in such a way that the trust of potential clients is earned. Mutual trust must develop if the process is to continue to step 3. In step 3, the firm focuses on the special technical and nontechnical needs of an existing and potential client and determines if the firm could meet those needs. In step 4, the client and firm enter into an agreement for provision of services. Under step 5, the last step, the firm closely monitors the client and all of its contracts with the client in the interest of achieving a mutually beneficial result and, for the consulting firm, additional contracts.

Feedback

Various "feedbacks" are suggested in Figure 14–3. Possible interpretations of the example feedbacks are as follows:

FEEDBACK IDENTIFICATION	POSSIBLE INTERPRETATION
A	Clients view the firm differently than the firm views the firm. Improved definition (step 1) and communication (step 2) are needed.

B	Client expresses need for particular service, so appropriate materials need to be identified or prepared and given to client (step 2).
C	Contract negotiations reveal that client expects firm to provide more services without an increase in fee. Firm needs to listen more carefully (step 3).
D	Client may soon be doing different type of project, but one that is within firm's service mix. Firm needs to find out more about the potential project (step 3).

Tools and Techniques

Many specific tools and techniques are available for implementing a marketing program. An effective set of tools and techniques must be selected for each of the previous presented five steps. Some available tools and techniques are summarized in Table 14–1. The five steps in the marketing program correspond to columns 1 through 5 in Table 14–1. A consulting firm could use this matrix, or a variation on it, to prepare a detailed marketing plan and then to implement the plan. The effort would, of course, be tailored to the firm's situation. Although the firm might use all five steps (the columns in the matrix), only the potentially most effective of the 40 listed tools and techniques (the rows) would be utilized. Note that only one of the 40 tools and techniques ("ask for contract") is "selling." Other kinds of organizations such as other businesses (service and manufacturing), educational institutions, and charitable organizations could use the matrix in Table 14–1 to guide the development and implementation of a marketing program tailored to their needs and aspirations.

Some of the listed marketing tools, such as resumes of staff members and statements of qualifications, should be kept on word processing systems—not in hard copy. Resist the temptation to have one resume or statement of qualifications for all purposes. Not only will this approach generally result in a document of excessive length, which will probably not be read, but it also will result in one that is not responsive to the current and specified needs of the potential client. Consider maintaining one complete and comprehensive resume for each member of the professional, and as appropriate, support staff on the word processing system and editing it down for each statement of qualifications, proposal, or other client-specific use. Do the same for the statement of qualifications for each specialty service. That is, respond to needs by carefully and quickly extracting sections of master resumes and project descriptions to create special short documents tailored to specific situations.

TABLE 14–1 MATRIX OF MARKETING STEPS VS MARKETING TOOLS AND TECHNIQUES

Marketing Tools and Techniques		Marketing Steps				
Name	Reference	1 Determine: Expertise Niche Strengths	2 Provide or Achieve: Information Positioning Publicity Earn: Trust	3 Listen with Empathy and Learn: Needs	4 Obtain: Contract Agreement	5 Do: Monitoring Follow-up
Conduct internal meeting workshop, retreat, etc.		*				
Assess recent fees by service type, clients, geographic area.	Dupies, 1979.	*				
Assess marketing costs—relative to total fees and fees by service area.	Dupies, 1979.	*				
Analyze demographic and economic data.	Dupies, 1979.	*				
Analyze existing and pending local, state, and federal laws and regulations.	Dupies, 1979.	*				
Interview/survey clients and community leaders with emphasis on services needed and how the service firm is viewed.	Dupies, 1979.	*				
Emphasize benefits, not features.	Dehne, 1991.	*	*			
Advertising.	Groob et al., 1987. Competitive Advantage, 1988.		*			
Audiovisuals.	Groob et al., 1987.		*			
Brochures.	Groob et al., 1987.		*			

Direct mail.	Groob et al., 1987. Devonshire Financial, 1988. Competitive Advantage, 1988.	*	
Trade show displays.	Groob et al., 1987. Devonshire Financial, 1988.	*	
Statement of qualifications.	Groob et al., 1987.	*	
Newsletter.	Groob et al., 1987.	*	
List name in professional directories.	Snyder, 1993, p. 29.	*	
Prepare press releases.	Groob et al., 1987.	*	
Participate in joint ventures.	Snyder, 1993, p. 31.	*	
Use office to make desired impression.	Competitive Advantage, 1988.	*	
Ask existing clients to call potential clients.		*	
Enter projects in award competition.		*	
Send business leads to private-sector potential clients.		*	*
Send articles.	Isphording, 1990.	*	*
Distribute copies of published papers.		*	*
Call “your” office anonymously; ask a friend to visit “your” office.	Groob et al., 1987. Townsend, 1970, p. 31.	*	
Entertain clients/potential clients.	Isphording, 1990.	*	*
Conduct seminars for clients/potential clients.	Isphording, 1990. Holtz, 1987. Severn et al., 1994.	*	*

(continued)

TABLE 14–1 MATRIX OF MARKETING STEPS VS MARKETING TOOLS AND TECHNIQUES (*CONTINUED*)

Marketing Tools and Techniques		Marketing Steps				
Name	Reference	1 Determine: Expertise Niche Strengths	2 Provide or achieve: Information Positioning Publicity Earn: Trust	3 Listen with Empathy and Learn: Needs	4 Obtain: Contract Agree-ment	5 Do: Monitoring Follow-up
Host client/potential client visits to service firm's office.	Isphording, 1990.		*	*		
Host client/potential client visits to service firm's projects and/or clients.			*	*		
Participate actively in civic and professional organizations.	Isphording, 1990.		*	*		
Visit clients/potential clients.	Isphording, 1990. Bakan, 1985.		*	*		
Ask open-ended questions.	Competitive Advan-tage, 1988.			*		
Assign backup partners/principals to each client.	Isphording, 1990.			*		*
Admit mistakes and suggest or take reme-dial actions.	Rabeler, 1991.		*			
Maintain resumes of staff members with emphasis on achievements.			*			
Recognize achievements/milestones of clients/potential clients.	Isphording, 1990.		*			
Conduct completed project postmortem with client.				*		*
Gather data on client/potential client.	Bakan, 1985.			*		
Follow up on all leads.	Groob et al., 1987.			*		
Ask for contract.					*	
Hand deliver reports.	Rabeler, 1991.					*

REFERENCES

BAKAN, L. H., "The Client Relationship: Effective Marketing Steps," *Journal of Management in Engineering—ASCE*, Vol. 1, No. 1 (January 1985), pp. 3–11.

COMPETITIVE ADVANTAGE INC., "The Competitive Advantage," sample issue, 1988.

COVEY, S. R., *The 7 Habits of Highly Effective People*, New York: Simon & Schuster, 1990.

DEHNE, G. C., "How Small Colleges Can Thrive in the '90s," *AGB Reports*, July/August 1991, pp. 6–11.

DEVONSHIRE FINANCIAL CORPORATION, "Sales and Marketing Digest," August 1988.

DUPIES, D. A., "Marketing for Engineering Organizations: Some Experiences," presented at the ASCE National Convention, Boston, Mass., April 2–6, 1979.

GROOB, J., K. SHOCKEY, L. WATTERS, and T. ALUISE, "Proven Tips for Marketing Professional Services," *Journal of Management in Engineering—ASCE*, Vol. 3, No. 1 (January 1987), pp. 28–37.

HENSEY, M. D., ed., "The Top 25," in Management Forum, *Journal of Management in Engineering—ASCE*, Vol. 3, No. 4 (October 1987), p. 263.

HOLTZ, H., *Expanding Your Consulting Practice with Seminars*, New York: John Wiley & Sons, Inc., 1987.

ISPHORDING, J., "Simple Marketing Techniques for the Service Profession," in Management Forum, *Journal of Management in Engineering—ASCE*, Vol. 6, No. 1 (January 1990), pp. 12–13.

KOLTER, P., and K. F. A. FOX, *Strategic Marketing for Educational Institutions*. Englewood Cliffs, N.J.: Prentice Hall, Inc., 1985.

RABELER, R. C., "Maintaining Existing Clients," *Journal of Management In Engineering—ASCE*, Vol. 7, No. 1 (January 1991), pp. 21–32.

SALWEN BUSINESS COMMUNICATIONS, "Successful A-E Firms List Service, Niche Marketing As Best Strategies for '90s, Survey Finds"; press release, New York, April 28, 1993.

SEVERN, S. R. T., S. C. GLADDEN, and K. S. NAKHJIRI, "Seminars: A Tool for Marketing Professional Services," *Journal of Management in Engineering—ASCE*, Vol. 10, No. 1 (January/February, 1994), pp. 14–18.

SHERIDAN, J. H., "Memos," *Industry Week*, Vol. 237, No. 1 (July 4, 1988), p. 5.

SMALLOWITZ, H., and D. MOLYNEUX, "Engineering a Marketing Plan," *Civil Engineering*, August 1987, pp. 70–72.

SNYDER, J., *Marketing Strategies for Engineers*. New York: American Society of Civil Engineers, 1993.

TOWNSEND, R., *Up the Organization*. New York: Alfred A. Kropf, 1970.

VERSAU, J. A., "Three Strategies for Sure Success: Positioning, Action, People Are Key," *Journal of Management in Engineering—ASCE*, Vol. 2, No. 3 (July 1986), pp. 191–199.

WAHBY, D., "Managing the A/E Firm in Turbulent Times," in Management Forum, *Journal of Management in Engineering—ASCE*, Vol. 9, No. 2 (April 1993), pp. 122–124.

SUPPLEMENTARY REFERENCES

ARDITI, D., and L. DAVIS, "Marketing of Construction Services," *Journal of Management in Engineering—ASCE*, Vol. 4., No. 4 (October 1988), pp. 297–315.

CREASON, G. W., "Redefining Marketing Activity," *Journal of Management in Engineering—ASCE*, Vol. 7, No. 3 (July 1991), pp. 295–298.

MANDINO, O., *The Greatest Salesmen in the World.* New York: Bantam Books, 1968.

PETERS, L. A., "Client-based Practice," *Journal of Management in Engineering—ASCE*, Vol. 3., No. 2 (April 1987), pp. 133–137.

EXERCISE

PREPARE MARKETING PLAN

Purpose

Demonstrate understanding of marketing basics by outlining a marketing plan for an actual or hypothetical consulting firm.

Tasks

1. Assume that you are responsible for the marketing function of an actual or hypothetical engineering or other technical services consulting firm. (An actual firm is preferred, although you may want to use a fictitious name. Perhaps you have learned about a consulting firm by working for them or as a result of interviewing.)
2. Prepare a written description of the firm. The description, which is not to exceed two double-spaced pages, should include, but not be limited to: size, office location(s), principal services offered, strengths (actual/perceived), and weaknesses (actual/perceived). Do not do any research. Simply use what you know or presume about a real consulting firm or develop a reasonable profile of a hypothetical firm.
3. After you have been director of marketing for about one year, your firm's Board of Directors indicates their desire to double the company's business over the next five years. Prepare a written marketing plan. Use the suggested working model presented in this chapter. Describe each of the five steps, and include the specific marketing tools and techniques you would use.
4. *Note:* This assignment will be weighted as twice a normal assignment.

15

The Future and You

> *In human affairs, the willed future always prevails over the logical future.*
>
> (Rene Dubois)
>
> *Chance favors only the prepared mind.*
>
> (Louis Pasteur)

What does the future hold for young engineers and other technical professionals? To what extent can the young professional hope to create the future of his or her career, the profession, and society at large? The answer to the second question is To a significant extent. Your future is something you can create—it does not have to be something that happens to you.

As suggested by the preceding questions and the answer to one of them, your future will be determined by a combination of changing external conditions, which you cannot control, and strategic individual actions, which you can control—if you choose. You can engineer your future. This chapter is intended to help you prepare for the future in three ways. First, changes likely to occur in the world of engineering and technical work are described. You should understand the changing stage on which you will play out your career. Second, the concept of paradigm pliancy, as opposed to paradigm paralysis, is explored as a means of helping you to shape your future. There always have been and always will be very different paths initially taken by a minute minority which have made all the difference for them and society. Third, the elements of leadership as an important part of the production, management, and leadership model introduced in Chapter 1 are discussed further to enable you to develop your leadership ability. Leadership is not an organizational position you work up into over a period of many years. On the contrary, leadership is a set of attitudes and skills that you, as a young professional, can begin to develop and apply now.

THE CHANGING WORLD OF ENGINEERING AND OTHER TECHNICAL WORK

Likely changes in the world of work as you and your contemporaries move into the twenty-first century can be seen by providing at least tentative answers to these three related questions:

- Who will be available to do engineering and other technical work?
- Who will the future engineers and other technical professionals serve?
- What kind of work will twenty-first century engineers and other technical professionals do?

Each of these questions is addressed in this section of the chapter based on material gathered and presented by Walesh (1992).

Who Will Be Available to Do Engineering and Other Technical Work?

About two-thirds of the Americans who will be working in the year 2000 are working now. But what about the new workers, more specifically, their gender? Almost two-thirds of the workers who have entered or will enter the workforce in the 1990s will be female (United Way, 1988). Women in the practice of engineering and other technical professions, relative to men, tend to be more supportive to fellow workers, to be more aware of and expressive of feelings as they relate to accomplishing project objectives, and tend to be more willing to share ideas. In contrast, as pointed out by Marcellino (1992), men tend toward task orientation, conceptualization, politics, and career building. Accordingly, women as a group have much to offer the engineering profession of the future.

Consider the ethnic profile of people entering the workforce in the 1990s. Over half of the new workers have been or will be nonwhite, female, or immigrant. Nearly one in three of all workers will be what are now called ethnic minorities by the year 2000. One in five Americans is 55 or older, and this group will grow (United Way, 1988). The growing group of older Americans, many of whom have elected to retire from their first or principal career early, provide a largely untapped resource for many organizations.

This concludes discussion of the first of the three questions, "Who will be available to do engineering and other technical work?" *Answer:* A much more heterogeneous workforce with greater participation by women, ethnic minorities, and seniors. There are many professional practice implications inherent in the ongoing and imminent changes in the workforce. Technical organizations need to further improve their recruitment and retention of women and ethnic minorities. Managers need to make fundamental changes in their language, metaphors, and style. There should be more appreciation of the gender-specific professional traits, with special recognition of female characteristics, and the wisdom of blending them within the engineering and business setting. New working paradigms are needed with some movement away from the traditional individualistic approach that pits individuals against each other in competition for promotions, bonuses, salary increases, and perquisites. Interdisciplinary awareness and teamwork should be encouraged. Finally, engineering and other technical organizations need to make more creative use of the knowledge, experience, and energy of senior citizens.

Who Will Future Engineers and Other Technical Professionals Serve?

Robert Reich (1991) writes about the future. Reich doesn't speak specifically to engineers and other technical professionals, but has much to say that is relevant to the technical community. He urges Americans to prepare for the global society with emphasis on the global market. He identifies a category of workers, which he refers to as *symbolic analysts,* who, among all workers, will have the most productive, secure, and satisfying careers. He defines symbolic analysts as those who "solve, identify, and broker problems by manipulating symbols . . . Their strength is not knowledge per se, but their ability to quickly and creatively use the knowledge. Symbolic analysts have four basic skills: abstraction, system thinking, experimentation, collaboration" (Reich, 1991, pp. 178, 229–233). If one accepts Reich's general thesis about a growing global economy and his specification of the symbolic analyst as the individual who will thrive in that economy, then engineering and other technical professions seem to have a promising future. The skills he identifies—abstraction, system thinking, experimentation, and collaboration—are the stuff that progressive engineering is made of.

Allen and Serards (1992) predict that the globalization of engineering practice, and presumably the practice of other technical professions, will result in major changes "in employers' and employees' attitudes toward international assignments." The traditional thinking that international assignments hurt careers because of the difficulty of reentry into the United States-based operations will give way to a focus on preparing the professional for and supporting the professional in a long series of international assignments. Arango (1991) argues that the United States can stop its backsliding by fundamental changes in the education and practice of engineering. Within engineering education, he calls for improvements in the direction of internationalization.

Harris (1992), looking to the future of the engineering profession, predicts that "The world of engineering will not be dominated by national security issues as in the past, but by the concerns for social security." He believes that problems to be solved by engineers will become more people-oriented and less thing-oriented.

This concludes discussion of the second of the three questions, "Who will these future engineers and other technical professionals serve?" *Answer:* People and organizations around the globe—national borders will be increasingly irrelevant. There are many implications for technical organizations and for young technical professionals like you beginning their careers. Probably, as part of your professional education, you've had an opportunity to make friends with students and faculty from other nations and to learn about their history and culture. Perhaps you've traveled or studied in other countries. If not, you should look for travel opportunities. In seeking employment, consider the international orientation or attitude of prospective employers. Find out if they have or seek international partners or assignments.

What Kind of Work Will Twenty-First-Century Engineers and Other Technical Professionals Do?

From the beginning of recorded history and all over the earth, people we would now label engineers, architects, and other technical professionals have met the basic needs of communal society (Walesh, 1990). Examples of these needs are providing shelter, water supply, wastewater disposal, irrigation, transportation, and environmental protection. Frankly, the fundamental work of these professionals hasn't changed. As discussed near the end of Chapter 1 in the section "Engineer as Builder," the common bond among engineers is building for the benefit of society. In the future, fundamental changes are not likely to occur. There certainly will be many changes in tools and techniques, but engineers and some other technical professionals will continue to be the builders.

Related to the topic of the kind of work future technical professionals will do is the topic of changes in how they will work. Driven primarily by the increasing availability of portable personal computers and various electronic communication devices, a growing minority of technical professionals will spend most of their work time in their residence or at other locations outside of the traditional office setting. As a result of factors such as longer expected life span, increased need for continuing education, particularly in science- and technology-driven disciplines, and the desire for personal renewal in the broader sense, life patterns may change markedly. The current education–work–retire pattern may give way, as illustrated in Figure 15–1, to a more varied life pattern. Long work periods will be interspersed with short periods of education, leisure, sabbaticals (paid), and leaves (unpaid) similar to those enjoyed by college faculty. In contrast with leisure, sabbaticals and leaves typically involve personal growth through study, travel, and temporary, different work.

This concludes the discussion of the third of the three questions, "What kind of work will the early twenty-first century engineers and other technical professionals do?" *Answer:* Meeting society's basic physical needs—as they have always done—no significant change in focus but some changes in workplace and pattern.

Recap of the Changing World of Technical Work

You should be preparing yourself for the way your profession will be practiced, not the way it is or was practiced. In the early twenty-first century, the workforce will be increasingly heterogenous with greater participation by women, ethnic minorities, and senior people. Future engineers and other technical professionals will increasingly serve people and organizations around the globe. The basic mission of technical professionals will not change, but will continue to focus on meeting society's basic physical needs.

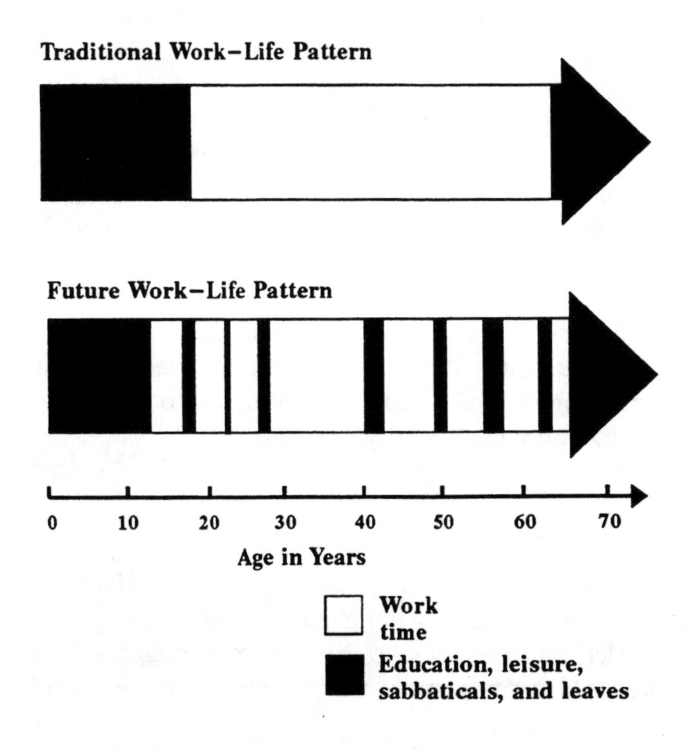

Figure 15–1. Changing work-life patterns (Source: Adapted with permission from United Way of America, *The Future World of Work: Looking Toward the Year 2000,* 1988.)

THE FUTURE—CAN YOU SPARE A PARADIGM?

> *. . . There is nothing more difficult to plan, more doubtful of success, nor more dangerous to manage than the creation of a new system. For the initiators have the enmity of all who would profit by the preservation of the old institutions and merely lukewarm defenders in those who would gain by the new one.*
>
> (Nicolo Machiavelli, 1537, pp. 49–50)

Considerable management and leadership knowledge and skill have been presented in the preceding chapters of this book. But management expert Peter Drucker (cited in Barker, 1989, p. 7) would probably argue that something is missing: the critical skill, particularly in turbulent times, of anticipation. Managers and leaders would like to anticipate what will happen so that they can identify and seize opportunities—or at least avoid problems.

Anticipative and Reactive Modes

The dynamics of the interactions between anticipation and reaction and opportunity identification and problem solving are illustrated in Figure 15–2. The least desirable domain is that of the lower left quadrant. Engineering examples of activities that would fall within this quadrant are flood fighting and rework or replacement of manufactured and shipped products. The next best domain is the upper left quadrant in which problems are anticipated and then solved. Examples include the planned evacuation of people from flood-prone areas, provision of a complaint counter in the retail organization, and inspection at the end of a production line. A better situation is that encompassed in the lower right quadrant, in which the focus is on avoiding problems in the first place. Examples include single-purpose flood-prevention facilities and the continuous improvement of processes in product and service organizations. The best domain is that in the upper right quadrant. The previously mentioned critical management skill of anticipation, coupled with opportunity identification, occurs in this quadrant. Examples include a comprehensive, multipurpose approach to the use of flood-prone areas and development of new concepts, products, and services.

Definition of Paradigm

Anticipating the future, which is one of the dimensions of the upper right quadrant of Figure 15–2, is at best very difficult. An important quality is being able to "take off the blinders," which naturally leads to the topic of paradigms. Covey (1990, p. 23) defines a paradigm as "the way we 'see' the world—not in terms of

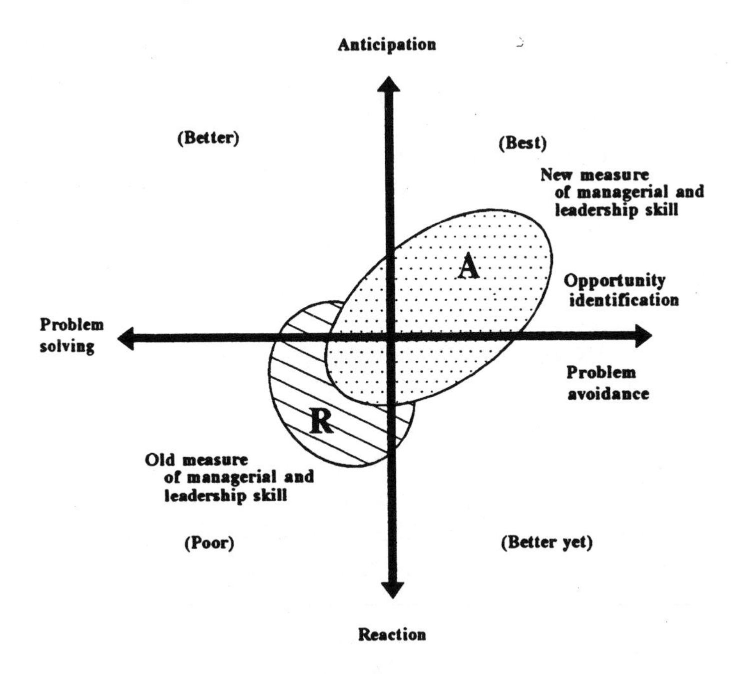

A: anticipative mode
R: reactive mode

Figure 15-2. Managerial domains (Source: Adapted with permission from J. A. Barker, *Paradigms: The Business of Discovering the Future.* New York, NY: Harper Business, 1992, p. 27.)

our visual sense of sight, but in terms of perceiving, understanding, interpreting." According to Barker (1989, p. 14), a paradigm is " . . . a set of rules and regulations that: (1) defines boundaries; and (2) tells you what to do to be successful within those boundaries."

Examples of Paradigms

Paradigms abound—they are all around us. Examples include:

- The traditional, pyramidal organizational structure of consulting organizations, construction firms, manufacturing operations, and academic institutions.
- The nine-month school year.
- Men and women participating in high school athletics.

- Japanese products being of high quality, the implication being that manufacturing organizations hoping to compete globally must match or exceed the quality of Japanese products.
- The 40-hour work week.
- The largely individual-based, competitive learning situation utilized in U. S. higher education.
- The philosophy that "buy American" helps the United States.

Some Characteristics of Paradigms

Barker (1989, pp. 67f) reiterates the idea that one of the characteristics of paradigms is that they are so common, that they are invoked or used implicitly with little thought. Paradigms are very useful, given the complexity of society. They almost always allow for more than one "right" answer. For example, there are many ways to interpret paradigms such as the nine-month school year and the 40-hour work week.

On a somewhat negative side, according to Barker (1989), paradigms tend to reverse the "seeing and believing" process. Intelligent, thoughtful people like to think that they are rational, that they "believe because they see." However, because of paradigms, people often "see because they believe." Consider, for example, some beliefs you hold and have held for a long time and you highly value. Are you not likely to find many examples of situations that tend to support your belief, and might you not actually be looking for them? And might you be "seeing because you believe" rather than "believing because you see?" As noted by Brown (1988, p. 106), "We do not see things as they are. We see things as we are."

Unfortunately, if paradigms are too strongly held, and they often are, the holders risk incurring paradigm paralysis. Paradigm pliancy is a much better strategy, according to Barker (1989), especially in turbulent times. Pliancy is a quality or state of yielding or changing. Fortunately, at least a handful of people in any organization can change their paradigms. Even for them, paradigm pliancy is at best difficult. The creators or advocates of new paradigms are often unwelcome. Bennis (1989, p. 124) asserts, "Most organizations would rather risk obsolescence than make room for the nonconformists in their midst." Discussions, debates, and disagreements over old versus new paradigms do not typically pit the incompetent against the competent, the weak against the strong, and the uncaring against the caring. Paradigm paralysis adherents differ from paradigm pliancy adherents mainly in their regard, or lack thereof, for the status quo. As explained by Heilmeier (1992, pp. 10–16), "History seems to indicate that breakthroughs are usually the result of a small group of capable people fending off a larger group of equally capable people with a stake in the status quo."

Another characteristic of paradigms is that they evolve as suggested by Figure 15–3. The vertical scale represents measures of effectiveness such as problems solved, things accomplished, people helped, clients served, and products sold per unit of time. Figure 15–3 illustrates the concept that the effectiveness of a paradigm is minimal early in its life and then accelerates significantly. Typically, with time, the effectiveness of a particular paradigm begins to decline. In other words, paradigms have a finite, but unknown life. A paradigm, according to Barker, follows the familiar S-shaped curve.

Bridges form from "old" to "new" paradigms, offering the possibility that a more effective paradigm can be found to replace an aging paradigm. The concept of paradigm bridging is presented in Figure 15–4. The series of paradigms might, for example, represent respectively various personal computers including the abacus, slide rule, electronic calculator, and digital computer. The most successful individuals and organizations are those that are among the first to recognize and cross the bridges to newly developing paradigms or, better yet, create and develop new paradigms and the bridges that other individuals and organizations will eventually adopt and use.

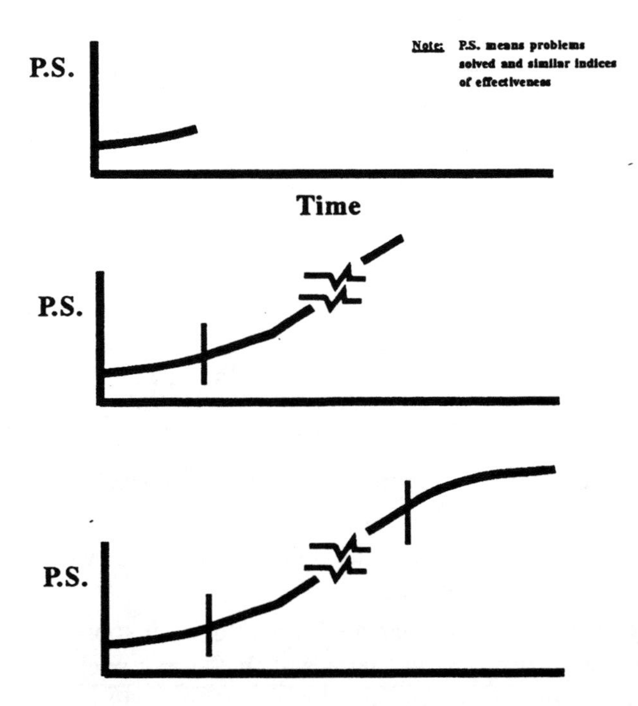

Figure 15–3. Evolution of a paradigm (Source: Adapted with permission from J. A. Barker, *Paradigms: The Business of Discovering the Future.* New York, NY: Harper Business, 1992, p. 46.)

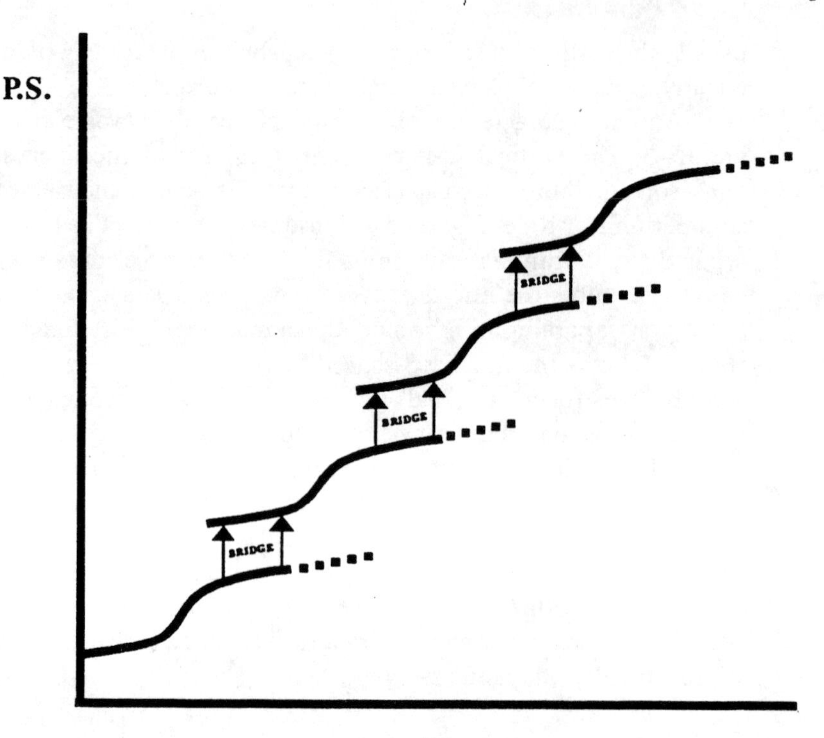

Figure 15–4. Paradigm bridging (Source: Adapted with permission from J. A. Barker, *Paradigms: The Business of Discovering the Future.* New York, NY: Harper Business, 1992, p. 72.)

Examples of Paradigm Shifts

As suggested by the abacus–slide-rule–electronic-calculator–digital-computer progression, paradigms do shift. Many examples can be found within technical professions and throughout society at large. Consider, for example, the following situations which existed about twenty years ago in the United States relative to today:

- Inter-high school and college sports were essentially exclusively for males (Barker, 1989, p. 98).
- The slide rule was the personal computer—a digital computer with its peripherals typically filled an entire room.
- There was a strong feeling that nuclear power would soon solve most energy problems throughout the United States and that many nuclear power plants would be built in successive decades (Barker, 1989, p. 98).

- Watches were mechanical, including complex gear and spring mechanisms.
- Really desirable cars had V-8 engines, some exceeding 300 horsepower.
- Letters routinely took days to be delivered.
- Worldwide conflicts between democracy and communism and capitalism and socialism would continue forever.
- Women did not study engineering.

Now consider some of the paradigm shifts that have occurred in the last few decades and the "shifters," "outsiders," and "odd balls," responsible for those shifts. For example:

- Fred Smith founded Federal Express (Barker, 1989, p. 26; Cypert, 1993, pp. 201–202) so that letters and packages, in the United States at least, could be routinely delivered overnight. As a Yale University student, Smith wrote a paper proposing overnight mail delivery in the United States using trucks and airplanes operating within a hub and spoke system. According to the professor who gave Mr. Smith a "C" on the paper, the idea was interesting, but would never work.
- Chester Carlson developed the process of electrostatic photography, offered it to 43 companies in the late 1940s and finally found one organization with foresight, resulting in the development of what is now called xerography. The problem at the time was that the photography paradigm consisted of film, developer, and a darkroom—there was no other way to do it (Barker, 1989, p. 55). At least 43 companies could not envision the now omnipresent copy machines.
- Swiss watch researchers in Neuchatel, Switzerland developed a novelty quartz watch with a digital display in 1967. At that time, Swiss mechanical watches enjoyed 60% of the global watch market. They displayed their unprotected novelty watch at an international watchmakers' conference. The idea was picked up by Japanese entrepreneurs and, as a result, the Swiss share of the global watch market has now diminished to less than 10% (Barker, 1989, pp. 57–60).
- The late American W. Edwards Deming, whose advice on what is now called Total Quality Management, and is discussed in Chapter 7, was ignored by American businesses. He assisted the Japanese who, in a matter of a few decades, set the world standard for manufactured products (Barker, 1989, pp. 61–66).
- College dropouts Stephen Jobs and Steve Wozniak, working in a garage in 1975, created the first Apple personal computer (Barker, 1989, pp. 91–96, Cypert, 1993, pp. 131–132).

Some Possible Future Paradigms

Barker (1989, pp. 99–100, 102, 116) suggests shifts to the following new paradigms:

- Patient diagnosis will occur at home.
- Babies being born at home will be the norm.
- Donations of organs will be mandatory.
- Zero-gravity manufacturing will be performed in outer space, permitting the production of products such as foamed metals.
- Vouchers for education will become widespread, thus forcing improvement of education through competition.

The Bottom Line

The most successful technical organization managers and leaders will not suffer paradigm paralysis—they will practice paradigm pliancy. They will create new paradigms and build bridges to them or at least recognize new paradigms when they are coming down the pike and see the business and professional opportunities within them. What paradigms will you contribute to your engineering or other technical field, or what paradigms created by others will you be receptive to and help advance?

ELEMENTS OF LEADERSHIP

Chapter 1 uses the metaphor of a three-legged stool to suggest an individual and organizational paradigm comprised of production, management, and leadership functions (Figure 1–4). With this paradigm each and every person in an organization is expected and enabled to fulfill, to widely varying degrees, all three functions. The emphasis of preceding chapters of this book is on the second of these three functions; that is, on managing or directing. Many and diverse elements of management are treated in great detail throughout the book and need not be further discussed in this chapter. However, leadership, which was introduced in Chapter 1 and carried through preceding chapters as a secondary theme, warrants further discussion in this final chapter. A goal of this book is to lead the reader into, or at least point the reader toward, the leadership function. There is, of course, no widely accepted list of elements of leadership. However, a review of books, papers, and articles on the topic and reflection on experience suggests seven important elements of leadership. These elements are presented in Figure 15–5 and discussed here for consideration by the entry-level engineer or other technical professional. In the following, the word *leader* is used, for brevity, to mean the leader in each of us or the leadership part in each of us.

Honest and Possessed of Integrity

When you tell the truth, you never have to worry about your lousy memory.

(Brown, 1988, p. 63)

Leaders are granted the privilege of leading by those who are prepared to be led. Honesty and integrity on your part are crucial to earning and retaining the privilege to lead. Although honesty and integrity are often used in a vague and even interchangeable manner, they warrant precise definitions. Covey (1990, pp. 195–196) says, "Honesty is telling the truth—in other words, conforming our words to reality. Integrity is conforming reality to our words—in other words, keeping promises and fulfilling expectations." Stated differently, honesty is retrospective and integrity is prospective; honesty is what you say about what you've done and integrity what you do about what you've said.

- **Honest and possessed of integrity**
- **Conscious of the mission: preaching it, teaching it, and reaching for it**

- **Setting goals, establishing strategies/tactics to achieve them, and following through**

- **Always learning**

- **Courageous**

- **Calm in crisis and comfortable with chaos**
- **Creative, synergistic, imaginative, innovative**

1 + 1 = 3

Figure 15–5. Elements of leadership

Conscious of the Mission: Preaching It, Teaching It, and Reaching for It

Leaders know where their organization is going; they believe in it, and much of what they say and do reflects and supports the vision. Each organization should have a strategic plan, and that plan should contain a brief, clear vision or mission statement. Examples follow.

- Our mission is to empower people and organizations to significantly increase their performance capability in order to achieve worthwhile purposes through understanding and living principle-centered leadership (mission statement of the Covey Leadership Center, Utah. Source: Covey, 1990, p. 355).
- The purpose of the College is to educate qualified and motivated individuals in a strong undergraduate environment who, upon graduation, are prepared for lifelong learning and the pursuit of professional excellence by ethically and creatively applying scientific knowledge to benefit society. (mission statement of the College of Engineering, Valparaiso University, Indiana. Source: College of Engineering, 1991, p. 3).

If the organization does not have such a statement and a plan, the leader takes action to get a statement articulated and a plan developed. But what if an individual's values and goals are in serious conflict with an organization's mission? The answer is obvious—try to effect change and, if not successful, move on. Life is too short to dissipate yourself in a hostile environment or prostitute yourself by feigning allegiance to alien values.

The leader interprets opportunities and problems in the context of the mission, always seeking ways to move the organization one more step in the direction of the vision. In the leader's mind, a constructive tension exists between many of the common, day-to-day occurrences and the organization's mission. That tension pulls those occurrences in the direction of the mission and is the force that enables the leader to take steps in keeping with the vision.

Phillips (1992, p. 169) describes President Abraham Lincoln as a visionary. His concluding Lincoln-based principles on vision include the following, which are quoted directly from Phillip's book:

- Provide a clear, concise statement of the direction of your organization, and justify the actions you take.
- Everywhere you go, at every conceivable opportunity, reaffirm, reassert,

and remind everyone of the basic principles upon which your organization was founded.

- Effective visions can't be forced on the masses. Rather, you must set them in motion by means of persuasion.
- When effecting renewal, call on the past, relate it to the present, and then use them both to provide a link to the future.

In summary, the leader enables others to understand, value, and commit to the mission of the organization and, as a result, direct their aspirations, talents, and skills toward the achievement of that mission.

Setting Goals, Establishing Strategies and Tactics to Achieve Them, and Following Through

The leader clearly understands and appreciates his or her area of responsibility, authority, and accountability. In the context of and in support of the organization's mission, leaders regularly set individual and subgroup goals and establish strategies and tactics to achieve those goals. Recall the discussion of personal goal-setting in Chapter 2 in the section titled "Time-Management Tips." Leaders do not wait for someone to tell them what to do, when to do it, and how to do it. Leaders act. Leaders view the future as something they make happen—not something that happens to them.

Drawing again on the example of Abraham Lincoln, Phillips (1992, p. 108) says that Lincoln had "an almost uncontrollable obsession" to achieve. Phillips (1992, p. 113) sets forth some principles at the conclusion of a chapter titled "Set Goals and be Results-Oriented." Some of these principles, quoted directly from Phillips, are

- Set specific short-term goals that can be focused on with intent and immediacy by subordinates.
- Sometimes it is better to plow around obstacles rather than to waste time going through them.
- Your war will not be won by strategy alone, but more by hard, desperate fighting.
- Your task will neither be done nor attempted unless you watch it every day and hour, and force it.
- Remember that half-finished work generally proves to be labor lost.

Always Learning

Leaders develop and maintain, through formal study, self-study, and experience, their special or unique set of skills and knowledge. Maintaining one's expertise is a leadership element that is particularly important in a technical organization because of the rapid changes in science and technology driving the services offered and products produced by such organizations.

Currency in skills and knowledge is crucial to an organization for three reasons. First, it contributes to what the organization can do or offer in serving its clients, customers, and constituents. Second, maintaining currency at the individual level sets a positive example for others in the organization, earns respect from them, and encourages them to develop and maintain their expertise. Third and finally, by being an expert, a person is much more likely to value expertise, people who possess it, and the wisdom of drawing on the proper mix of expertise and people in meeting needs of internal and external clients and customers.

Certainly the organization must support financially and in other ways the development and maintenance of individual skills and knowledge. However, the primary responsibility for maintaining expertise lies with the individual. The topic of managing personal professional assets is discussed in detail in Chapter 2.

A leader is a perpetual student of nontechnical topics, that is, of areas of concern and relevance outside of his or her area of expertise. Expertise implies depth of knowledge in contrast with breadth of knowledge and understanding of context, which are also necessary. President Harry S. Truman advocated and exemplified gaining breadth of knowledge through reading. He said it this way, "Readers of good books, particularly books of biography and history, are preparing themselves for leadership. Not all readers become leaders. But all leaders must be readers" (Poen, 1982, p. 139).

In writing about leadership traits for the future, Brown (1990, p. 34) says, "Leaders are going to have to become adept at learning instead of knowing . . . " The notion of a leader who has extensive and largely sufficient knowledge based on education and experience and who uses that knowledge to direct the efforts of narrowly and sufficiently trained subordinates is dimming. The modern leader must continuously seek and probe—and expects others to do likewise. Rather than claiming to know in a static and superior sense, leaders will increasingly focus on knowing how to learn and enabling others to do likewise. Continuous learning will increasingly characterize the world of work.

The perpetual student concept, so important to today's and tomorrow's leaders, is a common thread woven through early western writings. Ancient works of fiction and nonfiction portray the heroes and the elite as relentless pursuers of knowledge and ideas—even when the resulting revelations threaten the seeker. For example, although Socrates probably protested too much, his claim that he did not know the truth but instead diligently searched for it can serve as a model for leaders. Directed discussion, as used by Socrates and described by Plato, would seem to be the preferred modus operandi of viable organizations. Socrates convincingly expressed his faith in inquiry and discovery when he said (Plato, *Meno*, p. 76):

> *. . . but I would contend at all costs both in word and deed as far as I could that we will be better men, braver and less idle, if we believe that one must search for the things that one does not know, rather than if we believe that it is not possible to find out what we do know and that we must not look for it.*

Change will increasingly be the only certainty. Accordingly, leaders must create and support an intellectual environment in which discussion is directed toward identifying, interpreting, planning for, and, in some cases, influencing the direction and shape of change. Given the increasing complexity of the world, no one person can possess sufficient knowledge to accommodate change. Interdisciplinary teams employing Socratic-style directed discussions are a promising alternative approach.

However, unlike Socrates' era when information was minimal, data now abound—usually to excess—and must enter into an organization's dialogues. Used in a modern organization, the Socratic method promises to elicit informed and relative contributions from many experts and other individuals, build the confidence of participants, sharpen critical thinking, encourage synthesis, and occasionally lead to serendipity. On the negative side, directed dialogue requires time and patience, which are always precious resources in dynamic, action-oriented organizations. Dialogue may be difficult for some leaders, because the process may lead to challenges to their dearly held notions and operating principles. Paradigm paralysis may afflict them. A leader's search for knowledge must be credible—it cannot be conditioned on the expected positive or negative impact of the findings. The truth must be determined and dealt with.

Courageous

> *Do not follow where the path may lead. Go instead where there is no path and leave a trail.*
>
> (Anonymous)

Leadership requires courage—courage to hold people accountable for carrying out their responsibilities and keeping their promises, to confront individuals exhibiting unacceptable behavior, to walk away from a project or client on ethical grounds, to aim high and risk apparent great failure, to apologize and ask for a second chance, and to persist when all others have given up. But what constitutes courage and courageous people?

Aristotle offers a thoughtful and demanding perspective on courage. Aristotle defines courage as a precarious, difficult-to-prescribe balance among causes, motives, means, timing, and confidence. He says (Aristotle, p. 65):

> *The man, then, who faces and who fears the right things and from the right motive, in the right way and at the right time, and who feels confidence under the corresponding conditions, is brave; for the brave man feels and acts according to the merits of the case and in whatever way the rule directs.*

Aristotle goes on to say (Aristotle, pp. 66–67) that courage is a mean between cowardice and rashness, confidence and fear. In summary, he defines courage as a fully informed, carefully considered willingness to die for a noble cause (Aristotle, p. 64). Aristotle refutes the notion that courage is reactive or instinctive. You might be tempted to say that Aristotle was not totally serious about his definitions of courage and courageous people—at least with respect to the "willingness to die" aspect. After all, he must have intended death as a metaphor for a willingness to incur great loss. Perhaps this interpretation is acceptable, at least for purposes of this book.

Aristotle outlines in systematic and exhaustive fashion five kinds of false courage. These might be referred to as lesser degrees of courage. They encompass much of what passes for courage in our society (Aristotle, pp. 67–71) and help, by elimination, to define bona fide courage.

The first type of courage is coercion courage, or what Aristotle refers to as "the courage of the citizen-soldier." The possessor faces significant risks, but he or she has no choice. Leaders simply have to do many things—some of which are quite unpleasant and risky. Aristotle's coercion courage concept cautions the leader to maintain perspective and not to view these as courageous acts worthy of praise. These acts are part of the job—they come with the territory.

What might be called high information or calculated courage is the second type. Aristotle uses the example of the professional soldier who seems brave in battle, but in fact entered the fray with far superior information and other resources that virtually guaranteed victory. The modern leader may be tempted to feign courage, because he or she often has exclusive access to vital information.

The third type of courage is passion courage. These reactionary acts are in conflict with the choice and motive elements clearly evident in Aristotle's model of courage. While the emotional outburst or sharp retort is often viewed as courage, as in "You sure told him/her," these acts are often done without thought. Although passionate reactions may seem to immediately please onlookers, calm and reason in difficult circumstances may require more courage and lead to long-term benefits for all antagonists.

Sanguine, to use Aristotle's word, or what might be called overly optimistic courage, is the fourth type of counterfeit courage. A string of business or other successes can lead to unrealistic optimism or even complacency, which may be viewed by the participant or by others as courage. The United States global dominance in economic and military affairs during the four-decade post-World War II period is an illustration of Aristotle's sanguine courage. The modern leader must be alert and view expectations of continued success with suspicion. An earlier atmosphere of courage that enables an organization to achieve high levels of performance may gradually and unnoticeably give way to complacency.

Aristotle's fifth and last type of false courage is the ignorance variety. As he bluntly says, "People who are ignorant of the danger also appear brave." As we become an increasingly information-rich world, leaders must devote some resources to continuously sift through new knowledge to identify and assess opportunities and threats.

Informed by Aristotle's ideas, the leader of today is more likely to recognize his and others' bravado. There will always be some pretense of bravery, particularly by people in high and prestigious positions. Recognizing this, the leader should place a premium on his or her acts and the acts of others that, in the face of risk and calamity, are carefully considered and indicate a willingness to sacrifice for the corporate or community cause. Courageous acts don't have to be extreme acts. When leaders take extreme positions they may be less successful in defending a principle, advancing a cause, or achieving a worthy goal than when they assume courageous, but somewhat more moderate postures.

Calm in Crisis and Comfortable with Chaos

Whenever competent and committed people are involved in group efforts, often in competition with other organizations, difficult interpersonal and other serious conflicts and situations are inevitable. Confronted, usually unexpectedly, with such crises, the leader instills calm, seeks understanding, and does not make premature judgments. The leader should "Seek first to understand, then to be understood" (Covey, 1990, p. 239).

Having defused and perhaps even resolved the most recent crisis, the leader—regardless of his or her administrative level in an organization—is usually thrust back into the midst of a general chaotic situation typical of the dynamic organization striving to succeed in an ever-changing world. Engineering and other technical organizations must contend with rapid advances in science and technology; new environmental, personnel, and other laws and regulations; the globalization of business; client and customer demands or dissatisfaction;

new competition or old competitors offering new services in new ways; unexpected business or other opportunities and turnover of professional and other personnel. Typically, the changes cannot be predicted and defy quantification, but they must be continuously confronted. The leader in you is comfortable with chaos.

Creative, Synergistic, Imaginative, Innovative

Leaders seek ways to utilize the right hemisphere of the brain. As noted by Covey (1990, p. 130), decades of research have resulted in brain dominance theory. Covey explains:

> *Essentially, the left hemisphere is the more logical, verbal one and the right hemisphere the more intuitive, creative one. The left deals with words, the right with pictures; the left with parts and specifics, the right with wholes and the relationship between the parts. The left deals with analysis, which means to break apart; the right with synthesis, which means to put together. The left deals with sequential thinking; the right with simultaneous and holistic thinking. The left is time bound; the right is time free.*

Clearly, leadership qualities of creativity, synergism, imagination, and innovation are products of the right brain. Because of the nature of their college education, engineers and some other technical professionals are very likely to be dominated by the left hemisphere of the brain. Cultivation and development of the right hemisphere is, therefore, especially important for engineers and other similar professionals who want to be creative, synergistic, imaginative, and innovative.

According to Covey (1990, p. 130), leaders create things twice—first in their mind and then in physical reality. Leaders have strong visualization capabilities. Their mental images are vivid and all-encompassing. What will the final system, facility, product, organization, event, or thing look like? How will one feel to be in or around it? What will it smell like? How will it sound? The technical professional should resonate with the concept of creating things twice. This is the essence of first preparing plans and specifications—creating on paper—and then constructing—creating physically. Covey argues that leaders apply the process of creating twice to all aspects of their lives. Somewhat ominously, Covey (1990, p. 100) warns that " . . . there is a first creation to every part of our lives. We are either the second creation of our own proactive design, or we are the second creation of other people's agendas, of circumstances, or of past habits." Or, to repeat a conviction stated in the beginning of this chapter, leaders view the future as something they make happen—not something that happens to them.

Leaders leverage successes to produce even greater successes. They find ways to invest money or other resources at the margin so as to yield large returns at the margin (e.g., converting a successful project into a successful paper). Leaders are synergistic—they seek combinations such that the sum is greater than the parts. Leaders search for the silver lining in a black cloud. They believe—at least most of the time—that "This is not a problem; this is an opportunity."

REFERENCES

ALLEN, L., and J. SEWARDS, "Issues in Human Resources: Managing Talent in the 21st Century," *Journal of Management in Engineering—ASCE*, Vol. 8, No. 4 (October 1992), pp. 340–345.

ARANGO, I., "From U. S. Engineer to World Engineer," *Journal of Management in Engineering—ASCE*, Vol. 7, No. 4 (October 1991), pp. 412–427.

ARISTOTLE, *The Nicomachean Ethics*, translated by D. Ross and revised by J. L. Ackrill and J. O. Urmson. Oxford, England: Oxford University Press, 1987.

BARKER, J. A., *Discovering the Future: The Business of Paradigms.* St. Paul, Minn.: ILI Press, 1989.

BENNIS, W. G., *Why Leaders Can't Lead—The Unconscious Conspiracy Continues.* San Francisco, Calif.: Jossey-Bass Publishers, 1989.

BROWN, H. J., JR., *A Father's Book of Wisdom.* Nashville, Tenn.: Rutledge Hill Press, 1988.

BROWN, T., "Leaders for the '90s," *Industry Week*, March 5, 1990, p. 34.

COLLEGE OF ENGINEERING, "Strategic Plan for the College of Engineering at Valparaiso University 1991–1995," Valparaiso University, Valparaiso, Ind., August 1991.

COVEY, S. R., *The 7 Habits of Highly Effective People.* New York: Simon & Schuster, 1990.

CYPERT, S. A., *The Success Breakthrough.* New York: Avon Books, 1993.

HARRIS, J. G., "Engineering—The Bridge Between Two Cultures," *Newsletter*, ASEE Electrical Engineer Division and IEEE Education Society, Summer 1992.

HEILMEIER, G. H., "Some Reflections on Innovation and Invention," *The Bridge*, National Academy of Engineering, Vol. 22, No. 4 (Winter 1992).

MACHIAVELLI, N., *The Prince.* New York: New American Library, 1980, pp. 49–50. (Published in 1537; Luigi Ricci's translation published in 1903; this translation by E. R. P. Vincent was first published in 1935.)

MARCELLINO, P., "Management: The Study of People—Do Women and Men Really Have Different Styles?" *Woman Engineer,* April/May 1992, pp. 50–57.

PHILLIPS, D. T., *Lincoln on Leadership—Executive Strategies for Tough Times.* New York: Warner Books, 1992.

PLATO, *Five Dialogues: Euthyphro, Apology, Crito, Meno, Phaedo*, translation by G. M. A. Grube. Indianapolis, Ind.: Hackett Publishing Company, 1981.

POEN, M. M., ed., *Strictly Personal and Confidential—The Letters Harry Truman Never Mailed.* Boston, Mass.: Little, Brown and Company, 1982.

REICH, R. B., *The Work of Nations.* New York: Knopf, 1991.

UNITED WAY, *The Future World of Work: Looking Toward the Year 2000*, 1988.

WALESH, S. G., "Changing Demographics: Civil Engineering Implications," presented at "Challenges in a Changing World" the 1992 International Convention and Exposition of the American Society of Civil Engineers, New York, September 1992.

WALESH, S. G., "Water Resources Science and Technology: Global Origins," *Urban Stormwater Quality Enhancement*, Proceedings of an Engineering Foundation Conference, Davos, Switzerland, ASCE, 1990, pp. 1–27.

SUPPLEMENTAL REFERENCES

BROWN, T. L., "A Glimpse at the Future—Can Be an Intimidating Sight," *Industry Week*, July 1, 1991, p. 34.

HELGESEN, S., *The Female Advantage: Women's Ways of Leadership.* New York: Doubleday, 1990.

LABICH, K., "The Seven Keys to Business Leadership," *Fortune*, October 24, 1988, pp. 58–66.

NSPE, "A Vision for the Future of U. S. Engineers," Report of the 2000 Task Force, October 1991, Alexandria, Va.

ROESNER, L. A., and S. G. WALESH, "Urban Water Resources Issues in the 21st Century," *Journal of Professional Issues in Engineering—ASCE*, Vol. 114, No. 3 (July 1988), pp. 302–309.

Appendix A

SPECIAL FEATURES OF CIVIL ENGINEERING

The practice of civil engineering has some features that make it different, from a management and leadership perspective, from other areas of engineering. Accordingly, the civil engineering student or entry-level civil engineer should be aware of these differences and prepare accordingly (Roesner and Walesh, 1988).

ONE-OF-A-KIND CREATIONS

Unlike most other areas of engineering, civil engineers tend to design large, one-of-a-kind structures, facilities, and systems. Examples are flood control works; large, multistory buildings; water supply collection, transmission, treatment, and distribution systems; rapid transit systems; and airports. Not only are these systems unique, but the consequences of their failure can be catastrophic because of their precarious proximity to large concentrations of people. Furthermore, failure or substandard performance of structures, facilities, and systems like those commonly found within the realm of civil engineering can have a great, negative economic impact. To the credit of the civil engineering profession, catastrophic failures have been few and far between. However, the rare disasters are quickly and widely publicized.

GREATER SPAN OF FUNCTIONS AND EXPECTATIONS

About two-thirds of all civil engineers work in consulting firms or government. In contrast, most other engineers are employed in industry. Because of their place of employment, civil engineers tend to work for smaller organizations and, at almost any point in their early career, tend to have a wider range of functions to perform. For example, a new civil engineering graduate, during his or her first year of employment in a consulting firm, could expect to have the following kinds of varied assignments: write portions of a report, accompany a senior engineer to a meeting with a client and make part of the presentation, attend a seminar and report on it to colleagues in the office, coordinate the work of surveyors, do field reconnaissance at a potential construction site, and use one or more computer programs.

MORE DIRECT CONTACT WITH CLIENTS AND THE PUBLIC

Most engineers "wholesale" their services, that is, report to and interact primarily with professionals and other individuals within their organizations. In contrast, civil engineers tend to "retail" their services. For example, the civil engineer working either in consulting or government tends to have much more direct contact with the public, whereas other engineers working in large organizations tend to have more internal contact.

SOME IMPLICATIONS

As an entry-level civil engineer, you should be cognizant of the possibility of the negative impact of projects. You should also strive for an extra measure of written and other communication skills, and you should refine your interpersonal abilities so that you can function effectively within a highly varied group of coworkers, clients, and the public at large. Not only must the young civil engineer be prepared to interact with a wide variety of individuals and groups, but he or she must also recognize and be able to accommodate the largely nontechnical nature of most of the individuals and groups.

REFERENCE

ROESNER, L. A., and S. G. WALESH, "Urban Water Resources Issues in the 21st Century," *Journal of Professional Issues in Engineering—ASCE*, Vol. 114, No. 3 (July 1988), pp. 302–309.

Appendix B

ASCE Code of Ethics[a]

Effective January 1, 1977

FUNDAMENTAL PRINCIPLES[b]

Engineers uphold and advance the integrity, honor and dignity of the engineering profession by:

1. using their knowledge and skill for the enhancement of human welfare;
2. being honest and impartial and serving with fidelity the public, their employers and clients;
3. striving to increase the competence and prestige of the engineering profession; and
4. supporting the professional and technical societies of their disciplines.

FUNDAMENTAL CANONS

1. Engineers shall hold paramount the safety, health and welfare of the public in the performance of their professional duties.
2. Engineers shall perform services only in areas of their competence.

[a]As adopted September 25, 1976 and amended October 25, 1980.

[b]In April 1975, the American Society of Civil Engineers Board of Direction adopted THE FUNDAMENTAL PRINCIPLES of the ABET Code of Ethics of Engineers as accepted by the Accreditation Board for Engineering and Technology, Inc. (ABET).

(Source: Used with permission from ASCE.)

3. Engineers shall issue public statements only in an objective and truthful manner.
4. Engineers shall act in professional matters for each employer or client as faithful agents or trustees, and shall avoid conflicts of interest.
5. Engineers shall build their professional reputation on the merit of their services and shall not compete unfairly with others.
6. Engineers shall act in such a manner as to uphold and enhance the honor, integrity, and dignity of the engineering profession.
7. Engineers shall continue their professional development through their careers, and shall provide opportunities for the professional development of those engineers under their supervision.

GUIDELINES TO PRACTICE UNDER THE FUNDAMENTAL CANONS OF ETHICS

CANON 1. Engineers shall hold paramount the safety, health and welfare of the public in the performance of their professional duties.

a. Engineers shall recognize that the lives, safety, health and welfare of the general public are dependent upon engineering judgments, decisions and practices incorporated into structures, machines, products, processes and devices.

b. Engineers shall approve or seal only those design documents, reviewed or prepared by them, which are determined to be safe for public health and welfare in conformity with accepted engineering standards.

c. Engineers whose professional judgment is overruled under circumstances where the safety, health and welfare of the public are endangered, shall inform their clients or employers of the possible consequences.

d. Engineers who have knowledge or reason to believe that another person or firm may be in violation of any of the provisions of Canon 1 shall present such information to the proper authority in writing and shall cooperate with the proper authority in furnishing such further information or assistance as may be required.

e. Engineers should seek opportunities to be of constructive service in civic affairs and work for the advancement of the safety, health and well-being of their communities.

f. Engineers should be committed to improving the environment to enhance the quality of life.

CANON 2. Engineers shall perform services only in areas of their competence.

a. Engineers shall undertake to perform engineering assignments only when qualified by education or experience in the technical field of engineering involved.

b. Engineers may accept an assignment requiring education or experience outside of their own fields of competence, provided their services are restricted to those phases of the project in which they are qualified. All other phases of such project shall be performed by qualified associates, consultants, or employees.

c. Engineers shall not affix their signatures or seals to any engineering plan or document dealing with subject matter in which they lack competence by virtue of education or experience or to any such plan or document not reviewed or prepared under their supervisory control.

CANON 3. Engineers shall issue public statements only in an objective and truthful manner.

a. Engineers should endeavor to extend the public knowledge of engineering, and shall not participate in the dissemination of untrue, unfair or exaggerated statements regarding engineering.

b. Engineers shall be objective and truthful in professional reports, statements, or testimony. They shall include all relevant and pertinent information in such reports, statements, or testimony.

c. Engineers, when serving as expert witnesses, shall express an engineering opinion only when it is founded upon adequate knowledge of the facts, upon a background of technical competence, and upon honest conviction.

d. Engineers, shall issue no statements, criticisms, or arguments on engineering matters which are inspired or paid for by interested parties, unless they indicate on whose behalf the statements are made.

e. Engineers shall be dignified and modest in explaining their work and merit, and will avoid any act tending to promote their own interests at the expense of the integrity, honor and dignity of the profession.

CANON 4. Engineers shall act in professional matters for each employer or client as faithful agents or trustees, and shall avoid conflicts of interest.

a. Engineers shall avoid all known or potential conflicts of interest with their employers or clients and shall promptly inform their employers or clients of any business association, interests, or circumstances which could influence their judgment or the quality of their services.

b. Engineers shall not accept compensation from more than one party for services on the same project, or for services pertaining to the same project, unless the circumstances are fully disclosed to and agreed to, by all interested parties.

c. Engineers shall not solicit or accept gratuities, directly or indirectly, from contractors, their agents, or other parties dealing with their clients or employers in connection with work for which they are responsible.

d. Engineers in public service as members, advisors, or employees of a governmental body or department shall not participate in considerations or actions with respect to services solicited or provided by them or their organization in private or public engineering practice.

e. Engineers shall advise their employers or clients when, as a result of their studies, they believe a project will not be successful.

f. Engineers shall not use confidential information coming to them in the course of their assignments as a means of making personal profit if such action is adverse to the interests of their clients, employers or the public.

g. Engineers shall not accept professional employment outside of their regular work or interest without the knowledge of their employers.

CANON 5. Engineers shall build their professional reputation on the merit of their services and shall not compete unfairly with others.

a. Engineers shall not give, solicit or receive either directly or indirectly, any commission, political contribution, or a gift or other consideration in order to secure work, exclusive of securing salaried positions through employment agencies.

b. Engineers should negotiate contracts for professional services fairly and on the basis of demonstrated competence and qualifications for the type of professional service required.

c. Engineers shall not request, propose or accept professional commissions on a contingent basis under circumstances in which their professional judgments may be compromised.

d. Engineers shall not falsify or permit misrepresentation of their academic or professional qualifications or experience.

e. Engineers shall give proper credit for engineering work to those to whom credit is due, and shall recognize the proprietary interests of others. Whenever possible, they shall name the person or persons who may be responsible for designs, inventions, writings or other accomplishments.

f. Engineers may advertise professional services in a way that does not contain self-laudatory or misleading language or is in any other manner derogatory to the dignity of the profession. Examples of permissible advertising are as follows:

Professional cards in recognized, dignified publications, and listings in rosters or directories published by responsible organizations, provided that the cards or listing are consistent in size and content and are in a section of the publication regularly devoted to such professional cards.

Brochures which factually describe experience, facilities, personnel and capacity to render service, providing they are not misleading with respect to the engineer's participation in projects described.

Display advertising in recognized dignified business and professional publications, providing it is factual, contains no laudatory expression or implication and is not misleading with respect to the engineer's extent of participation in projects described.

A statement of the engineer's names or the name of the firm and statement of the type of service posted on projects for which they render services.

Preparation or authorization of descriptive articles for the lay or technical press, which are factual, dignified and free from laudatory implications. Such articles shall not imply anything more than direct participation in the project described.

Permission by engineers for their names to be used in commercial advertisements, such as may be published by contractors, material suppliers, etc., only by means of a modest, dignified notation acknowledging the engineers' participation in the project described. Such permission shall not include public endorsement of proprietary products.

g. Engineers shall not maliciously or falsely, directly or indirectly, injure the professional reputation, prospects, practice or employment of another engineer or indiscriminately criticize another's work.

h. Engineers shall not use equipment, laboratory or office facilities of their employers to carry on outside private practice without the consent of their employers.

CANON 6. Engineers shall act in such a manner as to uphold and enhance the honor, integrity, and dignity of the engineering profession.

a. Engineers shall not knowingly act in a manner which will be derogatory to the honor, integrity, or dignity of the engineering profession or knowingly engage in business or professional practices of a fraudulent, dishonest or unethical nature.

CANON 7. Engineers shall continue their professional development throughout their careers, and shall provide opportunities for the professional development of those engineers under their supervision.

a. Engineers should keep current in their specialty fields by engaging in professional practice, participating in continuing education courses, reading in the technical literature, and attending professional meetings and seminars.

b. Engineers should encourage their engineering employees to become registered at the earliest possible date.

c. Engineers should encourage engineering employees to attend and present papers at professional and technical society meetings.

d. Engineers shall uphold the principle of mutually satisfying relationships between employers and employees with respect to terms of employment including professional grade descriptions, salary ranges, and fringe benefits.

Appendix C

IEEE Code of Ethics[a]

We, the members of the IEEE, in recognition of the importance of our technologies in affecting the quality of life throughout the world, and in accepting a personal obligation to our profession, its members and the communities we serve, do hereby commit ourselves to the highest ethical and professional conduct and agree:

1. to accept responsibility in making engineering decisions consistent with the safety, health and welfare of the public, and to disclose promptly factors that might endanger the public or the environment;
2. to avoid real or perceived conflicts of interest whenever possible, and to disclose them to affected parties when they do exist;
3. to be honest and realistic in stating claims or estimates based on available data;
4. to reject bribery in all its forms;
5. to improve the understanding of technology, its appropriate application, and potential consequences;
6. to maintain and improve our technical competence and to undertake technological tasks for others only if qualified by training or experience, or after full disclosure of pertinent limitations;
7. to seek, accept, and offer honest criticism of technical work, to acknowledge and correct errors, and to credit properly the contributions of others;
8. to treat fairly all persons regardless of such factors as race, religion, gender, disability, age, or national origin;
9. to avoid injuring others, their property, reputation, or employment by false or malicious action;
10. to assist colleagues and co-workers in their professional development and to support them in following this code of ethics.

[a]Approved by the IEEE Board of Directors, August 1990.

(Source: Used with permission from IEEE.)

Appendix D

College Placement Council, Principles for Professional Conduct for Career Services and Employment Professionals

Career services and employment professionals are involved in an important process—helping students choose and attain personally rewarding careers, and helping employers develop effective college relations programs which contribute to effective candidate selections for their organizations. The impact of this process upon individuals and organizations requires commitment by practitioners to principles for professional conduct.

Career services and employment professionals are involved in this process in a partnership effort, with a common goal of achieving the best match between the individual student and the employing organization. The College Placement Council, Inc. (CPC), as the national professional association for career planning, placement, and recruitment, is also concerned with this process. The concern led CPC to the development and adoption of the *Principles for Professional Conduct.* The principles presented here are designed to provide practitioners with three basic precepts for career planning, placement, and recruitment:

- Maintain an open and free selection of employment opportunities in an atmosphere conducive to objective thought where job candidates can choose optimum long-term uses of their talents that are consistent with personal objectives and all relevant facts;
- Maintain a recruitment process that is fair and equitable to candidates and employing organizations;
- Support informed and responsible decision making by candidates.

Adherence to the guidelines will support the collaborative effort of career planning, placement, and recruitment professionals while reducing the potential

(Source: Reprinted from *Principles for Professional Conduct,* with permission of the College Placement Council, Inc. copyright holder.)

for abuses. The guidelines also apply to new technology or third-party recruiting relationships which may be substituted for the more traditional personal interaction among career service professionals, employer professionals, and students.

The principles are not all-inclusive; they are intended to serve as a framework within which the career planning, placement, and recruitment processes should function, and as a foundation upon which professionalism can be promoted.

As part of CPC's commitment to provide leadership in the ethics area and to facilitate the ongoing dialogue on ethics-related issues, the CPC Principles for Professional Conduct Committee has been established. The committee, made up of representatives of each of the seven regions—Western College Placement Association (WCPA), Rocky Mountain College Placement Association (RMCPA), Southwest Placement Association (SWPA), Southern College Placement Association (SCPA), Midwest College Placement Association (MCPA), Middle Atlantic Placement Association (MAPA), and Eastern College and Employer Network (ECEN)—will provide advisory opinions to members on the application of the *Principles,* act as an informational clearinghouse for various ethical issues arising within the regions, periodically review and recommend changes to this document, and resolve problems which may arise.

It is important to keep in mind one final point. The *Principles* do not address certain professional obligations to support state and regional associations, professional development programs, salary surveys, and other demographic trend surveys. Obligations such as these are recognized as vital to the continuing growth of our profession, but since they do not relate directly to the recruitment process, they are not addressed specifically in this document. However, the College Placement Council Board of Governors strongly encourages career services and employment professionals to support and participate in these activities.

The Board of Governors
The College Placement Council, Inc.
June 1994

PRINCIPLES FOR CAREER SERVICES PROFESSIONALS

1. Career services professionals, without imposing personal values or biases, will assist individuals in developing a career plan or making a career decision.

2. Career services professionals will know the career services field and the educational institution and students they represent, and will have appropriate counseling skills.

3. Career services professionals will provide students with information on a range of career opportunities and types of employing organizations. They will inform students of the means and resources to gain access to information which may influence their decisions about an employing organization. Career services professionals will also provide employing organizations with accurate information about the educational institution and its students and about the recruitment policies of the career services office.

4. Career services professionals will provide generally comparable services to all employers, regardless of whether the employers contribute services, gifts, or financial support to the educational institution or office and regardless of the level of such support.

5. Career services professionals will establish reasonable and fair guidelines for access to services by employers. When guidelines permit access to organizations recruiting on behalf of an employer and to international employers, the following principles will apply:

a. Organizations providing recruiting services for a fee will be required to inform career services of the specific employer they represent and the specific jobs for which they are recruiting, and will permit verification of the information. Third-party recruiters that charge fees to students will not be permitted access to career services;

b. Employers recruiting for work outside of the United States are expected to adhere to the (EEO) policy of the career services office. They will advise the career services office and the students of the realities of working in that country and of any cultural and foreign law differences.

6. Career services professionals will maintain EEO compliance and follow affirmative action principles in career services activities in a manner that includes the following:

a. Referring all interested students for employment opportunities without regard to race, color, national origin, religion, age, gender, sexual orientation, or disability and providing reasonable accommodations upon request;

b. Notifying employing organizations of any selection procedures that appear to have an adverse impact based upon the student's race, color, national origin, religion, age, gender, sexual orientation, or disability;

c. Assisting recruiters in accessing certain groups on campus to provide a more inclusive applicant pool;

d. Informing all students about employment opportunities, with particular emphasis on those employment opportunities in occupational areas where certain groups of students are underrepresented;

e. Developing awareness of, and sensitivity to, cultural differences and the diversity of students, and providing responsive services;

f. Responding to complaints of EEO noncompliance, working to resolve such complaints with the recruiter or employing organization, and, if necessary, referring such complaints to the appropriate campus department or agency.

7. Any disclosure of student information outside of the educational institution will be with prior consent of the student unless health and/or safety considerations necessitate the dissemination of such information. Career services professionals will exercise sound judgment and fairness in maintaining the confidentiality of student information, regardless of the source, including written records, reports, and computer data bases.

8. Only qualified personnel will evaluate or interpret tests of a career planning and placement nature. Students will be informed of the availability of testing, the purpose of such tests, and the disclosure policies regarding test result.

9. If the charging of fees for career services becomes necessary, such fees will be appropriate to the budgetary needs of the office and will not hinder student or employer access to services. Career services professionals are encouraged to counsel student and university organizations engaged in recruitment activities to follow this principle.

10. Career services professionals will advise students about their obligations in the recruitment process and establish mechanisms to encourage their compliance. Students' obligations include providing accurate information; adhering to schedules; accepting an offer of employment in good faith; notifying employers on a timely basis of an acceptance or nonacceptance and withdrawing from the recruiting process after accepting an offer of employment; interviewing only with employers for whom students are interested in working and whose eligibility requirements they meet; and requesting reimbursement of only reasonable and legitimate expenses incurred in the recruitment process.

11. Career services professionals will provide services to international students consistent with U.S. immigration laws; inform those students about these laws; represent the reality of the available job market in the United States; encourage pursuit of only those employment opportunities in the United States that meet the individual's work authorization; and encourage pursuit of eligible international employment opportunities.

12. Career services professionals will promote and encourage acceptance of these principles throughout their educational institution, and will respond to reports of noncompliance.

PRINCIPLES FOR EMPLOYMENT PROFESSIONALS

1. Employment professionals will refrain from any practice that improperly influences and affects job acceptances. Such practices may include undue time pressure for acceptance of employment offers and encouragement of revocation of another employment offer. Employment professionals will strive to communicate decisions to candidates within the agreed-upon time frame.

2. Employment professionals will know the recruitment and career development field as well as the industry and the employing organization that they represent, and work within a framework of professionally accepted recruiting, interviewing, and selection techniques.

3. Employment professionals will supply accurate information on their organization and employment opportunities. Employing organizations are responsible for information supplied and commitments made by their representatives. If conditions change and require the employing organization to revoke its commitment, the employing organization will pursue a course of action for the affected candidate that is fair and equitable.

4. Neither employment professionals nor their organizations will expect, or seek to extract, special favors or treatment which would influence the recruitment process as a result of support, or the level of support, to the educational institution or career services office in the form of contributed services, gifts, or other financial support.

5. Employment professionals are strongly discouraged from serving alcohol as part of the recruitment process. However, if alcohol is served, it will be limited and handled in a responsible manner in accordance with the law and the institution's and employer's policies.

6. Employment professionals will maintain equal employment opportunity (EEO) compliance and follow affirmative action principles in recruiting activities in a manner that includes the following:

a. Recruiting, interviewing, and hiring individuals without regard to race, color, national origin, religion, age, gender, sexual orientation, or disability, and providing reasonable accommodations upon request;

b. Reviewing selection criteria for adverse impact based upon the student's race, color, national origin, religion, age, gender, sexual orientation, or disability;

c. Avoiding use of inquiries that are considered unacceptable by EEO standards during the recruiting process;

d. Developing a sensitivity to, and awareness of, cultural differences and the diversity of the work force;

e. Informing campus constituencies of special activities that have been developed to achieve the employer's affirmative action goals;

f. Investigating complaints forwarded by the career services office regarding EEO noncompliance and seeking resolution of such complaints.

7. Employment professionals will maintain the confidentiality of student information, regardless of the source, including personal knowledge, written records/reports, and computer data bases. There shall be no disclosure of student information to another organization without the prior written consent of the student, unless necessitated by health and/or safety considerations.

8. Those engaged in administering, evaluating, and interpreting assessment tools, tests, and technology used in selection will be trained and qualified to do so. Employment professionals must advise the career services office of any test conducted on campus and eliminate such test if it violates campus policies. Employment professionals must advise students in a timely fashion of the type and purpose of any test that students will be required to take as part of the recruitment process and to whom the test results will be disclosed. All tests will be reviewed by the employing organization for disparate impact and job-relatedness.

9. When using organizations that provide recruiting services for a fee, employment professionals will respond to inquiries by the career services office regarding this relationship and the positions the organization was contracted to fill. This principle applies equally to any other form of recruiting that is used as a substitute for the traditional employer/student interaction.

10. When employment professionals conduct recruitment activities through student associations or academic departments, such activities will be conducted in accordance with the policies of the career services office.

11. Employment professionals will cooperate with the policies and procedures of the career services office, including certification of EEO compliance or exempt status under the Immigration Reform and Control Act, and will honor scheduling arrangements and recruitment commitments.

12. Employment professionals recruiting for international operations will do so according to EEO standards. Employment professionals will advise the career services office and students of the realities of working in that country and of any cultural or foreign law differences.

13. Employment professionals will educate and encourage acceptance of these principles throughout their employing institution and by third parties representing their employing organization on campus, and will respond to reports of noncompliance.

PRINCIPLES FOR THIRD-PARTY RECRUITERS

Preface: *These standards are designed to provide guidance to third-party recruiters who recruit college graduates through the college recruitment process. These standards are not to be construed as requiring or encouraging, prohibiting, or discouraging use of third-party recruiters by college or employer professionals.*

1. Definition of Third-Party Recruiter:

a. Third-party recruiters are agencies, organizations, or individuals recruiting candidates for temporary, part-time or full-time employment opportunities other than for their own needs. This includes entities that refer or recruit for profit or not for profit, and it includes agencies that collect student information to be disclosed to employers for purposes of recruitment and employment;

b. Third-party recruiting organizations charge for services using one of the following fee structures:

1. Applicant paid fee—The applicant pays the third-party recruiter a fee based upon the applicant's starting salary once the applicant is placed with an employer.

2. Employer paid fee—

a. Retainer—The employer pays a flat fee to the third-party recruiter for services performed in the recruiting of individuals to work for the employer.

b. Contingency fee—The employer pays to the third-party recruiter a percentage of the applicant's starting salary once the applicant is hired by the employer.

c. The above definition includes, but is not limited to, the following entities regardless of the fee structure used by the entity to charge for services:

1. Employment Agencies Organizations that list positions for a number of client organizations and receive payment when a referred candidate is hired. The fee for listing a position is paid either by the firm listing the opening (fee paid) or by the candidate who is hired.

2. Search Firms—Organizations that contract with clients to find and screen qualified persons to fill specific positions. The fees for this service are paid by the clients.

3. Contract Recruiter—Organizations that contract with an employer to act as the employer's agent in the recruiting and employment function.

4. Resume Referral Firms—Organizations that collect data on job seekers which is sent to prospective employers. Fees exist for the employer, job seeker, or both.

d. Temporary agencies—Temporary agencies are employers, not third-party recruiters, and will be expected to comply with the professional conduct principles set forth for employer professionals. These are organizations that contract to provide individuals qualified to perform specific tasks or complete specific projects for a client organization. Individuals perform work at the client organization, but are employed and paid by the agency.

2. Third-party recruiters will be versed in the recruitment field and work within a frame work of professionally accepted recruiting, interviewing, and selection techniques.

3. Third-party recruiters will follow EEO standards in recruiting activities in a manner that includes the following:

a. Referring qualified students to employers without regard to the student's race, color, national origin, age, gender, sexual orientation, or disability;

b. Reviewing selection criteria for adverse impact and screening students based upon job-related criteria only, not based upon the student's race, color, national origin, religion, age, gender, sexual orientation, or disability;

c. Refusing, in the case of resume referral entities, to permit employers to screen and select resumes based upon the student's race, color, national origin, religion, age, gender, sexual orientation, or disability;

d. Avoiding use of inquiries that are considered unacceptable by EEO standards during the recruitment process;

e. Affirming an awareness of and sensitivity to cultural differences and the diversity of the work force;

f. Investigating complaints forwarded by the career services office or the employer client regarding EEO noncompliance and seeking resolution of such complaints.

4. Third-party recruiters that directly charge students for services are not following accepted professional practices and will not be permitted on-campus recruiting privileges.

5. Third-party recruiters will disclose the following information to students and career services practitioners:

a. The client, or clients, that the third-party recruiter is representing and to whom the student's credentials will be disclosed. Career services will be permitted to verify this information by contacting the named client or clients. In the case of a resume referral entity, a list of clients that use the services of the entity must be made available.

b. The types of positions for which the third-party recruiter is recruiting. Resume referral firms do not have to disclose this information.

6. Third-party recruiters will not disclose to any employer, including the client-employer, any student information without obtaining prior written consent from the student. Under no circumstances can student information be disclosed for other than recruiting purposes nor can it be sold or provided to other entities.

7. Third-party recruiters attending career fairs will represent employers who have authorized them and will disclose to career services the names of the represented employers.

ADVISORY OPINIONS

A CPC member/regional association may request an advisory opinion regarding an interpretation of the *Principles* document at any time. The advisory opinion will apply to the situation as explained and will not be considered precedent for a subsequent complaint brought to CPC.

- The member/association will prepare a written statement detailing the conduct in question. Statements will include the section, or sections, of the *Principles* to be interpreted relative to the conduct in question.
- The information will be reviewed by the CPC Principles for Professional Conduct Committee and a response given to the member/association.

PROBLEM-SOLVING PROCEDURES

Questionable practices or problems involving recruiters and career services practitioners shall be resolved between the parties as quickly as possible. CPC recommends the following:

- Discuss the incident with all parties involved in the situation. Determine the specifics of the problem.
- Attempt to resolve the incident among the affected parties.
- Refer unresolved concerns to the supervisors of the involved individuals or to other appropriate officials.

- If informal resolution is not successful, the parties are encouraged to use the regional association's problem-solving mechanism.

If the problem remains unresolved, complaints or requests for advisory opinions may be presented to the CPC Principles for Professional Conduct Committee for ultimate determination by the CPC Board of Governors. Remedies for violations can include written warning, probation, suspension, and expulsion from CPC membership. For specific details for filing and processing complaints or for requesting an advisory opinion, contact:

EXECUTIVE DIRECTOR
College Placement Council, Inc.
62 Highland Avenue
Bethlehem, PA 18017
(610) 868–1421
(800) 544–5272

Appendix E

EXCERPTS FROM THE BOEING COMPANY'S BUSINESS CONDUCT POLICY AND GUIDELINES

1. Does management really expect employees to adhere to the ethical business conduct policy even if it would mean losing business or reducing profitability?

Yes. Maintaining high ethical standards is essential to staying in business and maintaining long-term profitability.

2. Will I get into trouble with my manager if I call the Corporate Office of Business Practices or my business ethics advisor about an ethics issue?

No. Boeing policy is to encourage employees to express concerns about ethical issues and to report any suspected violations. Boeing will make every effort to maintain the confidentiality of the caller. Managers are specifically precluded from retaliating against an employee for following this policy.

3. I have a friend who works for the DoD. Does Boeing policy prohibit me from giving her a Christmas gift?

No. Boeing policy requires employees to respect the standards of conduct imposed by our customers. Although the DoD standards prohibit gifts by vendors to DoD employees, this prohibition does not apply to the exchange of gifts between friends or family members when the exchange is exclusively the result of a personal and not a business relationship.

4. Why can't I give a pen and pencil set to an employee of the DoD or NASA when I could provide the same item to a commercial customer?

As currently defined by their respective regulations, DoD and NASA prohibit their employees from accepting advertising or promotional items of more

(Source: Used with permission of the Boeing Company, *Business Conduct Policy and Guidelines,* 1991.)

than $10.00 retail value. Boeing policy requires that we respect a customer's rules. Therefore, you are prohibited from providing a pen and pencil set to DoD or NASA employees. Most commercial customers permit their employees to accept promotional items such as an inexpensive pen and pencil set. In such cases, Boeing policy allows you to provide such a gift unless you are aware of a customer's policy that prohibits customer employees from accepting such business courtesies, or if you believe it would unduly influence the recipient and give Boeing an unfair competitive advantage.

5. I understand that certain government agencies prohibit Boeing from providing hospitality to government employees. I know that the lead government technical person on my program is a big sports fan. Can I take him to a football game and pay for it out of my own pocket?

No. Boeing policy does not permit an employee personally to offer business courtesies that the company is prohibited from offering. It is the act of providing the prohibited business courtesy, not the source of funds, that creates the problem.

6. A supplier representative knows that I am a big basketball fan. He invited me to accompany him to the All-Star game. His company was able to get tickets even though the game had been sold out for weeks. When I told him that Boeing policy would not allow me to accept his invitation, he offered to sell me the two tickets. Would purchasing the tickets violate Boeing policy?

Yes. The game is sold out and the tickets aren't generally available for purchase. The vendor's offer must be declined.

7. A representative of a supplier dropped by and left a pen and pencil set with me as a token of appreciation. May I accept it?

As long as the item is inexpensive, is widely available to others under similar circumstances, and your management allows you to accept such business courtesies, you may keep it for your personal use.

8. I am active in my local school district and do volunteer work to improve the quality of education for the residents of my district. An important levy is on the ballot and I am actively campaigning for it to be passed. May I use Company photocopy equipment to make copies of a flyer detailing the reasons that the levy should be passed?

Boeing encourages its employees to participate in the political process and to be active in the community. However, employees may not use Boeing resources in these activities unless approved by the office of the Corporate Vice President of Government Affairs.

9. My manager asked me to charge my time to an incorrect charge number. What should I do?

First, make sure that your manager knows what you are really working on, so that there is no misunderstanding about the work you are doing. If that doesn't resolve the problem, you should consider seeking advice from your next-level manager, your business ethics advisor, or the Corporate Office of Business Practices.

10. May I go out to a business meal with a representative of a supplier or vendor?

Ask your management whether your organization's policy permits the acceptance of business meals from those to whom you have the ability to allocate Company business. In some circumstances, modest and infrequent business meals may be accepted by Company employees. On other occasions, it may be more appropriate for Boeing to pay for the meal. However, whenever a vendor pays for a meal, always consider the specific circumstances and whether your impartiality could be affected or appear to others to be affected.

11. Because I was bumped on a Company trip, I was given a free ticket by the airline. May I use this ticket for a vacation?

No. Boeing policy requires that these benefits from Company business activities be returned to the Company. This is an important part of our effort to reduce overall costs.

12. Can I have Boeing Transportation drive two of our commercial customers from the airport to the hotel? What if these customers are from the U.S. Navy?

You can provide transportation for commercial customers except those whose company policy prohibits them from accepting this type of business courtesy. However, you cannot pick up your Navy guests at the airport without advance approval from the Navy's designated ethics official. You should contact your business ethics advisor or legal staff for guidance. The DoD Standards of Conduct prohibits DoD personnel from accepting this business courtesy, and Company policy requires Boeing employees to respect such restrictions. This restriction also applies to transportation in personal and rental cars, in addition to vehicles provided by Boeing Transportation.

Index

A

B

C